Family of Faith Library

W9-AON-354

FAMILY OF FAITH LIBRARY

GEOMETRY THE EASY WAY

By

LAWRENCE S. LEFF

Assistant Principal,
Mathematics Supervision
Franklin D. Roosevelt High School
Brooklyn, New York

BARRON'S EDUCATIONAL SERIES, INC.

New York • London • Toronto • Sydney

ACKNOWLEDGMENTS

A special word of thanks to Joan Cipriano and Donna DeBenedictis
of BARRON's Educational Series, Inc. for their untiring efforts and patience
during the preparation of this book.

© Copyright 1984 by Barron's Educational Series, Inc.
All rights reserved.
No part of this book may be reproduced in any form,
by photostat, microfilm, xerography, or any other
means, or incorporated into any information retrieval
system, electronic or mechanical, without the written
permission of the copyright owner.

All inquiries should be addressed to:
Barron's Educational Series, Inc.
250 Wireless Boulevard
Hauppauge, New York 11788

Library of Congress Catalog Card No. 84-9330

International Standard Book No. 0-8120-2718-3

Library of Congress Cataloging in Publication Data
Leff, Lawrence S.
Geometry the easy way.

Includes index.
1. Geometry, Plane. I. Title.
QA455.L35 1984   516.2'2   84-9330
ISBN 0-8120-2718-3

PRINTED IN THE UNITED STATES OF AMERICA
7 100 9 8 7

CONTENTS

PREFACE

Some geometry books "teach" geometry by quickly summarizing geometric principles and then providing a continuous stream of exercises. This can be a frustrating experience since it is often not clear *why* certain problems should be approached in a particular way or *why* the method to be used works. This book makes a special effort to anticipate and answer the "why" types of questions you might ask if the material were being explained by a teacher.

Why study geometry? The study of geometry provides us with the opportunity to investigate the properties and special relationships of familiar types of figures. If you enjoy following the line of reasoning used by detectives of the caliber of Sherlock Holmes and Hercule Poirot, then you should find the logical methods of reasoning illustrated in this book intriguing as well as challenging.

This book includes a number of features which should make it attractive whether you are studying the subject on your own or are enrolled in a formal course in plane geometry:

- There is an entire chapter (Chapter 16) devoted to developing *computer* solutions to problems in geometry using the BASIC programming language.

- Most concepts are *not* presented in "finished form," but are *developed* using an explanatory style complemented by numerous illustrations and figures. Emphasis throughout is on demonstrating how logical methods of reasoning can be used to discover and verify the validity of geometric concepts and relationships.

- A large number of fully explained demonstration problems are found within the development of each chapter. These examples are intended to illustrate and reinforce the concepts presented while providing you with an opportunity to test your understanding of the material *before* you proceed further.

- Each chapter concludes with an array of exercises that provides a comprehensive review of the geometric principles that were developed in the chapter.

- A solution key for *all* chapter review exercises is conveniently located at the rear of the book so that you may obtain feedback and guidance as you progress through the book.

- The presentation is modern in its use of notation and language. The coverage of topics parallels closely a typical full year's course in plane geometry at the high school level.

Lawrence S. Leff
June 1984

CHAPTER ONE
BUILDING A GEOMETRY VOCABULARY

1.1 THE BUILDING BLOCKS

Studying geometry is, in a sense, like building a house. Cement and bricks are often used to give a house a strong foundation. Instead of bricks, geometry uses the following types of building blocks:

1. UNDEFINED TERMS. These words are so fundamental that they cannot be satisfactorily defined. Point, line, and plane are undefined terms. Although these terms cannot be defined, they can be described.

2. DEFINED TERMS. These words are introduced in order to generate a common vocabulary making it easier to refer to geometric figures and relationships.

3. POSTULATES. These are statements which based on experience are assumed to be true.

4. THEOREMS. A theorem is a generalization which can be demonstrated to be true. A familiar theorem is that the sum of the measures of the three angles of a triangle is 180 degrees. Theorems and postulates may be considered to be opposites in the sense that a postulate is accepted as being true without proof, while a theorem cannot be assumed to be true unless it is first proved. Much of our work in geometry will be devoted to building on our current knowledge by suggesting and then proving theorems. These theorems, in turn, will be used to prove additional theorems. The result will be an expanding awareness of a variety of geometrical relationships.

We have just identified the four basic building blocks of geometry. Logical methods of reasoning will provide the cement that will bind the blocks together and allow us to constantly build on our foundation and "prove" new theorems. By prove we simply mean the logical manner in which undefined terms, definitions, postulates, and previously established theorems are systematically related in order to justify a new theorem. Although this may sound somewhat vague, the nature of a geometric proof is a central theme of this course, as will be made clear in the chapters to come.

UNDEFINED TERMS

Table 1.1 lists some undefined terms.

TABLE 1.1

UNDEFINED TERM	DESCRIPTION	NOTATION
Point	A point indicates position; it has no length, width, or depth.	• A A point is named by a single capital letter.
Line	A line is a set of continuous points that extend indefinitely in either direction. (NOTE: the term *line* will always be understood to mean straight line.)	A line is identified by naming two points on the line and writing a miniature line over the letters: $\overleftrightarrow{AB}$ Alternatively, a line may be named by using a single lowercase letter: line *l*
Plane	A plane is a set of points that forms a flat surface which extends indefinitely in all directions; a plane has no depth.	A plane is usually represented as a closed four-sided figure as illustrated above. Often a capital letter is placed at one of the vertices (corners). The plane is then referred to using this letter. This notation is not standardized.

Figure 1.1 illustrates that lines may lie in different planes or in the same plane. Line *l* and $\overleftrightarrow{AB}$ both lie in plane *Q*. Line *k* and $\overleftrightarrow{AB}$ lie in plane *P*. Lines *l* and *k* are contained in different planes while $\overleftrightarrow{AB}$ (the intersection of the two planes) is common to both planes. In order to simplify our discussions, we will always assume that we are working with figures which lie in the same plane. This branch of geometry takes a "flat," two-dimensional view of figures and is referred to as *plane geometry. Solid geometry* is concerned with figures and their spatial relationships as they actually exist in the world around us.

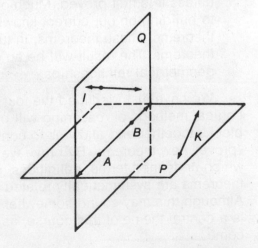

FIGURE 1.1

DEFINED TERMS

Table 1.2 lists some geometric terms and their definitions.

TABLE 1.2

TERM	DEFINITION	ILLUSTRATION
1. Line Segment	A *line segment* is a part of a line consisting of two points, called *end points*, and the set of all points between them.	A •——• B NOTATION: $\overline{AB}$
2. Ray	A *ray* is a part of a line consisting of a given point, called the *end point*, and the set of all points on one side of the end point.	M, L NOTATION: $\overrightarrow{LM}$ A ray is always named by using two points, the first of which must be the end point. The arrow on top must always point to the right.
3. Opposite Rays	*Opposite rays* are rays that have the same end point and which form a line.	X K B $\overrightarrow{KX}$ and $\overrightarrow{KB}$ are opposite rays.
4. Angle	An *angle* is the union of two rays having the same end point. The end point is called the *vertex* of the angle; the rays are called the *sides* of the angle.	J, K, L VERTEX: K SIDES: $\overrightarrow{KJ}$ and $\overrightarrow{KL}$

An angle may be named in one of three ways:

1. Using three letters, the center letter corresponding to the vertex of the angle and the other letters representing points on the sides of the angle. For example, in Figure 1.2, the name can be angle *RTB* or ∡*RTB* (*or* ∡*BTR*).

2. Placing a number at the vertex and in the *interior* of the angle. The angle may then be referred to by the number. For example, in Figure 1.3, the name can be ∡1 or ∡*RTB*.

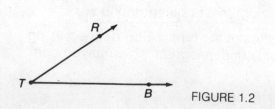

FIGURE 1.2

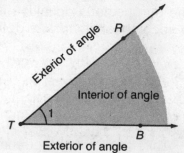

FIGURE 1.3

3. Naming the angle using a single letter that corresponds to the vertex, provided this causes no ambiguity. There is no question which angle on the diagram corresponds to angle *A* in Figure 1.4. Which angle on the diagram is angle *D*? Actually there are three angles formed at vertex *D*:

- Angle *ADB*
- Angle *CDB*
- Angle *ADC*

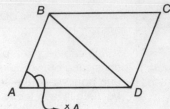

FIGURE 1.4

In order to uniquely identify the angle having *D* as its vertex, we must either name the angle using three letters or introduce a number into the diagram.

EXAMPLE 1.1

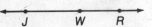

    a Name the accompanying line in three different ways.

    b Name three different segments.

    c Name four different rays.

    d Name a pair of opposite rays.

SOLUTION **a** $\overleftrightarrow{JW}$, $\overleftrightarrow{WR}$, and $\overleftrightarrow{JR}$

            b $\overline{JW}$, $\overline{WR}$, and $\overline{JR}$

    c $\overrightarrow{JR}$, $\overrightarrow{WR}$, $\overrightarrow{RJ}$ and $\overrightarrow{WJ}$

    d $\overrightarrow{WJ}$ and $\overrightarrow{WR}$

EXAMPLE 1.2 Use three letters to name each of the numbered angles in the accompanying diagram.

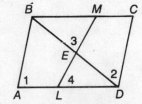

SOLUTION $\angle 1 = \angle BAD$ or $\angle DAB$ or $\angle LAB$

            $\angle 2 = \angle CDB$ or $\angle BDC$ (NOTE: Letter *E* may be used instead of letter *B*.)

            $\angle 3 = \angle BEM$ or $\angle MEB$

            $\angle 4 = \angle DLM$ or $\angle MLD$ (NOTE: Letter *E* may be used instead of letter *M*.)

1.2 DEFINITIONS AND POSTULATES

The purpose of a definition is to make the meaning of a term clear. In order to accomplish its task, a definition must

- Clearly identify the word (or expression) that is being defined
- State the distinguishing characteristics of the term being defined using only those terms which are commonly understood or which have been previously defined
- Take the form of a grammatically correct sentence

As an example, consider the term *collinear*. In Figure 1.5, points *A*, *B*, and *C* are collinear. In Figure 1.6, points *R*, *S*, and *T* are *not* collinear.

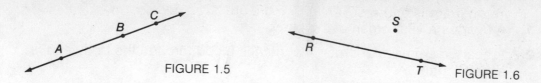

FIGURE 1.5

FIGURE 1.6

DEFINITION OF COLLINEAR POINTS

Collinear points are points which lie on the same line.

Notice that the definition begins by identifying the term being defined. The definition uses only those geometric terms (points and line) which have been previously discussed. Contrast this definition with the definition given for the term apothem: An *apothem* is a line segment drawn from the center of a regular polygon perpendicular to a side of the polygon.

Is this a good definition? No, it is not clear what an apothem is since there are several terms used in its definition which we have not explained, including the terms regular polygon and perpendicular.

Much of geometry involves using previously developed ideas to generate new concepts. For example, we can use our current knowledge of geometric terms to arrive at a definition of a triangle. How would you draw a triangle? If you start with three noncollinear points and connect them with line segments (Figure 1.7), a *triangle* is formed.

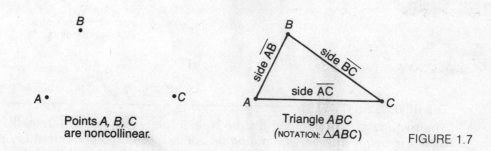

Points *A*, *B*, *C*
are noncollinear.

Triangle *ABC*
(NOTATION: △*ABC*)

FIGURE 1.7

DEFINITION OF TRIANGLE

A *triangle* is a figure formed by connecting three noncollinear points with line segments.

Notice that the definition capitalizes on our understanding of the term *collinear* points. Is it necessary to include that the three noncollinear points are connected by line segments? Observe (in Figure 1.8) that it is possible to join three noncollinear points without using line segments.

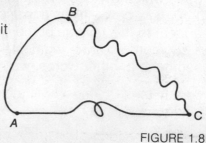

FIGURE 1.8

A good definition contains the *minimum* amount of information needed to accurately describe the term. To further illustrate this point, consider the following three alternative definitions for the word *heart*:

1. A heart is a vital organ of the body.

2. A heart is a muscular organ that pumps blood through the body and is a vital organ.

3. A heart is a muscular organ that pumps blood through the body.

The first definition offers too little information (there are other vital organs of the body), while the second definition offers too much detail (it is not necessary to add the phrase, "and is a vital organ.") The third definition is a good definition.

A good definition must be reversible as shown in the following table.

DEFINITION	REVERSE OF THE DEFINITION
Collinear points are points that lie on the same line.	Points that lie on the same line are collinear points.
A *right angle* is an angle whose measure is 90 degrees.	An angle whose measure is 90 degrees is a right angle.
A *line segment* is a set of points.	A set of points is a line segment.

The first two definitions are reversible since the reverse of the definition is a true statement. The reverse of the third "definition" is false since the points may be scattered as in Figure 1.9.

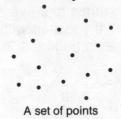

A set of points

FIGURE 1.9

THE REVERSIBILITY TEST

The reverse of a definition must be true. If the reverse of a statement which is being offered as a definition is false, then the statement is *not* a good definition.

The reverse of a definition will prove useful in our later work when attempting to establish geometric properties of lines, segments, angles, and figures. For example, a *midpoint* of a segment may be defined as a point which divides a segment into two segments of equal length. In Figure 1.10, how can we prove that point M is the midpoint of $\overline{AB}$? We must appeal to the *reverse* of the definition of a midpoint: a point which divides a segment into two segments of equal length is the midpoint of the segment. In other words, we must first show that $AM = MB$. Once this is

FIGURE 1.10

accomplished, we are entitled to conclude that point *M* is the midpoint of $\overline{AB}$. As another illustration, we may define an *even* integer as an integer which leaves a remainder of 0 when divided by 2. How can we prove that an integer is an even number? Simple—we use the reverse of the definition, show that when the number is divided by 2, the remainder is 0. If this is true, then the number must be even.

INITIAL POSTULATES

In addition to defined terms, a number of postulates (also referred to as *axioms*) will be introduced throughout this course. We begin by stating two postulates.

POSTULATE 1.1

Exactly one line contains two given points.

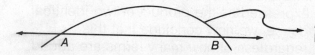

A second *line* cannot be drawn through points *A* and *B*.

POSTULATE 1.2

Exactly one plane contains three noncollinear points.

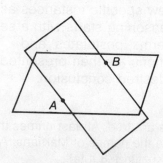

More than one plane may contain *two* points.

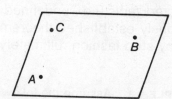

Exactly one plane contains three noncollinear points.

The previous two postulates are sometimes phrased as follows:

POSTULATE 1.1 (ALTERNATIVE)

Two points *determine* a line.

POSTULATE 1.2 (ALTERNATIVE)

Three noncollinear points *determine* a plane.

Postulate 1.1 implies that two points uniquely define a line, while Postulate 1.2 asserts that three noncollinear points uniquely define a plane.

1.3 INDUCTIVE VS. DEDUCTIVE REASONING

Consider the result of adding strings of consecutive odd integers beginning with 1.

INTEGER STRING	SUM
1 + 3	4
1 + 3 + 5	9
1 + 3 + 5 + 7	16
1 + 3 + 5 + 7 + 9	25
⋮	

Do you notice a pattern? It appears that the sum of consecutive odd integers, beginning with 1, will always be a *perfect square*. (A perfect square is a number that can be expressed as the product of two identical numbers.) If based upon this evidence, we now conclude that this relationship will always be true, regardless of how many terms are added, we have engaged in *inductive reasoning*. Inductive reasoning involves examining a few examples, observing a pattern, and then assuming that the pattern will always prevail. Inductive reasoning is *not* a valid method of proof, although it often suggests statements that can be proved by other methods.

Deductive reasoning may be considered to be the opposite of inductive reasoning. Rather than begin with a few specific instances as is common with inductive processes, deductive reasoning starts with a set of accepted *facts* (i.e., undefined terms, defined terms, postulates, and previously established theorems); logical assertions are then presented in a step-by-step fashion, ultimately leading to the desired conclusion.

EXAMPLE 1.3 Assume the following two statements are true. All last names that have seven letters with no vowels are the names of Martians. All Martians are 3 ft tall. Prove that Mr. Xhzftlr is 3 ft tall.

SOLUTION For illustrative purposes we shall adopt the two-column proof format that will be elaborated on in subsequent chapters.

PROOF	Statements	Reasons
	1. The name is Mr. Xhzftlr.	1. Given.
	2. Mr. Xhzftlr is a Martian.	2. All last names that have seven letters with no vowels are the names of Martians.
	3. Mr. Xhzftlr is 3 ft tall.	3. All Martians are 3 ft tall. (See Postulate 2.)

Notice that for each statement there is a corresponding factual type of supporting reason.

1.4 THE IF . . . THEN . . . SENTENCE STRUCTURE

"If I graduate from high school with a 90 or better grade average, then my parents will buy me a car." Will the student receive a car as a graduation present? Only if his or her average is 90 or better will the student receive a car. Notice that the statement in the "if clause" of the sentence represents a *condition*; if the condition is true, then the "then clause" specifies the action to be taken.

Statements in mathematics often take the "If . . . , then . . ." form. For example,

<u>If two lines are parallel</u>, then <u>they never meet</u>.

The statement in the if clause (the single underlined phrase) represents the hypothesis or given. The statement in the then clause (the double underlined phrase) represents the conclusion. If we assume the lines are parallel, then we may conclude that they never meet.

It is common for theorems to be expressed in this conditional form. When attempting to prove a *proposed* theorem, the statement in the if clause represents what we are given or entitled to assume; the phrase contained in the then clause corresponds to what has to be proved. For example, a theorem which will be proved in a later chapter is, "If two sides of a triangle are equal in length, then the angles opposite them are equal." What is the given? What is to be proved? The diagram that would be used in this proof together with a summary of the given and the prove follows.

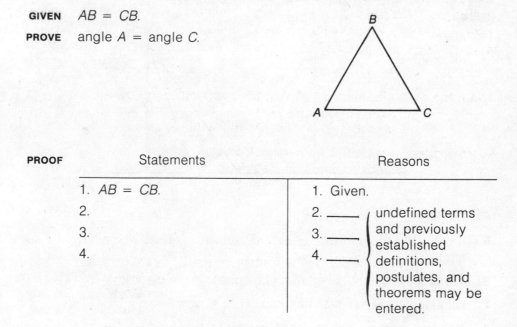

GIVEN $AB = CB$.

PROVE angle A = angle C.

PROOF	Statements		Reasons
	1. $AB = CB$.		1. Given.
	2.		2. ⎰ undefined terms
	3.		3. ⎱ and previously
	4.		4. established definitions, postulates, and theorems may be entered.

Once this proof is completed, the theorem is then taken as fact and can then be used to prove other theorems.

REVIEW EXERCISES FOR CHAPTER 1

1. In the accompanying diagram:

   a Name four rays having point *B* as an end point.

   b Name line *l* in three different ways.

   c Name line *m* in three different ways.

   d Name four angles that have the same vertex.

   e Name two pairs of opposite rays.

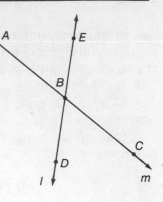

Use the following diagram for Exercises 2 and 3.

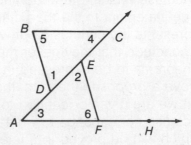

2. Name the vertex of angles: **a** 1   **b** 3   **c** 5.

3. Use three letters to name angles: **a** 2   **b** 4   **c** 6.

For Exercises 4 to 11, use the following diagram.

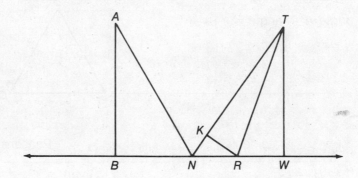

4. Name four collinear points.

5. If point *N* is the midpoint of $\overline{BW}$, name two segments which have the same length.

6. Name the different triangles which appear in the diagram.

7. Name each angle which has point *R* as its vertex.

8. Name an angle which is not an angle of a triangle.

9. Name two pairs of opposite rays.

10. Name a segment which is a side of two different triangles.

11. To prove *R* is the midpoint of $\overline{WN}$, which two segments must be demonstrated to have the same length?

12. Write the reverse of each of the following definitions.

   a An acute angle is an angle whose measure is less than 90 degrees.

   b An equilateral triangle is a triangle having three sides equal in length.

   c A bisector of an angle is the ray (or segment) which divides the angle into two congruent angles.

13. Identify each of the following as examples of inductive or deductive reasoning.

   a The sum of 1 and 3 is an even number; the sum of 3 and 5 is an even number; the sum of 5 and 7 is an even number; the sum of 7 and 31 is an even number; the sum of 19 and 29 is an even number. *Conclusion:* The sum of any two odd numbers is an even number.

   b All students in Mr. Euclid's geometry class are 15 years old. John is a member of Mr. Euclid's geometry class. *Conclusion:* John is 15 years old.

   c It has rained on Monday, Tuesday, Wednesday, Thursday, and Friday. *Conclusion:* It will rain on Saturday.

   d The sum of the measures of a pair of complementary angles is 90. Angle *A* and angle *B* are complementary. The measure of angle *A* is 50. *Conclusion:* The measure of angle *B* is 40.

14. A *median* of a triangle is a segment drawn from a vertex of a triangle to the midpoint of the opposite side of the triangle. Draw several large *right* (90-degree) triangles. See the diagram. For each triangle, locate the midpoint of the hypotenuse (the side opposite the 90-degree angle). Draw the median to the hypotenuse. Using a ruler, compare the lengths of the median and the hypotenuse in each triangle drawn. Use inductive reasoning to draw an appropriate conclusion. Note that *M* is the midpoint of $\overline{AB}$ if $\overline{AM}$ and $\overline{BM}$ measure the same length.

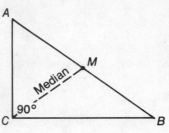

15. Draw several large triangles (not necessarily right triangles). In each triangle locate the midpoints of each side. Draw the three medians of each triangle. Use inductive reasoning to draw a conclusion related to where the medians intersect.

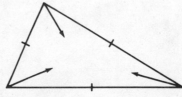

16. Use deductive reasoning to arrive at a conclusion based on the assumptions given.

   a ASSUMPTIONS 1. All Martians have green eyes.
   2. Henry is a Martian.

   b ASSUMPTIONS 1. The sum of the measures of the angles of a triangle is 180.
   2. In a particular triangle, the sum of the measures of two angles is 100.

17. A prime number is any whole number that is divisible only by itself and 1. For example, 7, 11, and 13 are examples of prime numbers. Evaluate the formula $n^2 + n + 17$ using all integer values of *n*, from 0 to 9, inclusive. Do you notice a pattern?

   a Using inductive reasoning, draw a conclusion.

   b Is your conclusion true for all values of *n*? Test *n* = 16.

CHAPTER TWO
MEASURE AND CONGRUENCE

2.1 MEASURING SEGMENTS AND ANGLES

We often describe the size of something by comparing it to something we are already familiar with. "She is as thin as a rail," conveys the message of a person being underweight, but it is not very precise. In geometry we must be precise. How could we determine the exact weight of a person? We might use a measurement instrument that is specifically designed for this purpose—the scale. To determine the length of a segment or the measure of an angle we must also use special measurement instruments—the *ruler* for measuring the length of a segment and the *protractor* for measuring an angle.

The units of measurement that we choose to express the length of a segment are not important, although they should be convenient. It would not be wise, for example, to try to measure and express the length of a postage stamp in terms of kilometers or miles. As illustrated in Figure 2.1, a segment is measured by lining up the end points of the segment with convenient markings of a ruler. In this example, the length or measure of line segment AB is 2 in. We abbreviate this by writing $m\overline{AB} = 2$, read as "the measure of line segment AB is 2." Alternatively, we could write $AB = 2$, read as "the distance between points A and B is 2." It is customary to use the expressions $m\overline{AB}$ and AB (no bar over the letters A and B) interchangeably and to interpret each as the length of line segment $\overline{AB}$. *Caution:* It is incorrect to write $\overline{AB} = 2$ as this implies the infinite set of points which makes up segment $\overline{AB}$ is equal to 2.

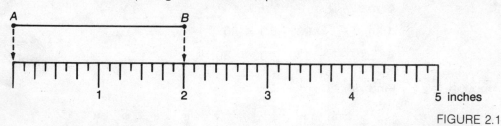

FIGURE 2.1

To measure an angle we use a protractor (see Figure 2.2) where the customary unit of measure is the degree.* In our example, the measure of

---

* Angles may be measured in terms of other units such as the radian. In this course, however, we shall assume that the measure of an angle corresponds to some number on the protractor, greater than 0 and less than *or* equal to 180.

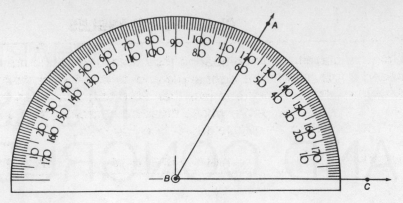

FIGURE 2.2

angle *ABC* is 60 degrees. We abbreviate this by writing $m\angle ABC = 60$, read as "the measure of angle *ABC* is 60." It is customary to omit the degree symbol (°); we *never* write $m\angle ABC = 60°$ or $\angle ABC = 60$ (omitting the "*m*").

EXAMPLE 2.1 Find the measures of these angles:

  a $m\angle APZ$

  b $m\angle FPZ$

  c $m\angle WPB$

  d $m\angle ZPB$

  e $m\angle SPZ$

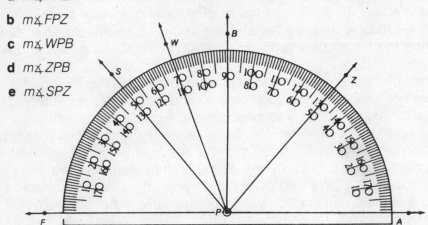

SOLUTION **a** $m\angle APZ = 50$ (read lower scale)

  b $m\angle FPZ = 130$ (read upper scale)

  c $m\angle WPB = 110 - 90 = 20$

  d $m\angle ZPB = 90 - 50 = 40$

  e $m\angle SPZ = 130 - 50 = 80$

EXAMPLE 2.2 Find $m\overline{RS}$.

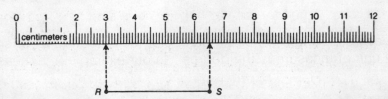

SOLUTION $m\overline{RS} = 6.5 - 3 = 3.5$ cm. NOTE: Length is a positive quantity so that we must always subtract the smaller reading on the ruler (called a coordinate) from the larger ruler coordinate.

CLASSIFYING ANGLES

Angles may be classified by comparing their measures to the number 90. An L-shaped angle is called a *right angle* and its measure is exactly equal to 90. An angle whose measure is less than 90 (but greater than 0) is called an *acute angle*. An angle whose measure is greater than 90 (but less than 180) is called an *obtuse angle*. See Figure 2.3.

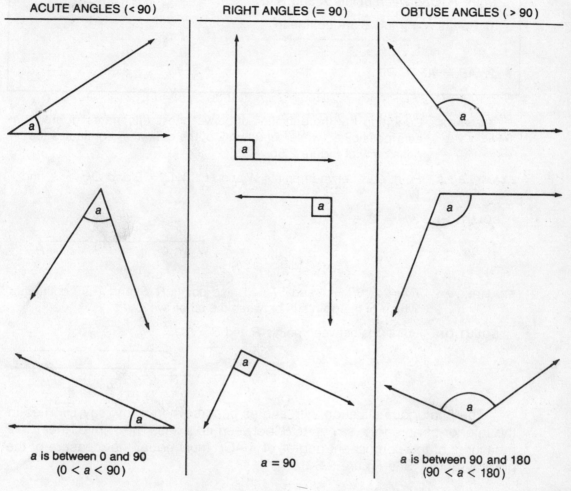

ACUTE ANGLES (< 90)	RIGHT ANGLES (= 90)	OBTUSE ANGLES (> 90)
a is between 0 and 90 (0 < *a* < 90)	*a* = 90	*a* is between 90 and 180 (90 < *a* < 180)

FIGURE 2.3

2.2 BETWEENNESS OF POINTS AND RAYS

Paul is standing on a line for theater tickets; he is standing *between* his friends Allan and Barbara. We represent this situation geometrically in Figure 2.4. We would like to be able to define this notion formally. The phrasing of the definition should eliminate the possibility that Paul may be standing "off" the ticket line or both Allan and Barbara are in front of Paul, or behind him, on the ticket line. (See Figure 2.5.)

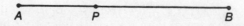

A   P   B

FIGURE 2.4

(a)

(Paul is not on line.)

(b)

(Paul is behind both Allan and Barbara.)

FIGURE 2.5

DEFINITION OF BETWEENNESS

Point *P* is *between* points *A* and *B* if:

1. Points *A*, *P*, and *B* are collinear

and

2. $AB = AP + PB$

REMARK Condition 1 of the definition of betweenness eliminates Figure 2.5*a* as a possibility, while condition 2 of the definition eliminates the possibility of Figure 2.5*b*.

EXAMPLE 2.3 Point *Q* is between points *W* and *H*. If $WQ = 2$ and $QH = 7$, find *WH*.

SOLUTION $WH = 2 + 7 = 9$

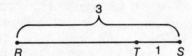

EXAMPLE 2.4 $RT = 2$, $RS = 4$ and $ST = 1$, and points *R*, *S*, and *T* are collinear. Which of the points is between the other two?

SOLUTION Point *T* is between points *R* and *S*.

The analogous situation with angles occurs when a ray, say $\overrightarrow{OP}$, lies in the interior of an angle, say $\angle AOB$, between its sides. The sum of the measures of the component angles of $\angle AOB$ must equal the measure of the original angle. See Figure 2.6.

NOTE: If $m\angle AOP = 40$ and $m\angle POB = 10$, then $m\angle AOB = 50$. This somewhat obvious relationship is given a special name: The Angle Addition Postulate.

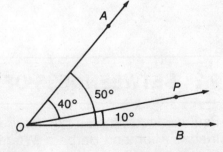

FIGURE 2.6

ANGLE ADDITION POSTULATE

If ray $\overrightarrow{OP}$ is in the interior of angle *AOB*, then

$$m\angle AOB = m\angle AOP + m\angle POB$$

REMARK   The Angle Addition Postulate may be expressed in the equivalent forms:

$$m\angle AOP = m\angle AOB - m\angle POB$$

and

$$m\angle POB = m\angle AOB - m\angle AOP$$

EXAMPLE 2.5 $\overrightarrow{BG}$ lies in the interior of $\angle ABC$. If $m\angle ABG = 25$ and $m\angle CBG = 35$, find $m\angle ABC$.

SOLUTION $m\angle ABC = 25 + 35 = 60$

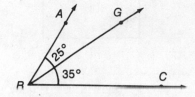

EXAMPLE 2.6 $\overrightarrow{KM}$ lies in the interior of $\angle JKL$. If $m\angle JKM = 20$ and $m\angle LKJ = 50$, find $m\angle MKL$.

SOLUTION $m\angle MKL = 50 - 20 = 30$

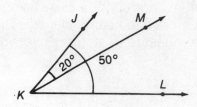

EXAMPLE 2.7   In the accompanying figure the Angle Addition Postulate is contradicted; the measure of the largest angle is *not* equal to the sum of the measures of the two smaller angles. Explain.

SOLUTION $\overrightarrow{BG}$ is not in the interior of angle *ABC*, thus violating the assumption (hypothesis) of the Angle Addition Postulate.

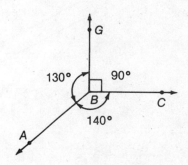

2.3 CONGRUENCE

If the spark plugs of a car are bad or the television set is not working, we usually do not buy a new car or television. We replace the broken parts. How do we know that the replacement parts will exactly fit where the broken part was removed? The new parts will fit because they have been designed to be interchangeable; they have been manufactured to be of exactly the *same size and shape*. Figures which have the same size and shape are said to be *congruent*.

Figures may agree in one or more dimensions, yet *not* be congruent. Diagrams *ABCD* and *JKLM* (Figure 2.7) each have four sides which are identical in length, but the figures are not congruent since their corresponding angles are not identical in measure. Figures are congruent only if they agree in *all* their dimensions.

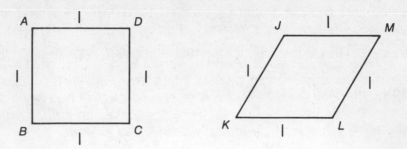

FIGURE 2.7

A line segment has a single dimension—its length. Two segments are congruent, therefore, if they have the same length. If line segments $\overline{AB}$ and $\overline{RS}$ have the same length, then they are congruent. We show that these segments are congruent by using the notation, $\overline{AB} \cong \overline{RS}$. Similarly, if two angles have the same measure, then they are congruent. If angle X has the same measure as angle Y, we write $\angle X \cong \angle Y$. See Figure 2.8.

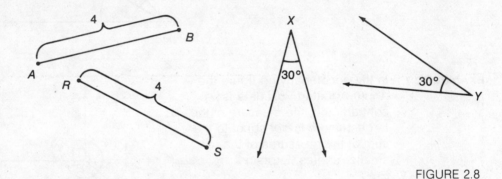

FIGURE 2.8

Congruence is one of the fundamental concepts of geometry. Solid objects that we can hold are congruent if they can be made to exactly coincide. Geometric figures that appear on a printed page cannot be "moved" or "held." The problem of establishing that two plane geometric figures, such as a pair of triangles, are congruent will be a primary concern in future chapters.

DEFINITION OF CONGRUENT SEGMENTS AND ANGLES

Segments (or angles) are said to be congruent if they have the same measure.

REMARK $\overline{AB} \cong \overline{RS}$ is read as "line segment AB is congruent to line segment RS.

2.4 MIDPOINT AND BISECTOR

Consider Figures 2.9 and 2.10. In Figure 2.9, $AM = MB = 3$. Since point M divides $\overline{AB}$ into two congruent segments ($\overline{AM} \cong \overline{MB}$), M is said to be the *midpoint* of $\overline{AB}$. Observe that the measure of each of the congruent segments is one-half the measure of the original segment, $\overline{AB}$. In Figure 2.10, $\overleftrightarrow{XY}$ intersects $\overline{AB}$ at point M, the midpoint of $\overline{AB}$. $\overleftrightarrow{XY}$ is said to *bisect* $\overline{AB}$; a line, ray, or segment that bisects a segment is called a *bisector*. Since an infinite number of lines, rays, or segments may be drawn through the midpoint of a segment, a line segment possesses an infinite number of bisectors. These terms may be formally defined as follows:

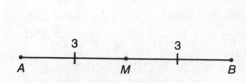

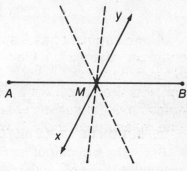

FIGURE 2.9   FIGURE 2.10

DEFINITION OF MIDPOINT
Point M is the *midpoint* of $\overline{AB}$ if

1. M is between A and B

    and

2. $AM = MB$

REMARK   If M is the midpoint of $\overline{AB}$, then M bisects $\overline{AB}$ and the following relationships involving the lengths of the segments thus formed are true:

$$AM = MB \qquad AM = \tfrac{1}{2}AB \qquad MB = \tfrac{1}{2}AB$$

$$\text{or} \qquad\qquad \text{or}$$

$$AB = 2AM \qquad AB = 2MB$$

DEFINITION OF A SEGMENT BISECTOR
A *bisector* of a line segment AB is *any* line, ray, or segment which passes through the midpoint of $\overline{AB}$.

REMARKS   1. We are assuming the segment has a midpoint.

2. It is common to use the expression "a bisector (midpoint) divides a segment into two congruent segments."

EXAMPLE 2.8 $\overleftrightarrow{RS}$ bisects $\overline{EF}$ at point P.

a If $EF = 12$, find PF.

b If $EP = 4$, find EF.

c If $EP = 4x - 3$ and $PF = 2x + 15$, find EF.

SOLUTION **a** $PF = \frac{1}{2}EF = \frac{1}{2}(12) = 6$.

b $EF = 2EP = 2(4) = 8$.

c Since $EP = PF$,

$$4x - 3 = 2x + 15$$
$$4x = 2x + 18$$
$$2x = 18 \text{ and } x = 9$$
$$EP = 4x - 3 = 4(9) - 3 = 36 - 3 = 33$$
$$EF = 2(EP) = 2(33) = 66$$

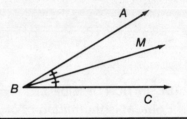

In a similar fashion, we define any ray that lies in the interior of an angle such that it divides the original angle into two congruent angles as the *bisector* of the angle.

DEFINITION OF ANGLE BISECTOR

$\overrightarrow{BM}$ is the bisector of $\angle ABC$ if M lies in the interior of $\angle ABC$ and $\angle ABM \cong \angle CBM$.

REMARKS **1.** An angle has exactly one bisector.

2. The measure of each of the angles formed by the bisector is one-half the measure of the original angles:

$$m\angle ABM = m\angle CBM \qquad m\angle ABM = \frac{1}{2}m\angle ABC$$
$$m\angle CBM = \frac{1}{2}m\angle ABC$$

3. It is common to use the expression "a bisector divides an angle into two congruent angles."

2.5 DIAGRAMS AND DRAWING CONCLUSIONS

Which line segment is longer, $\overline{AB}$ or $\overline{CD}$?

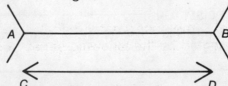

Actually both have the same length, although $\overline{AB}$ may give the illusion of being greater in length than $\overline{CD}$. When given a geometric diagram we must exercise extreme caution in drawing conclusions based on the diagram—pictures can be deceiving! In general, we may only assume collinearity and betweenness of points. We may not make any assumptions regarding the measures of segments or angles unless they are given to us. In Figure 2.11a although segments $\overline{AD}$ and $\overline{DC}$ appear to be equal in length, we may not conclude that $AD = DC$. The only assumption that we are entitled to make is that point D lies *between* points A and C. If we wished to start out with the fact that $AD = DC$, then we indicate this by writing the *given* information next to the figure. (See Figure 2.11b.)

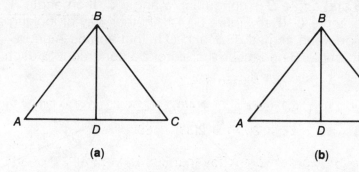

(a) **(b)**

FIGURE 2.11

We may then mark off the diagram to indicate the equal segments by drawing a single vertical bar through each segment. (See Figure 2.12.) An angle may only be assumed to be a right angle from the diagram if the angle contains the "corner" marking (⌐).

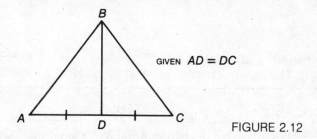

GIVEN $AD = DC$

FIGURE 2.12

2.6 PROPERTIES OF EQUALITY AND CONGRUENCE

John is taller than Kevin and Kevin is taller than Louis. How do the heights of John and Louis compare? We can analyze the situation with the aid of a simple diagram. (See Figure 2.13.) This leads us to conclude that John must be taller than Louis.

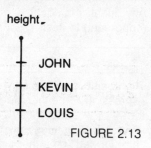

height

JOHN

KEVIN

LOUIS

FIGURE 2.13

Using the mathematical symbol for greater than, >, the height relationships can be represented by the following series of inequality statements:

If $J > K$

and $K > L$

then $J > L$

Without directly comparing John with Louis, we have used a *transitive property* to conclude that John's height is greater than Louis' height. The greater than relation is an example of a relation that possesses the transitive property. Is friendship a transitive relation? If Alice is Barbara's friend and Barbara is Carol's friend, does that mean that Alice and Carol must also be friends? Obviously, no. Some relations possess the transitive property, while others do not. The equality (=) and congruence (≅) relations possess the transitive property. For example, if angle *A* is congruent to angle *B* and angle *B* is congruent to angle *C*, then angle *A* must be congruent to angle *C*. (See Figure 2.14.) Another way of looking at this interrelationship between angles *A, B,* and *C* is that angles *A* and *C* are each congruent to angle *B* and must, therefore, be congruent to each other.

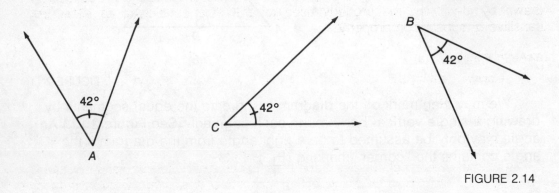

FIGURE 2.14

The equality and congruence relations also enjoy some additional properties. These are displayed and illustrated in Table 2.1.

TABLE 2.1

PROPERTY	EQUALITY EXAMPLE	CONGRUENCE EXAMPLE
REFLEXIVE		
The identical expression may be written on either side of the = or ≅ symbol. Any quantity is equal (congruent) to itself.	1. 9 = 9. 2. *AB* = *AB*.	∡*ABC* ≅ ∡*ABC*.
SYMMETRIC		
The positions of the expressions on either side of the = or ≅ symbol may be reversed. Quantities may be "flip-flopped" on either side of an = or ≅ sign.	1. If 4 = *x*, then *x* = 4. 2. If *AB* = *CD* then *CD* = *AB*.	If ∡*ABC* ≅ ∡*XYZ*, then ∡*XYZ* ≅ ∡*ABC*.
TRANSITIVE		
If two quantities are equal (congruent) to the same quantity, then they are equal (congruent) to each other.	If *AB* = *CD* and *CD* = *PQ*, then *AB* = *PQ*.	If ∡*X* ≅ ∡*Y* and ∡*Y* ≅ ∡*Z*, then ∡*X* ≅ ∡*Z*.

Another useful property of the equality relation is the *substitution property* of equality. If $AB = 2 + 3$, then an equivalent number may be substituted in place of the numerical expression on the right side of the equation. We may substitute 5 for $2 + 3$, and write $AB = 5$. The following diagram gives a geometric illustration of this often used property.

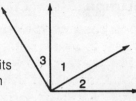

GIVEN $m\angle 1 + m\angle 2 = 90,$

$m\angle 2 = m\angle 3.$

CONCLUSION $m\angle 1 + m\angle 3 = 90.$

REASON Substitution property. The $m\angle 3$ replaces its equal ($m\angle 2$) in the first equation stated in the given.

In each of the following examples, the purpose is to justify the conclusion drawn by identifying the property used to draw the conclusion as either the transitive or substitution property.

EXAMPLE 2.9

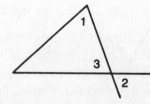

GIVEN $\angle 1 \cong \angle 2,$

$\angle 2 \cong \angle 3.$

CONCLUSION $\angle 1 \cong \angle 3.$

REASON ?

SOLUTION Since both angles 1 and 3 are congruent to the same angle, angle 2, they must be congruent to each other. This is the *transitive property of congruence*. Since we may only substitute equals in equations, we do *not* have a substitution property *of congruence*.

EXAMPLE 2.10

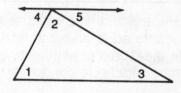

GIVEN $m\angle 1 = m\angle 4,$

$m\angle 3 = m\angle 5,$

$m\angle 4 + m\angle 2 + m\angle 5 = 180.$

CONCLUSION $m\angle 1 + m\angle 2 + m\angle 3 = 180.$

REASON ?

SOLUTION Substitution property. In the last equation stated in the given, the measures of angles 4 and 5 are replaced by their equals, the measures of angles 1 and 3, respectively.

EXAMPLE 2.11

GIVEN $RS = SM.$ (1)

$TW = SM.$ (2)

CONCLUSION $RS = TW.$

REASON ?

SOLUTION Since RS and TW are both equal to the same quantity, SM, they must be equal to each other. This is the transitive property.

or

In Equation (1), SM may be replaced by its equal, TW. We are using the information in Equation (2) to make a substitution in Equation (1). Hence, the conclusion can be also justified by using the substitution property.

EXAMPLE 2.12

GIVEN C is the midpoint of $\overline{AD}$,
            $AC = CE$.

CONCLUSION $CD = CE$.

REASON   ?

SOLUTION $AC = CD$, since point C is the midpoint of $\overline{AD}$.

We now have the set of relationships:

$$AC = CD \qquad (1)$$
$$AC = CE \qquad (2)$$

Since CD and CE are both equal to the same quantity (AC) they must be equal to each other. Hence, $CD = CE$ by the *transitive property of equality.*

or

We may replace AC by CE in Equation (1), also reaching the desired conclusion.

It should be clear from these examples that the transitive and substitution properties of equality, in certain situations, may be used interchangeably (see Examples 2.11 and 2.12 above). In these examples there are *two equations* that state that two quantities are each equal to the same quantity, thus leading to either the substitution or transitive properties of *equality.*

2.7   ADDITIONAL PROPERTIES OF EQUALITY

There are several properties of equality that have been previously encountered in elementary algebra which will prove useful when working with measures of segments and angles. Table 2.2 reviews these properties in their algebraic context.

TABLE 2.2

PROPERTY	ALGEBRAIC EXAMPLE	FORMAL STATEMENT
ADDITION (+)		
The same (or =) quantities may be added to both sides of an equation.	Solve for x: $x - 3 = 12$ $x - 3 + 3 = 12 + 3$ (same) $x = 15$	If equals are added to equals, their sums are equal. or If $a = b$, then $a + c = b + c$.
SUBTRACTION (−)		
The same (or =) quantities may be subtracted from both sides of an equation.	Solve for n: $n + 5 = 11$ $n + 5 - 5 = 11 - 5$ (same) $n = 6$	If equals are subtracted from equals, their differences are equal. or If $a = b$, then $a - c = b - c$.

TABLE 2.2 (*Continued*)

MULTIPLICATION (×)		
The same quantity may be used to multiply both sides of an equation.	Solve for y: $$\frac{y}{3} = 7$$ same $$3\left(\frac{y}{3}\right) = 3(7)$$ $$y = 21$$	If equals are multiplied by equals, their products are equal. or If $a = b$, then $ac = bc$.

DIVISION (÷)		
The same nonzero quantity may be used to divide both sides of an equation.	Solve for k: $$3k = 12$$ $$\frac{3k}{3} = \frac{12}{3}$$ same $$k = 4$$	If equals are divided by nonzero equals, their quotients are equal. or If $a = b$, then $\dfrac{a}{c} = \dfrac{b}{c}$, provided $c \neq 0$.

These equality properties may be summarized as follows: "Whatever you do to one side of an *equation*, be sure to do the same to the other side of the equation." The addition, subtraction, and multiplication/division properties may also be applied to geometric situations. The following examples illustrates how we may use these properties of equality in drawing conclusions about the measures of segments and angles.

USING THE ADDITION PROPERTY

a
GIVEN

$$\begin{aligned} AB &= AC \\ +BD &= +CE \end{aligned}$$

CONCLUSION

$$\underbrace{AB + BD}_{AD} = \underbrace{AC + CE}_{AE}$$

$$AD = AE$$
$$(7 = 7)$$

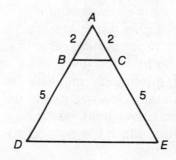

b
GIVEN

$$\begin{aligned} m\angle JXK &= m\angle MXL \\ +m\angle KXL &= +m\angle KXL \end{aligned}$$

CONCLUSION

$$m\angle JXL = m\angle KXM$$
$$(90 = 90)$$

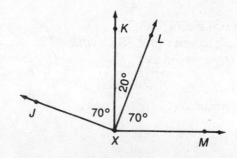

USING THE SUBTRACTION PROPERTY

a   GIVEN

$$m\angle BAD = m\angle DCB$$
$$- m\angle PAD = - m\angle BCQ$$

CONCLUSION
$$\overline{m\angle BAP = m\angle DCQ}$$
$$(40 = 40)$$

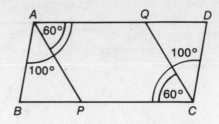

b   GIVEN

$$VI = NE$$
$$- EI = - EI$$

CONCLUSION
$$\overline{VI - EI = NE - EI}$$
$$\underbrace{}_{} \qquad \underbrace{}_{}$$
$$VE = NI$$
$$(3 = 3)$$

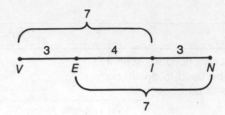

USING THE MULTIPLICATION/DIVISION PROPERTIES

GIVEN $AB = CB$,

$AR = \frac{1}{2}AB$,

$CT = \frac{1}{2}CB$.

CONCLUSION $AR = CT$. Why?

REASONING   Since we are multiplying equals
($AB = CB$) by the same number
($\frac{1}{2}$), their products must be equal:

$$\frac{1}{2}AB = \frac{1}{2}CB$$

By substitution,

$$AR = CT$$

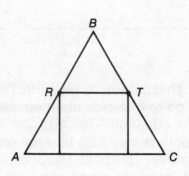

This chain of reasoning in which the multiplying factor is $\frac{1}{2}$ is used so often that we give it a special name, "*halves of equals are equal.*"

In order to help organize and fill in all the necessary logical steps in moving from the given to an appropriate conclusion, a two-column format is used; the first column lists the steps, while the second column justifies each step. Only the given, definitions, postulates, properties of equality and congruence, and theorems
may appear in the second column. See Table 2.3. Example 2.13 illustrates this format.

EXAMPLE 2.13

GIVEN $m\angle RST = m\angle WTS$,

$\overline{PS}$ bisects $\angle RST$,

$\overline{PT}$ bisects $\angle WTS$.

CONCLUSION $m\angle 1 = m\angle 2$.

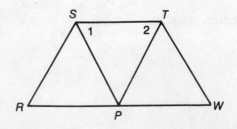

PROOF	Statements	Reasons
	1. $m\angle RST = m\angle WTS$.	1. Given.
	2. $\overline{PS}$ bisects $\angle RST$, $\overline{PT}$ bisects $\angle WTS$.	2. Given.
		3. Definition of angle bisector.
	3. $m\angle 1 = \frac{1}{2}m\angle RST$ $m\angle 2 = \frac{1}{2}m\angle WTS$	4. Halves of equals are equal.
	4. $m\angle 1 = m\angle 2$.	

TABLE 2.3

Pattern of . . .	Types of . . .
PROOF **Statements**	**Reasons**
• Start with the given information which is supplied in the original problem statement *or* with a conclusion that can be made based on the diagram (for example, betweenness of points). Label this statement number 1, and continue to number statements sequentially. The corresponding reason for each statement is assigned the same number.	• Given (Information is told to us in the statement of the problem.) • Definitions • Postulates • Previously proved theorems • Algebraic properties
• Develop a chain of reasoning, writing each logical step as a separate statement and writing its justification across the page in the adjacent "reasons" column.	
• Continue until you are able to write a statement which corresponds to the expression or equation that appears in the conclusion (or to prove) of the original problem statement.	

REVIEW EXERCISES FOR CHAPTER 2

1. In the accompanying diagram, classify each of the following angles as acute, right, obtuse, or straight:

   a ∡ *TOM*

   b ∡ *LOM*

   c ∡ *SOM*

   d ∡ *LOR*

   e ∡ *ROT*

   f ∡ *LOT*

   g ∡ *ROS*

   h ∡ *MOR*

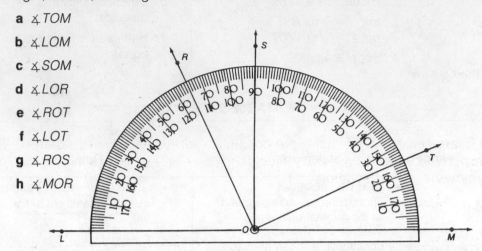

2. Point *P* is between points *H* and *G*. If *HP* = 3 and *PG* = 5, find *HG*.

3. If points *M*, *I*, and *Z* are collinear and *IZ* = 8, *MI* = 11, and *MZ* = 3, which point is between the other two?

4. $\overrightarrow{PL}$ lies in the interior of angle *RPH*. The *m*∡*RPL* = *x* − 5 and *m*∡*LPH* = 2*x* + 18. If *m*∡*HPR* = 58, find the measure of the smallest angle formed.

5. $\overline{XY}$ bisects $\overline{RS}$ at point *M*. If *RM* = 6, find the length of $\overline{RS}$.

6. $\overrightarrow{PQ}$ bisects ∡*HPJ*. If *m*∡*HPJ* = 84, find *m*∡*QPJ*.

7. $\overrightarrow{BP}$ bisects ∡*ABC*. If *m*∡*ABP* = 4*x* + 5 and *m*∡*CBP* = 3*x* + 15, classify angle *ABC* as acute, right, or obtuse.

8. If *R* is the midpoint of $\overline{XY}$ and *XR* = 3*a* + 1 and *YR* = 16 − 2*a*, find the length of $\overline{XY}$.

9. In the accompanying diagram, pairs of angles and segments are indicated as congruent. Use the letters in the diagram to write the appropriate congruence relation.

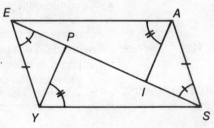

In Exercises 10 to 12, use the diagram and any information given to mark off the diagram with the given and draw the appropriate conclusion.

10.   **GIVEN** $\overline{BF}$ bisects $\overline{AC}$.

    CONCLUSION   ?

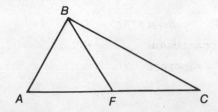

11. **GIVEN** $\overline{PT}$ bisects $\angle STO$.

   CONCLUSION ?

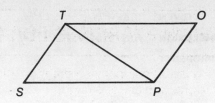

12. **GIVEN** $\overline{AC}$ bisects $\overline{BD}$,

   $\overline{BD}$ bisects $\angle ADC$.

   CONCLUSION ?

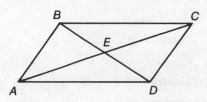

In Exercises 13 to 16, justify the conclusion drawn by identifying the property used to draw the conclusion as either reflexive, transitive, symmetric, or substitution.

13. **GIVEN** $\overline{LM} \cong \overline{GH}$,

   $\overline{GH} \cong \overline{FV}$.

   CONCLUSION $\overline{LM} \cong \overline{FV}$.

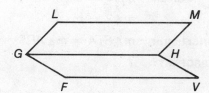

14. **GIVEN** Figure (quadrilateral) $ABCD$.

   CONCLUSION $\overline{AC} \cong \overline{AC}$.

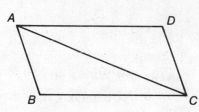

15. **GIVEN** $\overline{TW}$ bisects $\angle STV$,

   $\angle 1 \cong \angle 3$.

   CONCLUSION $\angle 2 \cong \angle 3$.

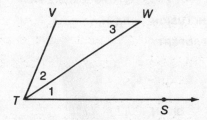

16. **GIVEN** $m\angle 1 + m\angle 2 = 90$,

   $m\angle 1 = m\angle 3$.

   CONCLUSION $m\angle 3 + m\angle 2 = 90$.

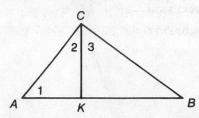

For Exercises 17 to 24, mark off the diagrams so that the corresponding pairs of equal or congruent parts are indicated. Then state the property of equality that could be used to justify each conclusion.

17. **GIVEN** $AC = BT$.

   CONCLUSION $AT = BC$.

   PROPERTY ?

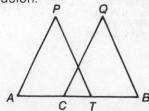

18.   **GIVEN** $m\angle KPN = m\angle LPM.$

      CONCLUSION $m\angle KPM = m\angle LPN.$

      PROPERTY   ?

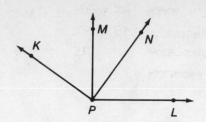

19.   **GIVEN** $AE = BE,$

                 $CE = DE.$

      CONCLUSION $AC = BD.$

      PROPERTY   ?

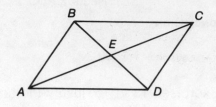

20.   **GIVEN** $m\angle 1 = m\angle 3,$

                 $m\angle 2 = m\angle 4.$

      CONCLUSION $m\angle STA = m\angle ARS.$

      PROPERTY   ?

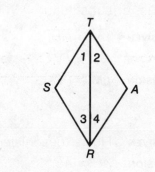

21.   **GIVEN** $m\angle WXY = m\angle ZYX,$

                 $\overline{HX}$ bisects $\angle WXY,$

                 $\overline{HY}$ bisects $\angle ZYX.$

      CONCLUSION $m\angle 1 = m\angle 2.$

      PROPERTY   ?

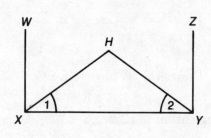

22.   **GIVEN** $JQ = LP.$

      CONCLUSION $JP = LQ.$

      PROPERTY   ?

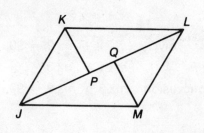

23.   **GIVEN** $m\angle SBO = m\angle TBH.$

      CONCLUSION $m\angle SBH = m\angle TBO.$

      PROPERTY   ?

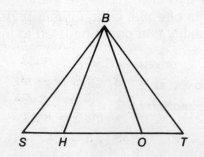

24.   **GIVEN** $AX = YS$,

$\qquad\qquad\quad XB = RY$.

CONCLUSION $AB = RS$.

PROPERTY   ?

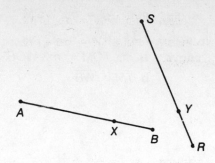

25.   **GIVEN** $\overline{PJ} \cong \overline{LR}$.

CONCLUSION $\overline{PR} \cong \overline{LJ}$.

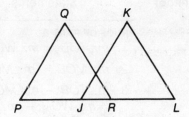

PROOF	Statements	Reasons
	1. $\overline{PJ} \cong \overline{LR}$	1. Given
	2. $PJ = LR$.	2. If two segments are congruent, then they are equal in length.
	3. $JR = JR$.	3. ?
	4. $PJ + JR = LR + JR$.	4. ?
	5. $PR = LJ$.	5. Substitution property of equality.
	6. $\overline{PR} \cong \overline{LJ}$.	6. If two segments are equal in measure, then they are congruent.

26.   **GIVEN** $\angle RLM \cong \angle ALM$,

$\qquad\qquad\quad \angle 1 \cong \angle 2$.

CONCLUSION $\angle 3 \cong \angle 4$.

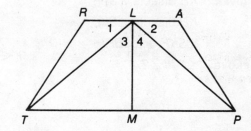

PROOF	Statements	Reasons
	1. $\angle RLM \cong \angle ALM$, $\angle 1 \cong \angle 2$.	1. ?
	2. $m\angle RLM = m\angle ALM$, $m\angle 1 = m\angle 2$.	2. ?
	3. $m\angle RLM - m\angle 1 = m\angle ALM - m\angle 2$.	3. ?
	4. $m\angle 3 = m\angle 4$.	4. ?
	5. $\angle 3 \cong \angle 4$.	5. ?

27. **GIVEN** $m\angle TOB = m\angle WOM,$
$\overline{TB} \cong \overline{WM}.$

PROVE **a** $m\angle TOM = m\angle WOB.$
b $\overline{TM} \cong \overline{WB}.$

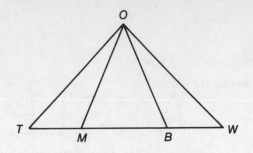

PROOF	Statements	Reasons
	PROOF OF PART a:	
	1. $m\angle TOB = m\angle WOM.$	1. ?
	2. $m\angle MOB = m\angle MOB.$	2. ?
	*3. $m\angle TOB - m\angle MOB$ $= m\angle WOM - m\angle MOB.$	3. ?
	*4. $m\angle TOM = m\angle WOB.$	4. Angle addition postulate (and substitution property of equality).
	PROOF OF PART b:	
	5. $\overline{TB} \cong \overline{WM}.$	5. ?
	6. $TB = WM.$	6. ?
	7. $MB = MB.$	7. ?
	†8. $TB - MB = WM - MB.$	8. ?
	†9. $TM = WB.$	9. Definition of betweenness (and substitution property of equality).
	10. $\overline{TM} \cong \overline{WB}.$	10. ?

* In later work, these steps are sometimes consolidated.
† In later work, these steps are sometimes consolidated.

28. **GIVEN** $\overline{KB}$ bisects $\angle SBF,$
$\overline{KB}$ bisects $\angle SKF,$
$\angle SKF \cong \angle SBF.$

PROVE $\angle 1 \cong \angle 2.$

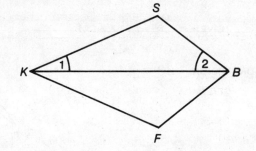

PROOF	Statements	Reasons
	1. $\angle SKF \cong \angle SBF.$	1. Given.
	2. $m\angle SKF = m\angle SBF.$	2. ?
	3. $\overline{KB}$ bisects $\angle SBF.$ $\overline{KB}$ bisects $\angle SKF.$	3. ?
	4. $m\angle 1 = \frac{1}{2}m\angle SKF.$ $m\angle 2 = \frac{1}{2}m\angle SBF.$	4. ?
	5. $m\angle 1 = m\angle 2.$	5. ?
	6. $\angle 1 \cong \angle 2.$	6. ?

CHAPTER THREE
ANGLE PAIRS AND PERPENDICULAR LINES

3.1 SUPPLEMENTARY AND COMPLEMENTARY ANGLE PAIRS

Important geometric conclusions may often be drawn based on whether a special relationship exists between a pair of angles. Supplementary and complementary angle pairs are of particular importance in the study of geometry.

> **DEFINITION OF SUPPLEMENTARY ANGLES**
> Two angles are *supplementary* if the sum of their measures is 180. If angle *A* is supplementary to angle *B*, then $m\angle A + m\angle B = 180$, and each angle is called the *supplement* of the other angle.

> **DEFINITION OF COMPLEMENTARY ANGLES**
> Two angles are *complementary* if the sum of their measures is 90. If angle *A* is complementary to angle *B*, then $m\angle A + m\angle B = 90$, and each angle is called the *complement* of the other angle.

EXAMPLE 3.1

GIVEN

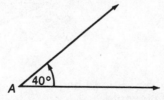

a Determine the measure of the supplement of angle *A*.

b Determine the measure of the complement of angle *A*.

SOLUTION **a** The supplement of angle *A* has measure 140.

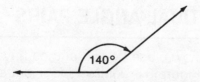

b The complement of angle *A* has measure 50.

EXAMPLE 3.2   In triangle ABC, angle A is complementary to angle B. Find the measure of angles A and B.

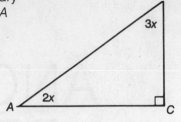

SOLUTION
$$2x + 3x = 90$$
$$5x = 90$$
$$x = 18$$
$$m\angle A = 2x = 2(18) = 36$$
$$m\angle B = 3x = 3(18) = 54$$

EXAMPLE 3.3   The measure of an angle and its supplement are in the ratio of $1:8$. Find the measure of the angle.

SOLUTION   *Method 1*:   Let x = measure of angle

then $180 - x$ = measure of the supplement of the angle

$$\frac{x}{180 - x} = \frac{1}{8}$$
$$180 - x = 8x$$
$$9x = 180$$
$$x = 20$$

Method 2:   Let x = measure of angle

then $8x$ = measure of the supplement of the angle

$$x + 8x = 180$$
$$9x = 180$$
$$x = 20$$

Therefore measure of angle = 20.

EXAMPLE 3.4   Determine the measure of an angle if it exceeds twice the measure of its complement by 30.

SOLUTION   Let x = measure of angle

then $90 - x$ = measure of the complement of the angle

$$x = 2(90 - x) + 30$$
$$x = 180 - 2x + 30$$
$$x = 210 - 2x$$
$$3x = 210$$
$$x = 70$$

Therefore, measure of angle = 70.

3.2   ADJACENT AND VERTICAL ANGLE PAIRS

Adjacent means lying next to. But how close do two angles have to be in order to be considered adjacent? Figure 3.1 contrasts four pairs of angles; only the first pair of angles is considered to be adjacent.

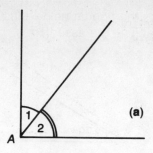
(a)

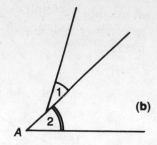
(b)

(a) Angles are adjacent.

(b) Angles do not share the same vertex.

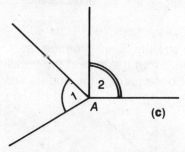
(c)

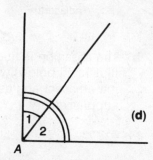
(d)

(c) Angles do not share a common side.

(d) Angles "overlap" — they have interior points in common.

FIGURE 3.1 Adjacent versus nonadjacent angle pairs.

Figure 3.1 suggests we make the following definition:

DEFINITION OF ADJACENT ANGLE PAIRS

Two angles are *adjacent* if they:

1. Have the same vertex;
2. Share a common side; and
3. Have no interior points in common.

REMARK In an angle pair the two sides which are not common to each of the angles are sometimes referred to as the *exterior* sides of the angles.

EXAMPLE 3.5 A beginning student of geometry wonders whether the following two assertions are true:

a If a pair of angles are supplementary, then they must be adjacent.

b If the exterior sides of a pair of adjacent angles form a straight line, then the angles are supplementary.

Comment on whether you think the statements are true or false. If you suspect that one or both are false, produce a diagram to help support your belief.

SOLUTION   **a**   A pair of supplementary angles, as the figures below illustrate, may be either adjacent or nonadjacent. The assertion is, therefore, false.

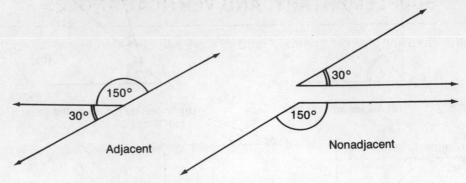

Adjacent   Nonadjacent

b   The assertion is true since a straight line is formed (as illustrated below) which implies that the sum of the measures of the adjacent angles is 180. Consequently, the angles are supplementary.

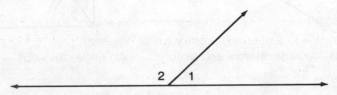

The assertion presented in Example 3.5*b* may be formally stated as a theorem.

THEOREM 3.1

If the exterior sides of a pair of adjacent angles form a straight line, then the angles are supplementary.

If two intersecting lines are drawn, four angles are formed, as shown below. Angles 1 and 3 are a vertical angle pair; angles 1 and 4 are *not* vertical angles since angles 1 and 4 are adjacent. Notice that vertical angles are "opposite" one another. Angles 2 and 4 are also a vertical angle pair.

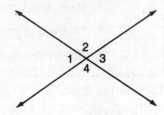

DEFINITION OF VERTICAL ANGLE PAIRS

Vertical angles are a pair of nonadjacent angles formed by two intersecting lines.

EXAMPLE 3.6   Name all pairs of vertical angles in the accompanying diagram.

SOLUTION   Angle pairs: 1 and 4; 2 and 5; 3 and 6.

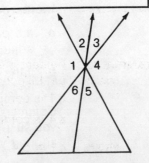

3.3 THEOREMS RELATING TO COMPLEMENTARY, SUPPLEMENTARY, AND VERTICAL ANGLES

In the diagram below, angles *A* and *B* are each complementary to angle *C*.

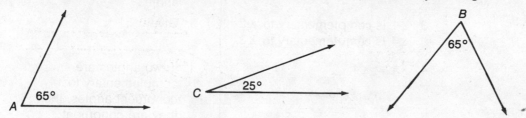

We may conclude that angle *A* must be equal in measure (or congruent to) angle *B*. Using the same reasoning, if angles *A* and *B* were each *supplementary* to angle *C*, they would necessarily be congruent to each other.

> **THEOREM 3.2**
>
> If two angles are complementary (or supplementary) to the same angle, then they are congruent.

An extension of Theorem 3.2 covers the situation where the original angles, say angle *A* and angle *B*, are each complementary (or supplementary) to a different angle, but the second pair of angles are congruent to each other. For example, suppose angle *A* is complementary to angle *C* and angle *B* is complementary to angle *D*; furthermore, angles *C* and *D* are congruent:

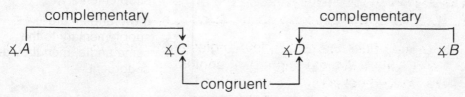

If $m\angle C$ = 20, then $m\angle A$ = 70. Since angles *C* and *D* are congruent, $m\angle D$ = 20 and $m\angle B$ = 70. Hence, angles *A* and *B* are congruent. If the original relationship specified that the angles were *supplementary* to a pair of congruent angles, the identical conclusion would result. This is summarized in Theorem 3.3.

> **THEOREM 3.3**
>
> If two angles are complementary (or supplementary) to congruent angles, then they are congruent.

EXAMPLE 3.7   Present a formal two-column proof.

GIVEN $\overline{LM}$ bisects $\angle$ *KMJ*,

         $\angle$1 is complementary to $\angle$2,

         $\angle$4 is complementary to $\angle$3.

PROVE $\angle$1 $\cong$ $\angle$4.

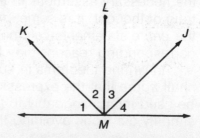

Statements	Reasons
1. $\overline{LM}$ bisects ∡KMJ.	1. Given
2. ∡2 ≅ ∡3.	2. A bisector divides an angle into two congruent angles.
3. ∡1 is complementary to ∡2. ∡4 is complementary to ∡3.	3. Given
4. ∡1 ≅ ∡4.	4. If two angles are complementary to congruent angles, then they are congruent.

EXAMPLE 3.8 Present a formal two-column proof.
GIVEN Lines *l* and *m* intersect at point *P*.
PROVE ∡1 ≅ ∡3.

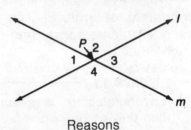

SOLUTION
PROOF

Statements	Reasons
1. Lines *l* and *m* intersect at point *P*.	1. Given
2. ∡1 is supplementary to ∡2, ∡3 is supplementary to ∡2.	2. If the exterior sides of a pair of adjacent angles form a straight line, then the angles are supplementary.
3. ∡1 ≅ ∡3.	3. If two angles are supplementary to the same angle, then they are congruent.

Notice that in the formal proofs presented in Examples 3.7 and 3.8, assertions are placed in their logical sequence and numbered accordingly in the "statements" column. The reason used to support each assertion receives a corresponding number and is written in the reasons column. In addition to the given, the "reasons" column may include only previously stated definitions, postulates, and theorems. Keep in mind that once a theorem is proved, it may then be included in the repertoire of statements which may be used in the reasons column of subsequent proofs.

In approaching a proof, the beginning student is urged to focus on developing the logical flow of the proof by planning and then organizing the necessary assertions in the statements column. In order to maintain a train of thought, it is sometimes helpful to first concentrate on completing the *entire* statements column; afterwards, in order to complete the proof, the corresponding reasons may be entered.

In writing theorems in support of statements in the reasons column we shall agree that the expressions "congruent" and "equal in measure" can be used interchangeably in the phrasing of a theorem. This will sometimes simplify our work, avoiding the need to change to measure and then back again to congruence. For example, Theorem 3.2 may, if convenient, be used in the following form:

> If two angles are complementary (or supplementary) to the same angle, then they are *equal in measure*.

Let us return to Example 3.8. Observe that angles 1 and 3 are vertical angles and congruent. Using the same approach we could establish that vertical angles 2 and 4 are congruent. This leads to Theorem 3.4.

THEOREM 3.4

Vertical angles are congruent.

EXAMPLE 3.9   **a** Find the value of x.

b Find the measures of angles AEC, DEB, DEA, and BEC.

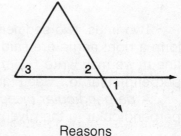

SOLUTION   **a** By Theorem 3.4:

$$3x - 18 = 2x + 5$$
$$3x = 2x + 23$$
$$x = 23$$

b $m\angle AEC = m\angle DEB = 3x - 18$
$$= 3(23) - 18$$
$$= 51$$

Since angles AEC and DEA are supplementary,

$$m\angle DEA = 180 - 51 = 129$$
$$m\angle DEA = m\angle BEC = 129$$

EXAMPLE 3.10   Present a formal two-column proof.

GIVEN $\angle 2 \cong \angle 3$.

PROVE $\angle 1 \cong \angle 3$.

SOLUTION
PROOF

Statements	Reasons
1. $\angle 1 \cong \angle 2$.	1. Vertical angles are congruent.
2. $\angle 2 \cong \angle 3$.	2. Given.
3. $\angle 1 \cong \angle 3$.	3. Transitive property of congruence.

SUMMARY

Two angles are congruent if they are

- Vertical angles formed by two intersecting lines.

- Complements of the same or of congruent angles.

- Supplements of the same or of congruent angles.

3.4 DEFINITIONS AND THEOREMS RELATING TO RIGHT ANGLES AND PERPENDICULARS

Recall that a *right angle* is an angle of measure 90. The following theorems are useful in proving other theorems:

> **THEOREM 3.5**
> All right angles are congruent.

* INFORMAL PROOF   Let angles 1 and 2 be right angles. We must show that $\angle 1 \cong \angle 2$.

Since angles 1 and 2 are right angles, $m\angle 1 = 90$ and $m\angle 2 = 90$. Therefore

$$m\angle 1 = m\angle 2 \xrightarrow[\text{implies}]{} \angle 1 \cong \angle 2$$

> **THEOREM 3.6**
> If two angles are congruent and supplementary, then each is a right angle.

INFORMAL PROOF   Let angles 1 and 2 be congruent and supplementary. We must show that angles 1 and 2 are right angles.

Since they are supplementary, $m\angle 1 + m\angle 2 = 180$

By substitution, $\qquad\qquad\qquad m\angle 1 + m\angle 1 = 2m\angle 1 = 180$

Therefore $\qquad\qquad\qquad\qquad\qquad m\angle 1 = 90 = m\angle 2$

Hence, angles 1 and 2 are right angles.

Two lines, two segments, or a line and a segment which intersect to form a right angle are said to be *perpendicular*. If line *l* is perpendicular to line *m* we may write $l \perp m$ where the symbol $\perp$ is read as "is perpendicular to." See Figure 3.2.

A *perpendicular bisector* of a line segment is a line or segment that is perpendicular to the given segment at its midpoint. See Figure 3.3.

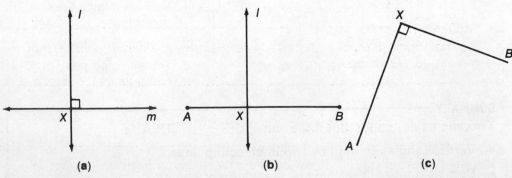

(a)   **(b)**   **(c)**

FIGURE 3.2   Perpendiculars.

* Sometimes it will be preferable to give an outline description of the proof. Section 3.5 briefly discusses alternative types of proofs.

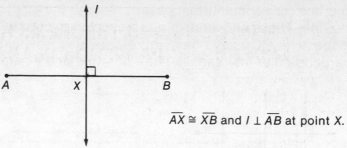

$\overline{AX} \cong \overline{XB}$ and $l \perp \overline{AB}$ at point X.

FIGURE 3.3   A perpendicular bisector.

DEFINITION OF PERPENDICULAR LINES

Perpendicular lines are a pair of lines which intersect to form right angles. If a line is perpendicular to a segment and intersects the segment at its midpoint, then the line is called the *perpendicular bisector* of the segment.

EXAMPLE 3.11

GIVEN $\overline{AC} \perp \overline{BC}$, angles 1 and 2 are adjacent.

PROVE $m\angle 1 + m\angle 2 = 90$.

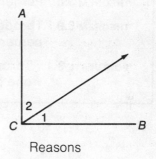

SOLUTION

PROOF

Statements	Reasons
1. $\overline{AC} \perp \overline{BC}$.	1. Given
2. $\angle ACB$ is a right angle.	2. Perpendicular segments intersect to form a right angle.
3. $m\angle ACB = 90$.	3. A right angle has measure 90.
4. $m\angle ACB = m\angle 1 + m\angle 2$.	4. Angle Addition Postulate.
5. $m\angle 1 + m\angle 2 = 90$.	5. Substitution.

Since the sum of the measures of angles 1 and 2 is 90, it follows that angles 1 and 2 are complementary. This example represents the proof of the following theorem:

THEOREM 3.7

If the exterior sides of a pair of adjacent angles are perpendicular, then the angles are complementary.

EXISTENCE OF PERPENDICULARS

How many perpendiculars can be drawn to a given line? A line has an infinite number of perpendiculars; however, it can be proven that through a particular point *on* the line, there exists exactly one perpendicular to the given line. At a point *not* on a line, we postulate that there exists exactly one line (or segment) that passes through the point and is perpendicular to the line. These situations are represented in Figure 3.4*a*, *b*, and *c*, respectively. Observe that Figure 3.4*c* also illustrates that when a perpendicular intersects a line, *four* right angles are formed.

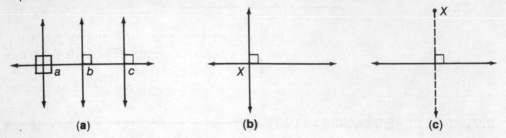

(a)   **(b)**   **(c)**

FIGURE 3.4   Existence of perpendiculars.

We may formally state these results.

FACTS ABOUT PERPENDICULARS

THEOREM 3.8   Perpendicular lines intersect to form four right angles.

THEOREM 3.9   Through a given point on a line, there exists exactly one perpendicular to the given line.

POSTULATE 3.1   Through a given point *not* on a line, there exists exactly one perpendicular to the given line.

DISTANCE

The term *distance* in geometry is always interpreted as the *shortest* path between two points. In Figure 3.5 the distance between points *P* and *Q* is the length of segment $\overline{PQ}$. A point that is exactly the same distance from two other points is said to be *equidistant* from the two points. The midpoint of a segment, for example, is equidistant from the endpoints of the segment. See Figure 3.6.

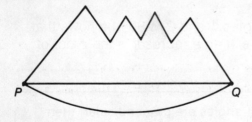

FIGURE 3.5   Representing the distance between two points. The shortest path is represented by the segment joining the points rather than any zigzag or circular type of route.

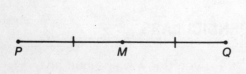

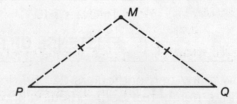

FIGURE 3.6   In each figure, point *M* is equidistant from points *P* and *Q*.

The shortest distance from a point not on a line to the line is measured by the length of the segment drawn from the point perpendicular to the line.

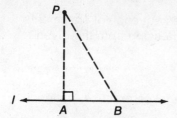

The distance from point *P* to line *l* is the perpendicular segment $\overline{PA}$ rather than any other segment such as $\overline{PB}$.

In summary, distance is defined according to whether it is being determined between two points or between a point and a line.

DEFINITIONS OF DISTANCE

1. The distance *between two points* is the length of the segment joining the points.

2. The distance *between a line and a point not on the line* is the length of the perpendicular segment drawn from the point to the line.

METHODS FOR PROVING LINES ARE PERPENDICULAR

By using the definition of perpendicularity, we may conclude that two lines are perpendicular if they intersect to form a right angle. It will sometimes be convenient in subsequent work with perpendicular lines to use an alternative method. In Figure 3.7 it seems intuitively clear that if we continue to rotate line *l* in a clockwise fashion, then eventually lines *l* and *m* will be perpendicular, and this will be true when angles 1 and 2 are congruent.

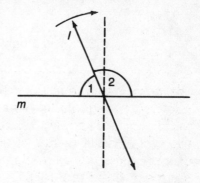

FIGURE 3.7

This leads to the following theorem:

THEOREM 3.10

If two lines intersect to form congruent adjacent angles, then the lines are perpendicular.

EXAMPLE 3.12 Prove Theorem 3.10.

SOLUTION In approaching the proof of a theorem that is stated verbally, we must

1. Identify the given from the information contained in the "if" clause of the statement of the theorem

2. Identify what is to be proved from the relationship proposed in the "then" clause of the theorem

3. Use the information obtained in steps 1 and 2 to draw an appropriate diagram

4. Organize our thoughts by planning the sequence of statements that must be followed in logically progressing from the given to the final conclusion

5. Write the formal proof

Do not despair—writing an original proof takes practice, patience, persistence, and learning by example.

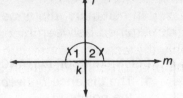

GIVEN Lines *l* and *m* intersect at *k*, ∢1 ≅ ∢2.

PROVE *l* ⊥ *m*.

PROOF

Statements	Reasons
1. Lines *l* and *m* intersect at *k* such that ∢1 ≅ ∢2.	1. Given.
2. ∢1 is supplementary to ∢2.	2. If the exterior sides of a pair of adjacent angles form a straight line, then the angles are supplementary.
3. ∢1 is a right angle.	3. If two angles are congruent and supplementary, then each is a right angle.
4. *l* ⊥ *m*.	4. If two lines intersect to form a right angle, then the lines are perpendicular. (This follows from the reverse of the definition of perpendicularity.)

TO PROVE THAT TWO LINES ARE PERPENDICULAR:

1. Show that the lines intersect to form a right angle.

2. Show that the lines intersect to form a pair of congruent adjacent angles (Theorem 3.10).

3.5 A WORD ABOUT PROOFS IN MATHEMATICS

Although the formal, two-column deductive proof is the dominant mode of proof found in introductory courses in geometry, it is not the only type of proof used in mathematics (including geometry). In general, proofs in mathematics may proceed in either direct or indirect fashion. The two-column proof uses a direct approach, progressing from the Given to the Prove in a finite number of sequential steps. Using an indirect method of proof, all possible outcomes are identified and proven false or contradictory *except one,* implying that the remaining possibility must be true by the process of elimination.

A mathematical proof may appeal to a deductive approach in which previously introduced definitions and principles are systematically used to establish a new conclusion. In advanced branches of mathematics the principle of *mathematical* induction is sometimes used in order to prove generalizations. The format of the proof may also vary. The two-column deductive proof is usually referred to as a "formal" proof since each assertion is listed with a corresponding justification following a very structured format. An "informal" proof usually takes paragraph form with only the essential ingredients of the proof explained in outline form. Informal proofs were given for Theorems 3.5 and 3.6 of this chapter.

EXAMPLE 3.13 Using the postulate "Exactly one line may be drawn between two given points," use an indirect method of proof to establish the following theorem:

THEOREM 3.11

If two lines intersect, then they intersect in exactly one point.

SOLUTION *An Informal Proof Using the Indirect Method of Proof:*

Let lines *l* and *m* represent the intersecting lines. There are two possibilities: (1) the lines intersect in exactly one point, or (2) they intersect in more than one point. Assume the latter is true and the lines intersect in *two* points, say points *P* and *Q.* This implies that both lines *l* and *m* pass through points *P* and *Q* which contradicts the postulate which asserts that one and only one line may be drawn between two points. Therefore, the remaining conclusion that there is exactly one point of intersection must be true.

Yet another type of proof is "proof by counterexample." In order to prove a statement is *false,* all we need do is show that it fails in one specific instance. For example, consider the assertion, "all pairs of supplementary angles are right angles." We can prove that this statement is false merely by producing an angle pair that satisfies the hypothesis (the angles are supplementary) but contradicts the conclusion (namely that the angles are right angles). The angle pair 100 and 80 is one such pair. The assertion is therefore false.

You may now be wondering which type of proof is "best" or which one should be used. The answer is very simple—it depends; it depends on the nature of the problem, the sophistication of the student, and the demands placed on the student by his/her teacher. In beginning geometry courses it is traditional to stress the two-column deductive proof. Whenever appropriate, however, informal proofs of theorems will be offered. In a subsequent chapter, more will be said about the indirect method of proof as it is an extremely powerful approach that is used extensively in all branches of mathematics, including geometry.

REVIEW EXERCISES FOR CHAPTER 3

1. In the accompanying diagram, list all pairs of vertical and adjacent angles.

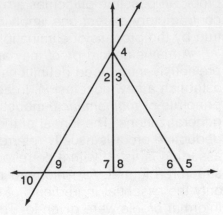

2. The measure of an angle is 5 times as great as the measure of its complement. Find the measure of the angle.

3. The measure of an angle exceeds 3 times its supplement by 4. Find the measure of the angle.

4. The measure of the supplement of an angle is three times as great as the measure of the angle's complement. Find the measure of the angle.

5. The difference between the measures of an angle and its complement is 14. Find the measure of the angle and its complement.

6. Find the measure of an angle if it is 12 less than twice the measure of its complement.

7. Find the value of *x*:

a

$4x + 5$

$x + 32$

b

$x + 2$

$3x + 18$

c

$2x - 5$

$3x + 5$

8. $\overleftrightarrow{XY}$ and $\overleftrightarrow{AB}$ intersect at point *C*. If $m \angle XCB = 4x - 9$ and $m \angle ACY = 3x + 29$, find $m \angle XCB$.

State whether each of the following statements is true or false. Prove a statement is false by providing a counterexample.

9. Complements of vertical angles are congruent.

10. If two lines intersect, then the bisectors of a pair of adjacent angles are perpendicular to each other.

11. If two lines intersect to form congruent angles, then the lines are perpendicular.

12. If an angle is congruent to its supplement, then it is a right angle.

13. If a point *C* is equidistant from points *A* and *B*, then point *C* is the midpoint of the segment which joins *A* and *B*.

14.   **GIVEN** $\angle 1 \cong \angle 4$.

       PROVE $\angle 2 \cong \angle 3$.

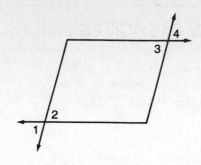

15.   **GIVEN** $\overline{BD}$ bisects $\angle ABC$.

       PROVE $\angle 1 \cong \angle 2$.

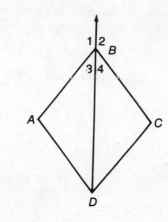

16.   **GIVEN** $\angle 3$ is complementary to $\angle 1$,
                     $\angle 4$ is complementary to $\angle 2$.

       PROVE $\angle 3 \cong \angle 4$.

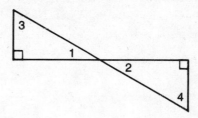

17.   **GIVEN** $\overline{AB} \perp \overline{BD}$, $\overline{CD} \perp \overline{BD}$,
                     $\angle 2 \cong \angle 4$.

       PROVE $\angle 1 \cong \angle 3$.

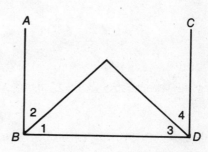

18.   **GIVEN** $\overline{KL} \perp \overline{JM}$,
                     $\overline{KL}$ bisects $\angle PLQ$.

       PROVE $\angle 1 \cong \angle 4$.

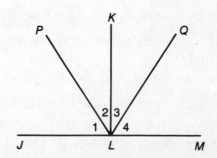

19. GIVEN $\overline{NW} \perp \overline{WT}$,
$\overline{WB} \perp \overline{NT}$,
$\angle 4 \cong \angle 6$.

PROVE $\angle 2 \cong \angle 5$.

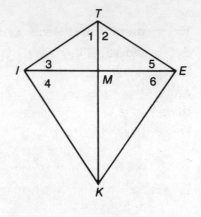

20. GIVEN $\overline{MT}$ bisects $\angle ETI$,
$\overline{KI} \perp \overline{TI}$, $\overline{KE} \perp \overline{TE}$,
$\angle 3 \cong \angle 1$, $\angle 5 \cong \angle 2$.

PROVE $\angle 4 \cong \angle 6$.

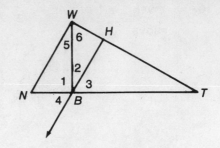

CHAPTER FOUR
PARALLEL LINES

4.1 PLANES AND LINES

Any flat surface such as a chalkboard or floor may be thought of as a *plane*. In a strict geometric context, a plane is characterized by the following properties:

1. It has length and width, but not depth or thickness.

2. Its length and width may, in theory, be extended indefinitely in each direction.

Figure 4.1 illustrates that two lines may lie in the same plane or in different planes. Lines, segments, rays, or points which lie in the same plane are said to be *coplanar;* if they do not lie in the same plane, they are referred to as non-coplanar. Throughout the discussion in this chapter we assume that lines are coplanar, unless otherwise stated.

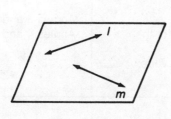

coplanar lines

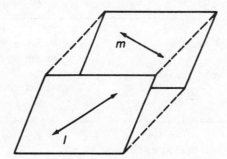

noncoplanar lines (called *skew* lines)

FIGURE 4.1

When a pair of lines are drawn, the plane is divided into distinct regions. The region bounded by both lines is referred to as the *interior* region; the remaining outside areas are *exterior* regions. See Figure 4.2.

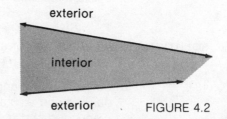

FIGURE 4.2

A line that intersects two or more lines in different points is called a *transversal*. In Figure 4.3 line *t* represents a transversal since it intersects lines *l* and *m* at two distinct points. Notice that at each point of intersection there are four angles formed. Those angles that lie in the interior region—angles 1, 2, 5, and 6—are referred to as *interior angles*. Angles 3, 4, 7, and 8 lie in the exterior region and are called *exterior angles*. It will be helpful to further classify angle pairs according to their position relative to the transversal. See Table 4.1.

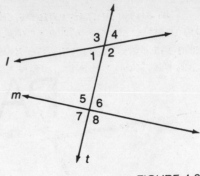

FIGURE 4.3

TABLE 4.1

TYPE OF ANGLE PAIR	DISTINGUISHING FEATURES	*EXAMPLES
Alternate Interior Angles	Angles are interior angles. Angles are on opposite sides of the transversal. Angles do not have the same vertex.	Angles 1 and 6; Angles 2 and 5.
Corresponding Angles	One angle is an interior angle; the other angle is an exterior angle. Angles are on the same side of the transversal. Angles do not have the same vertex.	Angles 3 and 5; Angles 4 and 6; Angles 1 and 7; Angles 2 and 8.
Alternate Exterior Angles	Angles are exterior angles. Angles are on opposite sides of the transversal. Angles do not have the same vertex.	Angles 3 and 8; Angles 4 and 7.

*Angles named refer to Figure 4.3.

EXAMPLE 4.1 Name all pairs of:
  a Alternate interior angles
  b Corresponding angles
  c Alternate exterior angles

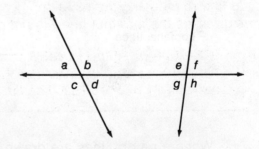

SOLUTION **a** Alternate interior angle pairs: *b* and *g*; *d* and *e*.

  b Corresponding angle pairs: *a* and *e*; *b* and *f*; *c* and *g*; *d* and *h*

  c Alternate exterior angle pairs: *a* and *h*; *c* and *f*

In analyzing diagrams, alternate interior angle pairs may be identified by their Z shape while corresponding angles form an F shape. The Z and F shapes, however, may be rotated so that the letter may appear reversed or upside down. See Figure 4.4.

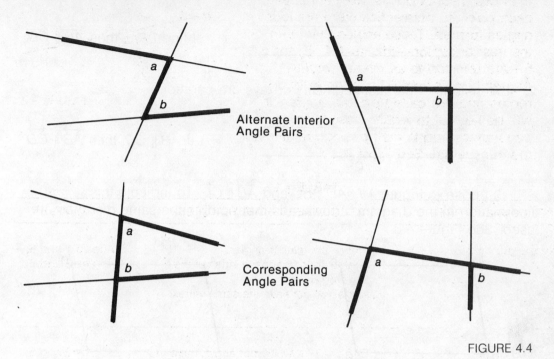

Alternate Interior Angle Pairs

Corresponding Angle Pairs

FIGURE 4.4

4.2 PROPERTIES OF PARALLEL LINES

Wires strung on telephone poles, bars on a prison cell, and lines drawn on a sheet of note paper are examples of *parallel* lines. We postulate that a pair of coplanar lines, if extended indefinitely, will eventually either intersect or never meet. Lines which never intersect are defined to be parallel.

In Figure 4.5, we may concisely express the fact that line *l* *is parallel to* line *m* by using the notation: *l* ∥ *m*; in Figure 4.6, the notation *l* ∦ *m* expresses the fact that line *l* is *not* parallel to line *m*.

DEFINITION OF PARALLEL LINES ————————————————
Parallel lines are coplanar lines that never intersect.
NOTATION: *l* ∥ *m* is read as "line *l* is parallel to line *m*."

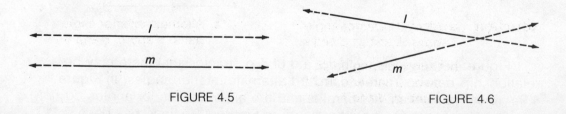

FIGURE 4.5   FIGURE 4.6

The following two observations are an immediate consequence of the previous definition:

1. Portions (segment or rays) of parallel lines are parallel.

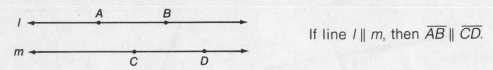

If line $l \parallel m$, then $\overline{AB} \parallel \overline{CD}$.

2. Extensions of parallel segments or rays are parallel.

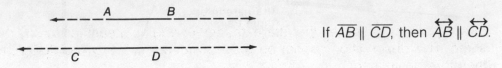

If $\overline{AB} \parallel \overline{CD}$, then $\overleftrightarrow{AB} \parallel \overleftrightarrow{CD}$.

Suppose in Figure 4.7, $\overline{AD} \parallel \overline{BC}$ and $\overline{AB} \parallel \overline{CD}$. To indicate this information in the diagram, arrowheads that point in the same direction are used. See Figure 4.8.

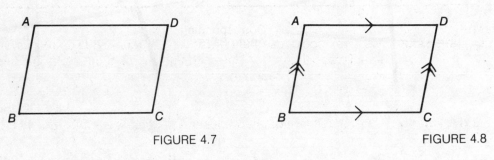

FIGURE 4.7   FIGURE 4.8

EXAMPLE 4.2   Mark off the diagram with the information supplied.

GIVEN $\overline{SR} \parallel \overline{TW}$,
$\overline{RL} \parallel \overline{TM}$,
$\overline{ST} \cong \overline{RW}$.

SOLUTION

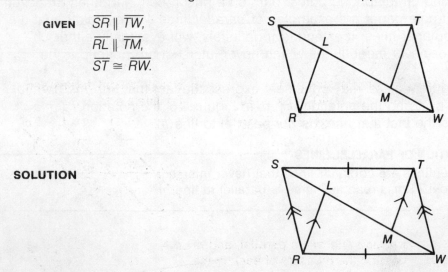

Figure 4.9 suggests, with the aid of a protractor, that there may be a relationship between parallelism and alternate interior angles. In Figure 4.9a, lines l and m are *not* parallel and the alternate interior angles do *not* measure the same. On the other hand, in Figure 4.9b line l has been

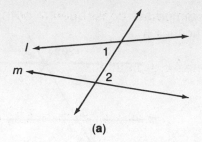
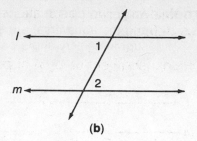

(a)

(b)

FIGURE 4.9

drawn parallel to line *m*; the alternate interior angles measure exactly the same. This observation cannot be proved with our existing knowledge; it is therefore postulated.

ALTERNATE INTERIOR ANGLE POSTULATE (4.1)

If two lines are parallel, then their alternate interior angles are congruent.

REMARK   Many textbooks use the phrase "If two parallel lines are *cut by a transversal*, then. . . ." However, this seems a bit redundant since the existence of alternate interior angles implies the presence of a transversal.

EXAMPLE 4.3   Given that the indicated lines are parallel, determine the value of *x*.

a

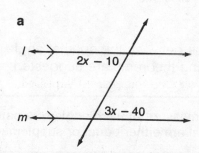

b

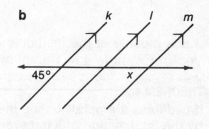

SOLUTION   **a** From Postulate 4.1,
$$3x - 40 = 2x - 10$$
$$3x = 2x + 30$$
$$x = 30.$$

b Using vertical angles and Postulate 4.1:

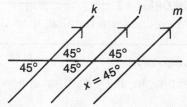

EXAMPLE 4.4   If lines *l* and *m* are parallel, and $m\angle a = 60$, find the measure of each of the numbered angles.

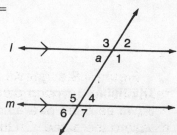

SOLUTION Angles formed at the intersection of line *l* and the transversal are either vertical or supplementary angle pairs.

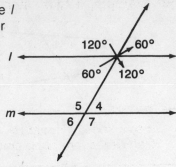

By Postulate 4.1, alternate interior angles are equal in measure:

$$m\angle a = m\angle 4 = 60$$

$$m\angle 1 = m\angle 5 = 120$$

And by vertical angles,

$$m\angle 6 = m\angle 4 = 60$$

$$m\angle 7 = m\angle 5 = 120$$

The completed diagram is

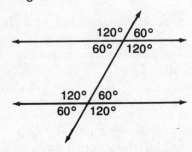

Look closely at the relationships between the special angle pairs derived in Example 4.4. The following theorems are suggested:

THEOREM 4.1

If two lines are parallel, then the measure of each angle of a pair formed by the intersection of a transversal are either equal or supplementary.

THEOREM 4.2

If two lines are parallel, then their corresponding angles are congruent.

For example, if *l* ∥ *m*, then $\angle a \cong \angle b$.

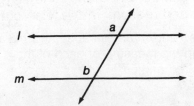

THEOREM 4.3

If two lines are parallel, then their alternate exterior angles are congruent.

For example, if $l \parallel m$, then $\angle c \cong \angle d$.

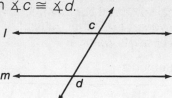

THEOREM 4.4

If two lines are parallel, then interior angles on the same side of the transversal are supplementary.

For example, if $l \parallel m$, then $\angle e$ and $\angle f$ are supplementary.

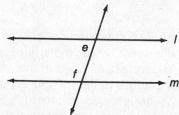

EXAMPLE 4.5 Prove Theorem 4.2.

    GIVEN $l \parallel m$.

    PROVE $\angle a \cong \angle b$.

SOLUTION

    PLAN   Introduce $\angle c$ in the diagram so that angles b and c are alternate interior angles. Then use Postulate 4.1 and vertical angles to establish the conclusion.

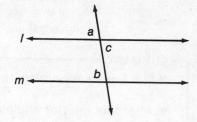

PROOF

Statements	Reasons
1. $l \parallel m$.	1. Given.
2. $\angle a \cong \angle c$.	2. Vertical angles are congruent.
3. $\angle c \cong \angle b$.	3. If two lines are parallel, then their alternate interior angles are congruent.
4. $\angle a \cong \angle b$.	4. Transitive property.

Note that we have introduced the notion of including a formal statement of a plan of attack. An indication of a "plan" and the corresponding annotation of the diagram is suggested before you begin to write the statements and reasons in the two-column proof.

EXAMPLE 4.6

    GIVEN $\overline{RS} \parallel \overline{TW}$,

                $\overline{TS} \parallel \overline{WX}$.

    PROVE $\angle S \cong \angle W$.

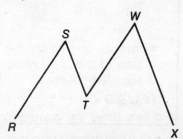

PLAN   Using alternate interior angles, $\angle S \cong \angle T$ and $\angle T \cong \angle W$. Therefore, $\angle S \cong \angle W$.

PROOF

Statements	Reasons
1. $\overline{RS} \parallel \overline{TW}$.	1. Given.
2. $\angle S \cong \angle T$.	2. If two lines are parallel, then their alternate interior angles are congruent.
3. $\overline{TS} \parallel \overline{WX}$.	3. Given.
4. $\angle T \cong \angle W$.	4. Same as reason 2.
5. $\angle S \cong \angle W$.	5. Transitive property of congruence.

CAUTION   Alternate interior, corresponding, and alternate exterior angle pairs are formed whenever two or more lines are intersected by a transversal, regardless of whether the original set of lines are parallel. These angle pairs, therefore, may or may *not* be congruent. We are entitled to conclude that alternate interior, corresponding, and alternate exterior angle pairs are congruent only if it is first determined that the lines intersected by the transversal are parallel.

SUMMARY

If two lines are parallel, then

- Alternate interior angles are congruent.

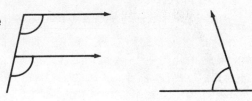

- Corresponding angles are congruent.

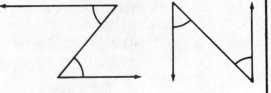

- Alternate exterior angles are congruent.

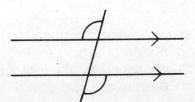

- Interior angles on the same side of the transversal are supplementary.

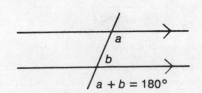

4.3 CONVERSES AND METHODS OF PROVING LINES PARALLEL

Geometry depends on preciseness of language as well as on logical rigor. Often the meaning of a sentence can change dramatically simply by switching the positions of words or phrases in the sentence. As an example, consider the statement:

If I am 12 years old, then I am not eligible to vote.

A new statement may be formed by interchanging the phrase appearing in the if clause (the single underlined phrase) with the statement appearing in the then clause (the double underlined phrase):

If I am not eligible to vote, then I am 12 years old.

The new statement formed is called the *converse* of the original statement. The relationship between the original statement and its converse is illustrated below.

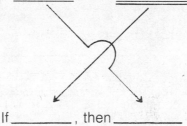

Original statement: If condition, then conclusion

Converse: If _____ , then _____

Our example illustrates that the converse of a true statement may not be assumed to be true. If I am not eligible to vote, I can be 13 years old, 14 years old, or perhaps not a citizen.

As another example, form the converse of the statement, "If yesterday was Saturday, then today is Sunday." The converse is: "If today is Sunday, then yesterday was Saturday." In this instance the original statement and its converse are true. The only generalization that can be made regarding the truth or falsity of the converse is that it may be either true or false.

Why then do we mention converses? Converses are important in mathematics since they provide *clues*—they suggest avenues to explore that may lead to the discovery of true relationships. In the previous section we studied how to prove certain types of angle relationships were true, *given* that lines were parallel. In this section we examine the converse situation. How can we prove lines are parallel, *given* that certain types of angle relationships are true?

A logical starting point is to investigate the validity of the statement derived by taking the converse of Postulate 4.1:

POSTULATE 4.1	CONVERSE
If two lines are parallel, then their alternate interior angles are congruent.	If a pair of alternate interior angles are congruent, then the lines are parallel.

NOTE: In forming the converse, the phrase "a pair of" was inserted; if one pair of alternate interior angles are congruent, so must the other pair be congruent.

The truth of the converse is consistent with experience—if it is known that a pair of alternate interior angles are congruent, then the lines must be parallel. See Figure 4.10.

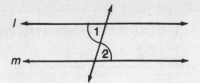

FIGURE 4.10

Based on our current knowledge, the converse cannot be proven. We therefore offer it as a postulate.

CONVERSE OF ALTERNATE INTERIOR ANGLE POSTULATE (4.2)

If a pair of alternate interior angles are congruent, then the lines are parallel.

EXAMPLE 4.7

GIVEN $\overline{BC} \parallel \overline{AD}$,

$\angle 2 \cong \angle 3$.

PROVE $\overline{AB} \parallel \overline{CD}$.

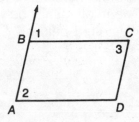

SOLUTION

PLAN To show $\overline{AB} \parallel \overline{CD}$ first prove $\angle 1 \cong \angle 2$ and then use the given to conclude that $\angle 1 \cong \angle 3$. By Postulate 4.2 the line segments must be parallel.

PROOF

Statements	Reasons
1. $\overline{BC} \parallel \overline{AD}$.	1. Given.
2. $\angle 1 \cong \angle 2$.	2. If two lines are parallel, then their corresponding angles are congruent.
3. $\angle 2 \cong \angle 3$.	3. Given.
4. $\angle 1 \cong \angle 3$.	4. Transitive property of congruence.
5. $\overline{AB} \parallel \overline{CD}$.	5. If a pair of alternate interior angles are congruent, then the lines are parallel. (Postulate 4.2).

Additional methods of proving lines parallel may be derived from forming and then proving the converses of Theorems 4.2, 4.3, and 4.4. The following theorems can be proved using Postulate 4.2:

THEOREM 4.5

If a pair of corresponding angles are congruent, then the lines are parallel.

THEOREM 4.6

If a pair of alternate exterior angles are congruent, then the lines are parallel.

THEOREM 4.7

If a pair of interior angles on the same side of the transversal are supplementary, then the lines are parallel.

EXAMPLE 4.8

GIVEN $\overleftrightarrow{AB} \perp t$,
$\overleftrightarrow{CD} \perp t$.

PROVE $l \parallel m$.

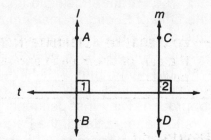

SOLUTION

PLAN Perpendicular lines intersect to form right angles, all right angles are congruent. Therefore, the lines are parallel by either Theorem 4.5 or by Postulate 4.2. (Theorem 4.7 may also be used.)

PROOF

Statements	Reasons
1. $\overleftrightarrow{AB} \perp t$ and $\overleftrightarrow{CD} \perp t$.	1. Given.
2. Angles 1 and 2 are right angles.	2. Perpendicular lines intersect to form right angles.
3. $\angle 1 \cong \angle 2$.	3. All right angles are congruent.

4.4 THE PARALLEL POSTULATE

One of the most controversial postulates relates to how many parallel lines can be drawn to a line through a point not on the line. See Figure 4.11. Once again we must appeal to intuition and experience. Since we cannot prove it, we postulate that there is exactly one such line. Since this postulate affirms the existence of parallel lines, it is commonly referred to as the Parallel Postulate.

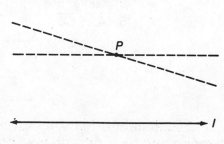

FIGURE 4.11

THE PARALLEL POSTULATE (4.3)

Through a given point not on a line, exactly one line may be drawn parallel to the line.

The Parallel Postulate is of great historic interest. Euclid was a Greek mathematician who lived in approximately 300 B.C. It is Euclid who is credited with organizing the various aspects of the subject of geometry into a coherent discipline and then writing the first major treatise on the subject, called the *Elements*. In the *Elements* Euclid presents 10 postulates on which the development of the subject is based. The Parallel Postulate corresponds to the fifth postulate listed in the *Elements*; as a result it is also referred to as Euclid's Fifth Postulate. Actually, the statement of Postulate 4.3 most closely resembles the version used by the mathematician John Playfair (1748–1819).

The Parallel Postulate implies the existence of parallel lines and from the time of Euclid on mathematicians have been somewhat skeptical that this assumption represents a self-evident truth. For 2000 years mathematicians have attempted to provide a satisfactory proof of Euclid's Fifth Postulate; all efforts have proved unsuccessful. In the 1800's two mathematicians, Bolyai and Lobachevsky, used a slightly different approach. They independently investigated how the subject of geometry would be impacted if the Parallel Postulate was relaxed and it was assumed that there may be more than one line that could be drawn through a point not on a line and parallel to the line. In the mid 1800s, Riemann took yet another approach. He postulated that through a given point not on a line, *no* line could be drawn parallel to the line. Geometries which tamper with Euclid's Fifth Postulate are collectively referred to as non-Euclidean geometries. Although interesting, their investigation goes well beyond the scope of this book.

EXAMPLE 4.9

  GIVEN $l \parallel m$.

  FIND $m\angle ABC$.

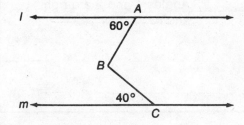

SOLUTION Using the Parallel Postulate, through point B draw a line parallel to line l (and, therefore, line m).

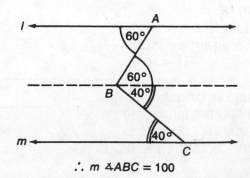

$\therefore m \angle ABC = 100$

REVIEW EXERCISES FOR CHAPTER 4

Classify the indicated pair of angles labeled *a* and *b* in Exercises 1 to 3 as alternate interior, corresponding, or alternate exterior. (*Suggestion*: First identify the transversal and then look for a Z or F shape; remember that the angle pair may be neither alternate interior nor corresponding.)

1.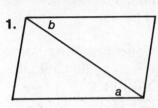

2.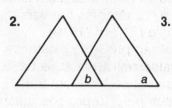

3.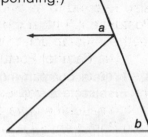

4. Find the value of *x* and *y*.

a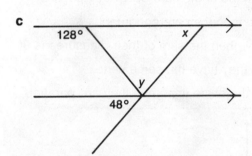

b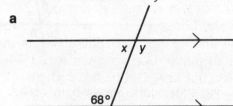

c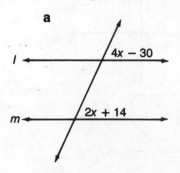

d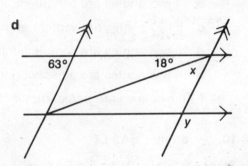

5. Given *l* ∥ *m*, find the value of *x*.

a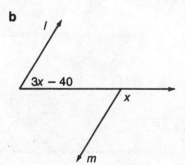

b

c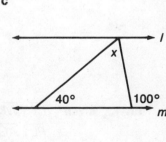

6. Two parallel lines are cut by a transversal. Find the measures of the angles if a pair of interior angles on the same side of the transversal:

a Are represented by $5x - 32$ and $x + 8$.

b Have measures such that one angle is 4 times the measure of the other.

7. Find *x*.

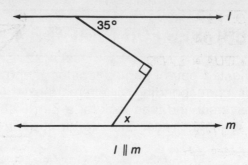

l ∥ *m*

8. In the accompanying diagram, $\overleftrightarrow{AB} \parallel \overleftrightarrow{CD}$ and $\overline{EF}$ bisects ∡*AFG*.

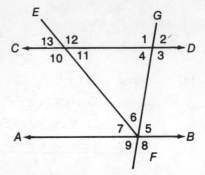

   a If *m*∡1 = 100, find the measure of each of the numbered angles.

   b If *m*∡3 = 4*x* − 9 and *m*∡5 = *x* + 19, find the measure of each of the numbered angles.

9. Form the converse of each of the following statements. State whether the converse is true or false.

   a If I live in New York, then I live in the United States.

   b If two angles are congruent, then the angles are equal in measure.

   c If two angles are vertical angles, then they are congruent.

   d If two angles are complementary, then the sum of their measures is 90.

   e If two angles are adjacent, then they have the same vertex.

   f If two lines are perpendicular to the same line, then they are parallel.

10.   **GIVEN** $\overline{BA} \parallel \overline{CF}$,
                $\overline{BC} \parallel \overline{ED}$.

      PROVE ∡1 ≅ ∡2.

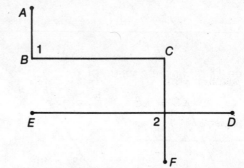

11.   **GIVEN** $\overline{LT} \parallel \overline{WK} \parallel \overline{AP}$,
                $\overline{PL} \parallel \overline{AG}$.

      PROVE ∡1 ≅ ∡2.

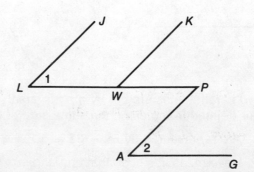

12. **GIVEN** $\overline{QD} \parallel \overline{UA}$,
$\overline{QU} \parallel \overline{DA}$.

PROVE $\angle QUA \cong \angle ADQ$.

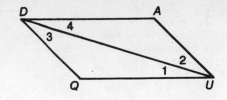

13. **GIVEN** $\overline{AT} \parallel \overline{MH}$,
$\angle M \cong \angle H$.

PROVE $\angle A \cong \angle T$.

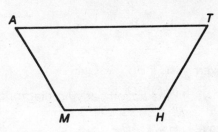

14. **GIVEN** $\overline{IB} \parallel \overline{ET}$,
$\overline{IS}$ bisects $\angle EIB$,
$\overline{EC}$ bisects $\angle TEI$.

PROVE $\angle BIS \cong \angle TEC$.

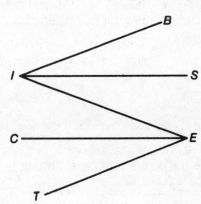

15. **GIVEN** $\angle B \cong \angle D$,
$\overline{BA} \parallel \overline{DC}$.

PROVE $\overline{BC} \parallel \overline{DE}$.

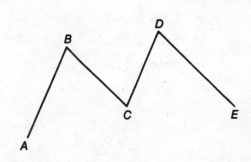

16. **GIVEN** $k \parallel l$,
$\angle 5 \cong \angle 8$.

PROVE $j \parallel l$.

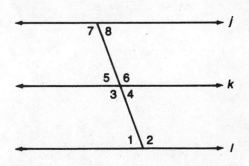

17. **GIVEN** $\angle K \cong \angle P$,
$m\angle J + m\angle P = 180$.

PROVE $\overline{KL} \parallel \overline{JP}$.

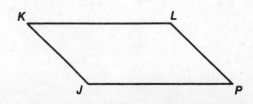

18. **GIVEN** $\overline{AB} \perp \overline{BC}$,

           $\angle ACB$ is complementary
           to $\angle ABE$.

    PROVE $\overleftrightarrow{AC} \parallel \overleftrightarrow{EBD}$.

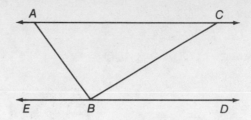

19. **GIVEN** $\overline{AG} \parallel \overline{BC}$,

           $\overline{KH} \parallel \overline{BC}$,

           $\angle 1 \cong \angle 2$.

    PROVE $\overline{HK} \perp \overline{AB}$.

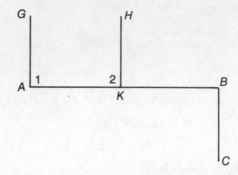

20. Prove if two lines are each parallel to a third line, then the lines are parallel to each other.

21. Prove if the rays that bisect a pair of alternate interior angles are parallel, then the lines are parallel.

22. Prove if two lines are parallel, then the rays which bisect a pair of corresponding angles are parallel.

CHAPTER FIVE
ANGLES OF A POLYGON

5.1 THE ANATOMY OF A POLYGON

Until now we have focused on points, segments, lines, rays, and angles. In this chapter we begin our study of geometric figures. A geometric figure whose sides are line segments is called a *polygon*, provided each segment, that is, side, intersects another segment at its end point. Perhaps it will help to clarify matters by presenting figures which are *not* polygons, as shown in Table 5.1.

A polygon is named by choosing any vertex and then writing the letters corresponding to each vertex consecutively, proceeding in either a clockwise or counterclockwise fashion. For example, the previous polygon (*ABCD*) may be named by starting at any vertex, say vertex *C*, and proceeding in a clockwise direction: *CDAB*; if we traveled counterclockwise from point *C*, the polygon would be named *CBAD*.

In Figure 5.1 both figures are polygons. However, in Figure 5.1*a*, angle *AED* is greater than 180; in Figure 5.1*b*, angle *AED* is less than 180. Figure 5.1*b* is an example of a *convex* polygon. Nonconvex polygons such as the one illustrated in Figure 5.1*a* will not be studied in this course. Instead, we shall restrict our attention to examining the properties of both general as well as special types of convex polygons.

DEFINITION OF CONVEX POLYGON

A *convex* polygon is a polygon in which the measure of each angle is less than 180.

REMARK   Whenever the term polygon is used, it will be understood that it refers to a convex polygon.

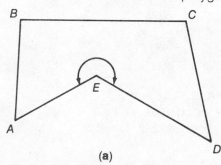

(a)

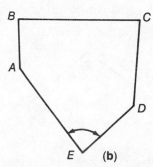

(b)

FIGURE 5.1   Convex vs. Nonconvex Polygons.

TABLE 5.1

DIAGRAM	COMMENT
	ABCD is *not* a polygon since points *A* and *D* are not joined by a segment.
	ABCDE is *not* a polygon since $\overline{AB}$ and $\overline{ED}$ do not intersect a side at its end point; a polygon must be closed.
	JKLMN is *not* a polygon since sides $\overline{JN}$ and $\overline{KM}$ intersect at *L* which is not the end point of a side.
	ABCD is a polygon. The intersection of any two sides is called a *vertex* of the polygon.

PARTS OF A POLYGON

There are several important terms related to a polygon: interior angle, side, vertex, and diagonal. See Figure 5.2.

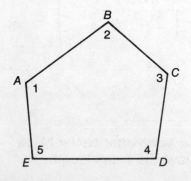

VERTICES: *A, B, C, D,* and *E*
SIDES: $\overline{AB}$, $\overline{BC}$, $\overline{CD}$, $\overline{DE}$, and $\overline{AD}$
INTERIOR ANGLES: 1, 2, 3, 4, and 5

FIGURE 5.2   Some Parts of a Polygon.

A diagonal is defined as a segment joining a pair of *nonadjacent* vertices. See Figure 5.3. Clearly, there exists a relationship between the number of sides of a polygon and the number of diagonals which can be drawn.

DIAGONALS: $\overline{AC}$, $\overline{BD}$, $\overline{EC}$, $\overline{EB}$, and $\overline{AD}$

FIGURE 5.3

THEOREM 5.1

If a polygon has N sides, then $\frac{1}{2}N(N - 3)$ diagonals can be drawn.

INFORMAL PROOF

Consider a polygon having N sides. Draw segments from any vertex, say A, to each of the other $N - 1$ vertices, resulting in $N - 1$ segments. However, this number also includes the segments which coincide with sides $\overline{AB}$ and $\overline{AC}$. Hence, subtracting the two segments, $N - 3$ segments remain which correspond to diagonals. This analysis may be repeated with each of the N vertices, yielding a total of $N(N - 3)$ diagonals. But, each diagonal has been counted twice, since its end points are a vertex of the polygon. Therefore, the total number of distinct diagonals is $\frac{1}{2}N(N - 3)$.

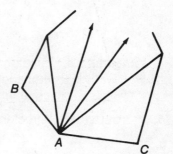

EXAMPLE 5.1 How many diagonals can be drawn in a polygon having seven sides?

SOLUTION Number of Diagonals = $\frac{1}{2}N(N - 3)$
For $N = 7$,

$$\text{Diagonals} = \tfrac{1}{2}(7)(7 - 3) = \tfrac{1}{2}(7)(4) = 14$$

CLASSIFYING POLYGONS

Polygons may be classified according to their number of sides. The most commonly referred to polygons are as follows:

POLYGON NAME	NUMBER OF SIDES
Triangle	3
Quadrilateral	4
Pentagon	5
Hexagon	6
Octagon	8
Decagon	10
Duodecagon	12

Polygons may also be categorized according to whether each of their angles are equal in measure and/or their sides have the same length.

DEFINITIONS OF POLYGONS HAVING PARTS EQUAL IN MEASURE

1. An *equiangular* polygon is a polygon in which each angle has the same measure.

2. An *equilateral* polygon is a polygon in which each side has the same length.

3. A *regular* polygon is a polygon that is both equiangular and equilateral.

REMARK A polygon may *not* be assumed to be equiangular, equilateral, or regular unless specifically given.

EXAMPLE 5.2 Give an example of a regular quadrilateral.

SOLUTION A square is both equiangular and equilateral. A square is, therefore, a regular quadrilateral.

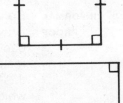

EXAMPLE 5.3 Give an example of a quadrilateral that is equiangular but *not* necessarily equilateral.

SOLUTION A rectangle.

5.2 ANGLES OF A TRIANGLE

One of the most familiar geometric relationships is that the sum of the measures of the angles of a triangle is 180. See Figure 5.4. An experiment to test the general validity of this relationship can be performed by drawing any triangle, tearing off angles 1 and 3 (see Figure 5.5), and then aligning the edges (sides) of the angle about angle 2 so that a straight line is formed.

FIGURE 5.4 $m\angle 1 + m\angle 2 + m\angle 3 = 180.$

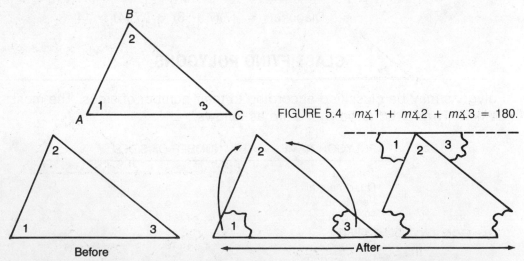

FIGURE 5.5 "Proving" The Sum of the Measures of the Angles of a Triangle is 180. Angles 1, 2, and 3 can be aligned to form a straight angle. Since a straight angle has measure 180, the sum of the measures of the angles of the triangle must be 180.

A formal geometric proof is presented below. Since we cannot "tear up" a triangle, the proof introduces the concept of drawing an extra (auxiliary) line in the original diagram.

THEOREM 5.2 SUM OF THE ANGLES OF A TRIANGLE THEOREM

The sum of the measures of the angles of a triangle is 180.

GIVEN △ABC (read as "triangle ABC").

PROVE $m\angle 1 + m\angle 2 + m\angle 3 = 180$.

PLAN Draw an auxiliary line *l* through *B* and parallel to $\overline{AC}$. The sum of the measures of angles 4, 2, and 5 is 180. Using alternate interior angles and substitution, the desired relationship is obtained.

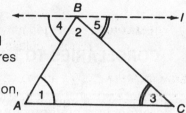

PROOF

Statements	Reasons
1. Through point *B*, draw line *l* parallel to $\overline{AC}$.	1. Through a point not on a line exactly one line may be drawn parallel to the line.
2. $m\angle 4 + m\angle 2 + m\angle 5 = 180$.	2. A straight angle has measure 180.
3. $m\angle 1 = m\angle 4$, $m\angle 3 = m\angle 5$.	3. If two lines are parallel, then their alternate interior angles are equal in measure.
4. $m\angle 1 + m\angle 2 + m\angle 3 = 180$.	4. Substitution.

EXAMPLE 5.4 The measures of the angles of a triangle are in the ratio of 2:3:4. Find the measure of each angle.

SOLUTION

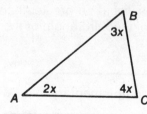

$2x + 3x + 4x = 180$ $m\angle A = 2x = 40$

$9x = 180$ $m\angle B = 3x = 60$

$x = 20$ $m\angle C = 4x = 80$

EXAMPLE 5.5 Find the value of *x*.

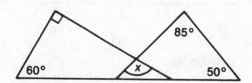

SOLUTION $m\angle BCA = 30$,

$m\angle EDF = 45 \implies$

$m\angle DGC = 105$,

so $x = 105$.

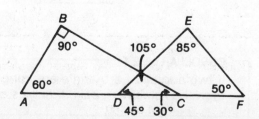

EXAMPLE 5.6 Given $l \parallel m$ and $\overline{CD} \perp \overline{AB}$, find x.

SOLUTION $m\angle A = 35$, $m\angle ADC = 90 \Longrightarrow x = 55$

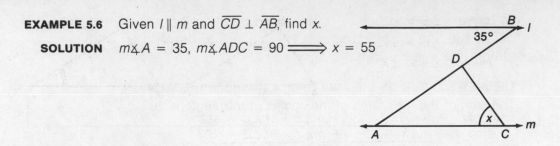

COROLLARIES TO THE SUM OF THE ANGLES OF A TRIANGLE THEOREM

A *corollary* is a statement which is a direct consequence of a previously proved theorem. The results follow immediately from Theorem 5.2.

> **COROLLARY 5.2.1**
> The measure of each angle of an equiangular triangle is 60.

COMMENT 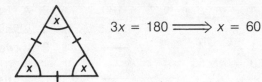 $3x = 180 \Longrightarrow x = 60$

> **COROLLARY 5.2.2**
> A triangle may have at most one right or one obtuse angle.

COMMENT Use an indirect method of proof—suppose it had two such angles. Then the sum of their measures would be greater than or equal to 180. What about the measure of the third angle of the triangle?

> **COROLLARY 5.2.3**
> The acute angles of a right triangle are complementary.

COMMENT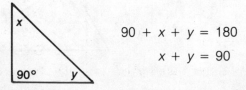
$$90 + x + y = 180$$
$$x + y = 90$$

> **COROLLARY 5.2.4**
> If two angles of a triangle are congruent to two angles of another triangle, then the remaining pair of angles are congruent.

GIVEN $\angle B \cong \angle Y$,
         $\angle C \cong \angle Z$.
PROVE $\angle A \cong \angle X$.

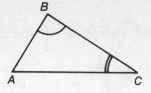

OUTLINE OF $m\angle A$ $+ m\angle B + m\angle C = 180$
PROOF $- m\angle X$ $+ m\angle Y + m\angle Z = 180$
             $\overline{m\angle A - m\angle X + 0 \quad\quad + 0 \quad\quad = 0}$

Here, $m\angle A = m\angle X$.

The preceding four corollaries could have been labeled theorems. Again we stress that the use of the term corollary serves to emphasize that the statement is intimately related to a particular theorem.

EXAMPLE 5.7   Use Corollary 5.2.4 to help construct a formal proof for the following problem:
GIVEN $\overline{BD}$ bisects $\angle ABC$,
         $\overline{BD} \perp \overline{AC}$.
PROVE $\angle A \cong \angle C$.

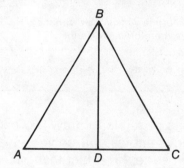

SOLUTION
PLAN   **1.** Mark off the diagram with the given information.
NOTE: In order to facilitate reference to an angle in the proof, angles have been numbered.

2. Since two angles of $\triangle ADB$ are congruent to two angles of $\triangle CDB$, the third pair of angles must be congruent. Hence $\angle A \cong \angle C$.

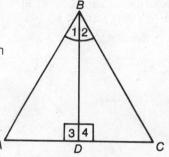

PROOF

Statements	Reasons
1. $\overline{BD}$ bisects $\angle ABC$.	1. Given.
2. $\angle 1 \cong \angle 2$.	2. A bisector divides an angle into two congruent angles.
3. $\overline{BD} \perp \overline{AC}$.	3. Given.
4. Angles 3 and 4 are right angles.	4. Perpendicular lines intersect to form right angles.
5. $\angle A \cong \angle C$.	5. If two angles of a triangle are congruent to two angles of another triangle, then the remaining pair of angles are congruent.

5.3 EXTERIOR ANGLES OF A TRIANGLE

Consider Figure 5.6. At each vertex of the triangle the sides have been extended forming additional angles at each vertex. Angles which are adjacent and supplementary to the interior angle at each vertex are called *exterior* angles of the triangle. Notice that at each vertex there are two such exterior angles:

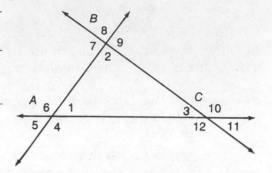

VERTEX	INTERIOR ANGLE	EXTERIOR ANGLES*
A	1	4 and 6
B	2	7 and 9
C	3	10 and 12

* Angles 5, 8, and 11 form a vertical pair with the interior angle and are therefore *not* exterior angles.

FIGURE 5.6   Exterior Angles of a Triangle.

Although it is possible to draw two exterior angles at each vertex, when we refer to exterior angles of a polygon we shall normally mean drawing only one at each vertex. The fact that this may be done in several different ways is unimportant. Notice in Figure 5.7 that the exterior angle and the interior angle at a vertex form an adjacent pair of angles such that their exterior sides form a straight line.

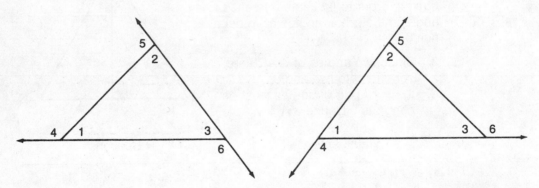

FIGURE 5.7

DEFINITION OF EXTERIOR ANGLE OF A POLYGON

An *exterior angle* of a polygon is an angle adjacent to an interior angle such that their exterior sides form a straight line.

EXAMPLE 5.8   The measure of two angles of a triangle are 80 and 60. Find the sum of the measures of the exterior angles (one exterior angle at a vertex) of the triangle.

SOLUTION   First note that the third angle of the triangle has measure 40. Using the fact that at a vertex, interior and exterior angles are supplementary, we deduce that

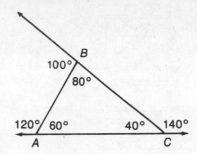

and sum = 120 + 100 + 140 = 360. In the next section we shall state and prove that the sum of the measures of the exterior angles of any polygon, regardless of the number of sides, is 360.

In Example 5.8 observe that the measure of each exterior angle is equal to the sum of the measures of the two nonadjacent (remote) interior angles.

THEOREM 5.3   EXTERIOR ANGLE OF A TRIANGLE THEOREM
The measure of an exterior angle of a triangle is equal to the sum of the measures of the two nonadjacent interior angles.

EXAMPLE 5.9   Find the value of x.

a

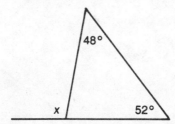

b

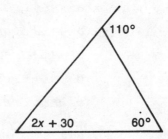

c

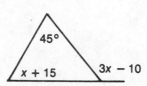

d

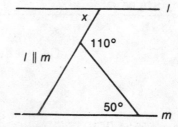

SOLUTION   **a** $x = 48 + 52 = 100$

b $110 = 2x + 30 + 60$
$110 = 2x + 90$
$20 = 2x$
$x = 10$

c $3x - 10 = x + 15 + 45$
$3x - 10 = x + 60$
$3x = x + 70$
$2x = 70$
$x = 35$

d $x = 60$

We have already established that $m\angle 1 = m\angle b + m\angle c$, as shown in the figure at the right. Since $m\angle b$ and $m\angle c$ represent positive numbers, if we delete one of these quantities from the equation, we no longer have an equality. The following inequalities result:

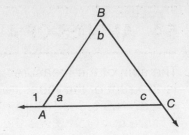

$$m\angle 1 = m\angle b + m\angle c \Longrightarrow m\angle 1 > m\angle b$$

$$m\angle 1 = m\angle b + m\angle c \Longrightarrow m\angle 1 > m\angle c$$

The symbol $>$ is read "is greater than" (and the symbol $<$ is read "is less than"). We now state the corresponding theorem that will prove useful in future work involving inequality relationships in triangles.

THEOREM 5.4 EXTERIOR ANGLE INEQUALITY THEOREM

The measure of an exterior angle of a triangle is greater than the measure of either nonadjacent interior angle.

EXAMPLE 5.10 Inside the box $\square$ insert the correct symbol (either $>$, $<$, or $=$) so that the resulting statement is true. If it is not possible to draw a conclusion based on the given diagram, then write "cannot be determined."

a $m\angle 4 \ \square \ m\angle 1$

b $m\angle 2 \ \square \ m\angle 6$

c $m\angle 3 \ \square \ m\angle 2$

d $m\angle 7 \ \square \ m\angle 1$

e $m\angle 5 \ \square \ m\angle 2$

f $m\angle 4 \ \square \ m\angle 1 + m\angle 2$

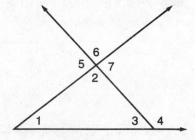

SOLUTION

a $>$ **d** $>$

b $=$ **e** Cannot be determined

c Cannot be determined   **f** $=$

SUMMARY: ANGLE RELATIONSHIPS IN A TRIANGLE

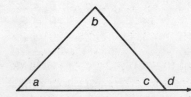

$a + b + c = 180$

$d = a + b$

$d > a$ and $d > b$

5.4 ANGLES OF A POLYGON

The sum of the measures of the angles of a polygon of 3 sides is 180. What is the sum of the measures of the angles of a polygon having 4 sides? 5 sides? 100 sides? Our approach to answering these questions is to derive a formula which gives the relationship between the sum of the angles of a polygon and the number of its sides. To accomplish this we must fall back on the fact that the sum of the angles of a triangle is 180 and attempt to resolve any polygon into an equivalent set of triangles. See Figure 5.8.

Figure 5.8 illustrates that there is a definite relationship between the number of sides of a polygon and the number of triangles it can be divided into. It can be proven that a polygon of n sides can be separated into $n - 2$ triangles. Since there are 180 degrees in each such triangle, it follows that the sum of the angles can be found by multiplying $(n - 2)$ by 180.

THEOREM 5.5   SUM OF THE INTERIOR ANGLES OF A POLYGON

The sum of the measures of the interior angles of a polygon having n sides is $180(n - 2)$.

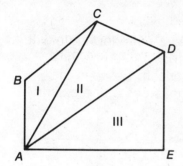 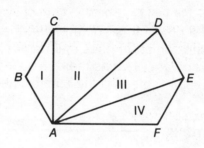 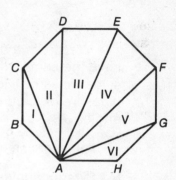

FIGURE 5.8   Separating a Polygon Into Triangles.

EXAMPLE 5.11   Find the sum of the measures of the interior angles of an octagon.

SOLUTION   For $n = 8$, sum = $180(n - 2) = 180(8 - 2) = 180(6) = 1,080$.

EXAMPLE 5.12   If the sum of the measures of the angles of a polygon is 900, determine the number of sides.

SOLUTION   For sum = 900,

$$900 = 180(n - 2)$$

$$\frac{900}{180} = n - 2$$

$$5 = n - 2$$

$$n = 7 \text{ sides}$$

We have noted that the sum of the measures of the *exterior* angles of a triangle is 360. We are now prepared to state and prove that this sum remains constant, regardless of the number of sides in the polygon.

THEOREM 5.6 SUM OF THE EXTERIOR ANGLES OF A POLYGON

The sum of the measures of the exterior angles of any polygon (one exterior angle at a vertex) is 360.

INFORMAL PROOF An n-sided polygon has n vertices. At each vertex the sum of the interior and exterior angles is 180:

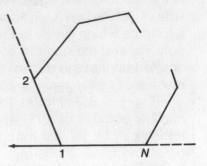

At vertex 1: Interior angle 1 + exterior angle 1 = 180
At vertex 2: Interior angle 2 + exterior angle 2 = 180

$\qquad\vdots\qquad\qquad\vdots\qquad\qquad\vdots$

At vertex n: Interior angle n + exterior angle n = 180

$$\begin{aligned}
\textit{Adding:} \quad \text{Sum of interior angles} + \text{sum of exterior angles} &= 180n \\
\textit{Substituting:} \quad 180(n-2) + \text{sum of exterior angles} &= 180n \\
\textit{Simplifying:} \quad 180n - 360 + \text{sum of exterior angles} &= 180n \\
-360 + \text{sum of exterior angles} &= 0
\end{aligned}$$

$$\text{Sum of exterior angles} = 360$$

We stress that if a polygon is equiangular, then each interior angle will have the same measure which implies that each exterior angle has the same measure. For *regular* polygons there is a relationship between the measure of each exterior angle and the sum of the measures of the exterior angles:

FOR REGULAR POLYGONS

$$\text{Exterior angle} = \frac{360}{n}$$

EXAMPLE 5.13 Find the measure of each interior angle and each exterior angle of a regular decagon.

SOLUTION *Method 1*:
$$\begin{aligned}
\text{Sum} &= 180(n-2) \\
&= 180(10-2) \\
&= 180(8) \\
&= 1{,}440
\end{aligned}$$

Since there are 10 interior angles, each of which is identical in measure,

$$\text{Interior angle} = \frac{1{,}440}{10} = 144$$

Since interior and exterior angles are supplementary,

$$\text{Exterior angle} = 180 - 144 = 36$$

Method 2: First determine the measure of an exterior angle:

$$\text{Exterior angle} = \frac{360}{10} = 36$$

$$\text{Interior angle} = 180 - 36 = 144$$

EXAMPLE 5.14 The measure of each interior angle of a regular polygon is 150. Find the number of sides.

SOLUTION We use a method similar to the approach illustrated in Method 2 of Example 5.13. Since the measure of each interior angle is 150, the measure of an exterior angle is $180 - 150$, or 30. Therefore,

$$30 = \frac{360}{n} \Longrightarrow n = 12$$

SUMMARY OF GENERAL PRINCIPLES

1. An n-sided polygon has n vertices and n interior angles. At each vertex an exterior angle may be drawn by extending one of the sides.

2. If the polygon is regular, then the measures of the interior angles are equal and the measures of the exterior angles are equal.

3. The sum of the measures of the *interior* angles of *any* polygon is given by the formula: Sum $= 180(n - 2)$. The sum of the measures of the *exterior* angles of any polygon, regardless of the number of sides, is 360.

4. To find the measures of the interior and exterior angles of a *regular* polygon, given the number of sides (or vice versa), use the following relationships:

$$\text{Exterior angle} = \frac{360}{n}$$

and

$$\text{Interior angle} = 180 - \text{exterior angle}$$

REVIEW EXERCISES FOR CHAPTER 5

1. Find the value of x:

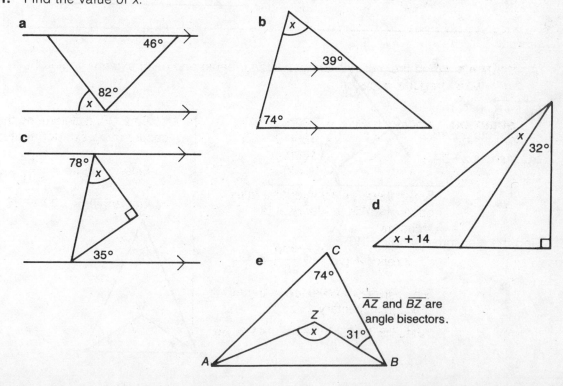

a 46°, 82°, x

b x, 39°, 74°

c 78°, x, 35°

d x, 32°, x + 14

e C 74°, Z x, 31°, A, B
$\overline{AZ}$ and $\overline{BZ}$ are angle bisectors.

2. Given that $l \parallel m$, $m \angle 2 = 110$ and $m \angle 6 = 70$, find each of the remaining numbered angles.

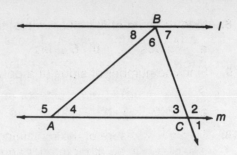

3. A triangle may be classified according to the measure of its largest angle. A triangle which contains an obtuse angle is called an *obtuse triangle*; if it contains a right angle it is referred to as a *right triangle*; an *acute triangle* is a triangle in which each angle is an acute angle.

For each of the following, the measures of the angles of $\triangle ABC$ are represented in terms of x. Find the value of x and classify the triangle as acute, right, or obtuse.

a $m \angle A = 3x + 8$
$m \angle B = x + 10$
$m \angle C = 5x$

b $m \angle A = x + 24$
$m \angle B = 4x + 17$
$m \angle C = 2x - 15$

c $m \angle A = 3x - 5$
$m \angle B = x + 14$
$m \angle C = 2x - 9$

For Exercises 4 to 6 find x.

4.

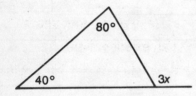

5.

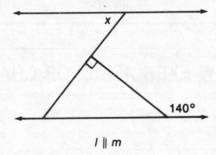

$l \parallel m$

6.

$\overleftrightarrow{EA} \parallel \overleftrightarrow{FB}$,
$\overline{AC}$ bisects $\angle EAB$,
$\overline{BC}$ bisects $\angle FBA$.

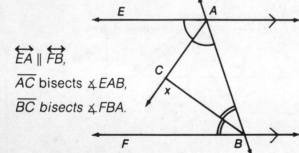

7. Find the measure of angle RWT.

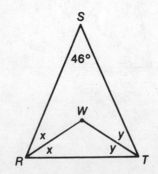

8. Find the sum of the measures of the interior angles of a polygon having:

   a 4 sides   **b** 6 sides   **c** 9 sides   **d** 13 sides

9. Find the number of sides of a polygon if the sum of the measures of the interior angles is:

   a 1,800   **b** 2,700   **c** 540   **d** 2,160

10. Find the measure of the remaining angle of each of the following figures, given the measures of the other interior angles.

   a Quadrilateral: 42, 75, and 118

   b Pentagon: 116, 138, 94, 88

   c Hexagon: 95, 154, 80, 145, 76

11. Find the measure of each interior angle of a regular polygon having:

   a 5 sides   **b** 24 sides   **c** 8 sides   **d** 15 sides

12. Find the number of sides of a regular polygon if the measure of an interior angle is:

   a 162   **b** 144   **c** 140   **d** 168

13. Which of the following cannot represent the measure of an exterior angle of a regular polygon?

   a 72   **b** 15   **c** 27   **d** 45

14. Find the number of sides in a polygon if the sum of the measures of the interior angles is 4 times as great as the sum of the measures of the exterior angles.

15. Find the number of sides in a regular polygon if:

   a The measure of an interior angle is three times the measure of an exterior angle.

   b The measure of an interior angle equals the measure of an exterior angle.

   c The measure of an interior angle exceeds 6 times the measure of an exterior angle by 12.

16.   **GIVEN** $\overline{AB} \perp \overline{BD}$, $\overline{ED} \perp \overline{BD}$.

       PROVE $\angle A \cong \angle E$.

17.   **GIVEN** $\overline{AC} \perp \overline{CB}$, $\overline{DE} \perp \overline{AB}$.

       PROVE $\angle 1 \cong \angle 2$.

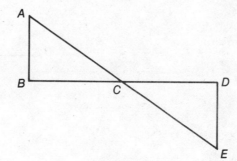

18. **GIVEN** $\overline{JP} \perp \overline{PR}$,
$\overline{LK} \perp \overline{KM}$,
$\overline{KM} \parallel \overline{PR}$.

PROVE $\overline{KL} \parallel \overline{PJ}$.

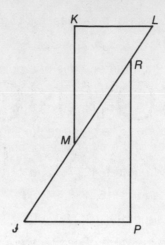

19. **GIVEN** $\overline{AB} \perp \overline{BC}$,
$\overline{DC} \perp \overline{BC}$,
$\overline{DE} \perp \overline{AC}$.

PROVE $\angle 1 \cong \angle 2$.

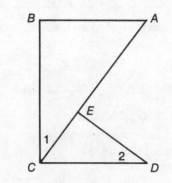

20. **GIVEN** $\overleftrightarrow{AX} \parallel \overleftrightarrow{CY}$,
$\overline{BA}$ bisects $\angle CAX$,
$\overline{BC}$ bisects $\angle ACY$.

PROVE $\angle ABC$ is a right angle.

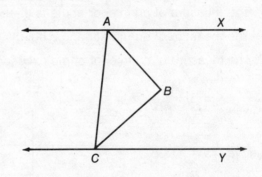

CHAPTER SIX
PROVING TRIANGLES ARE CONGRUENT

6.1 CORRESPONDENCES AND CONGRUENT TRIANGLES

On a triple blind date, Steve, Bob, and Charles accompanied Jane, Lisa, and Kris. To indicate which young man from the set of gentlemen is being paired with which young lady, the following notation may be used:

Steve ↔ Jane   Bob ↔ Lisa   Charles ↔ Kris

Such a pairing of the members of one group with the members of another group is called a *correspondence.* A correspondence may also be established between the vertices of two triangles, as shown in Figure 6.1.

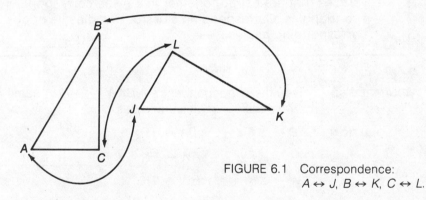

FIGURE 6.1 Correspondence:
$A \leftrightarrow J, B \leftrightarrow K, C \leftrightarrow L.$

We may concisely express this correspondence using the notation: △ABC ↔ △JKL. The order in which the vertices are written is critical since it defines the pairing of vertices:

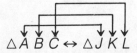

We stress that this correspondence may be written in more than one way. For example, △CAB ↔ △LJK defines the same correspondence as △ABC ↔ △JKL since the same pair of vertices are matched together. On the other hand, △ABC ↔ △JKL and △ACB ↔ △LKJ define two different correspondences.

    A specified correspondence also serves to define a set of corresponding angles and a set of corresponding sides. Given the correspondence $\triangle RST \leftrightarrow \triangle XYZ$, for example, angle pairs R and X, S and Y, and T and Z are corresponding angles. Segments which join corresponding vertices determine the corresponding sides:

$$\triangle R\ S\ T \leftrightarrow \triangle X\ Y\ Z \qquad \triangle R\ S\ T \leftrightarrow \triangle X\ Y\ Z \qquad \triangle R\ S\ T \leftrightarrow \triangle X\ Y\ Z$$
$$\overline{RS} \leftrightarrow \overline{XY} \qquad\qquad \overline{ST} \leftrightarrow \overline{YZ} \qquad\qquad \overline{RT} \leftrightarrow \overline{XZ}$$

Corresponding sides are $\overline{RS}$ and $\overline{XY}$, $\overline{ST}$ and $\overline{YZ}$, and $\overline{RT}$ and $\overline{XZ}$. Drawing the actual triangles in Figure 6.2, we see that *corresponding sides lie opposite corresponding angles.*

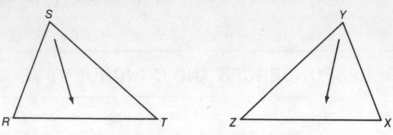

FIGURE 6.2   Corresponding sides lie opposite corresponding angles.

DEFINITION OF CORRESPONDING PARTS OF TRIANGLES

1. Corresponding *angles* are pairs of angles whose vertices are paired together in a given correspondence between two triangles.

2. Corresponding *sides* are pairs of segments whose end points are vertices that are paired together in a given correspondence between two triangles. Corresponding sides of triangles lie opposite corresponding angles.

EXAMPLE 6.1   Given the correspondence $\triangle BAW \leftrightarrow \triangle TFK$, name all pairs of corresponding angles and sides.

SOLUTION $\angle B \leftrightarrow \angle T \qquad \overline{BA} \leftrightarrow \overline{TF}$

$\angle A \leftrightarrow \angle F \qquad \overline{AW} \leftrightarrow \overline{FK}$

$\angle W \leftrightarrow \angle K \qquad \overline{BW} \leftrightarrow \overline{TK}$

EXAMPLE 6.2   Rewrite the correspondence given in Example 6.1 so that the same pairing of elements is maintained.

SOLUTION   There is more than one way of writing the same correspondence, including $\triangle WAB \leftrightarrow \triangle KFT$.

CONGRUENT TRIANGLES

Congruence means "can be made to coincide" or "exactly fits." Two segments, for example, can be made to coincide (assuming, of course, that we could move the segments) if they have the same size—that is, if they have the same length. Two angles can be made to coincide if they have the same shape—that is, if they have the same measure. In order for geometric *figures* to be congruent, they must have the same size *and* the same shape.

┌─ **DEFINITION OF CONGRUENT TRIANGLES** ────────────

Two triangles are *congruent* if their vertices can be paired in a correspondence so that:

1. All pairs of corresponding angles are congruent.

and

2. All pairs of corresponding sides are congruent.

In order for two triangles to be congruent the definition states that *six* pairs of parts must be congruent: three pairs of angles and three pairs of sides. A similar definition may be used to define congruent polygons of any number of sides.

EXAMPLE 6.3 Are the following two triangles congruent? If they are, write the correspondence between the triangles that establishes the congruence.

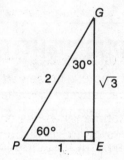

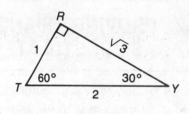

SOLUTION Yes. △ *PEG* ≅ △ *TRY.*

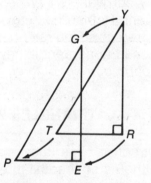

INCLUDED ANGLES AND SIDES

In subsequent work with congruent triangles the terms *included side* and *included angle* will be used.

With respect to angles *A* and *C* in Figure 6.3, side $\overline{AC}$ is called an included side; or side $\overline{AC}$ is said to be included between angle *A* and angle *C*. With respect to angles *B* and *C*, $\overline{BC}$ is the included side. What is the included side between angles *A* and *B*? ($\overline{AB}$)

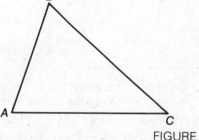

FIGURE 6.3

Similarly, with respect to sides $\overline{AB}$ and $\overline{CB}$, angle *B* is called an included angle; or angle *B* is said to be included between side $\overline{AB}$ and side $\overline{CB}$. With respect to sides $\overline{BA}$ and $\overline{CA}$, angle *A* is the included angle.

What angle is included between sides $\overline{BC}$ and AC? If you've said "angle C," then you have the idea.

EXAMPLE 6.4 For figure, use three letters to name the angle included between sides:

a $\overline{RS}$ and $\overline{WS}$

b $\overline{SW}$ and $\overline{TW}$

name the segment that is included between:

c Angles *WST* and *WTS*

d Angles *SWR* and *SRW*

SOLUTION   **a** ∡*RSW*

b ∡*TWS*

c $\overline{TS}$

d $\overline{RW}$

6.2 PROVING TRIANGLES CONGRUENT: SSS, SAS, AND ASA POSTULATES

Much of our work in geometry will be devoted to trying to prove that two triangles are congruent, given certain facts about the triangles. In order to simplify our work, we present several shortcut methods for proving triangles congruent. Rather than proving triangles congruent by demonstrating that they agree in *six* pairs of parts, it is possible to conclude that a pair of triangles is congruent if they agree in *three* pairs of parts, *provided they are a particular set of three pairs of congruent parts*.

TRIANGLES WHICH AGREE IN THREE SIDES

An instructive experiment may be performed by using a ruler and drawing two triangles so that the three sides of the first triangle have the same length as the corresponding sides in the second triangle. For example, see Figure 6.4.

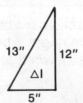
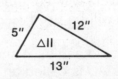

FIGURE 6.4

After the triangles are drawn, use a pair of scissors and cut out △II. You should be able to demonstrate that the two triangles may be made to coincide. That is, if three sides of one triangle are congruent to three sides of another triangle, that will "force" the corresponding angles of the triangles to be congruent and, therefore, make the triangles congruent. This is postulated as follows:

┌─ **SIDE-SIDE-SIDE (SSS) POSTULATE** ──────────────────────┐
│ If the vertices of two triangles can be paired so that three sides of one │
│ triangle are congruent to the corresponding sides of the second triangle, │
│ then the two triangles are congruent. │
└──┘

EXAMPLE 6.5

GIVEN $\overline{AB} \cong \overline{BC}$,
M is the midpoint of $\overline{AC}$.

PROVE $\triangle AMB \cong \triangle CMB$.

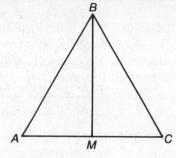

SOLUTION

PLAN   **1.** Mark off the diagram with the given. Note that an "x" is used to indicate that side $\overline{BM}$ is congruent to itself.

2. After the three pairs of congruent sides are identified, write up the formal proof.

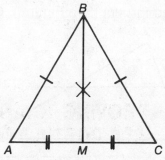

PROOF

Statements	Reasons
1. $\overline{AB} \cong \overline{BC}$. (Side)	1. Given.
2. M is the midpoint of $\overline{AC}$.	2. Given.
3. $\overline{AM} \cong \overline{CM}$. (Side)	3. A midpoint divides a segment into two congruent segments.
4. $\overline{BM} \cong \overline{BM}$. (Side)	4. Reflexive property of congruence.
5. $\triangle AMB \cong \triangle CMB$.	5. SSS Postulate.

THE SIDE-ANGLE-SIDE (SAS) AND THE ANGLE-SIDE-ANGLE (ASA) POSTULATES

In addition to the SSS Postulate, there are several other abbreviated methods that can be used to prove triangles congruent. The Side-Angle-Side Postulate is illustrated in Figure 6.5. If it is known that two sides and

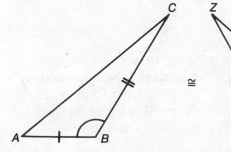

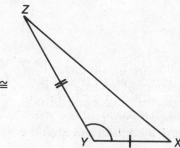

$$\overline{AB} \cong \overline{XY} \quad \boxed{S}$$

$$\angle B \cong \angle Y \quad \boxed{A}$$

$$\overline{BC} \cong \overline{YZ} \quad \boxed{S}$$

FIGURE 6 5

the included angle of one triangle are congruent to the corresponding parts of another triangle, then we may conclude that the triangles are congruent. Because the congruent angle must be sandwiched in between the congruent pairs of sides, this method is commonly referred to as the SAS Postulate.

SIDE-ANGLE-SIDE (SAS) POSTULATE

If the vertices of two triangles can be paired so that two sides and the included angle of one triangle are congruent to the corresponding parts of the second triangle, then the two triangles are congruent.

Similarly, if two angles and the included side (ASA) of one triangle are congruent to the corresponding parts of another triangle, then the triangles are congruent. See Figure 6.6.

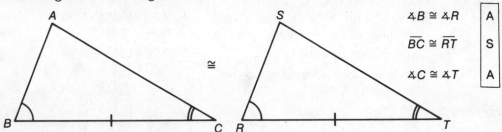

FIGURE 6.6

ANGLE-SIDE-ANGLE (ASA) POSTULATE

If the vertices of two triangles can be paired so that two angles and the included side of one triangle are congruent to the corresponding parts of the second triangle, then the two triangles are congruent.

It should be noted that both the SAS and ASA postulates may be proved using the SSS Postulate. However, these proofs are difficult, and such rigor is beyond the scope of this presentation.

EXAMPLE 6.6

GIVEN C is the midpoint of $\overline{BE}$,
$\angle B \cong \angle E$.

PROVE $\triangle ABC \cong \triangle DEC$.

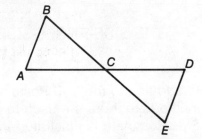

SOLUTION

PLAN **1.** Mark off the diagram with the given *and* any additional information that can be deduced to be congruent based on the diagram. In this instance, mark off the vertical angle pair.

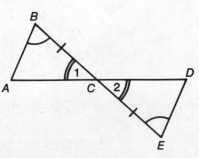

2. In this section we have presented three methods for proving triangles congruent: SSS, SAS, and ASA. Based on the diagram that has been marked off in Step 1, decide on which method to use. In this case, ASA is the appropriate method.

3. Write up the formal proof.

PROOF	Statements	Reasons
	1. $\angle B \cong \angle E$. (Angle)	1. Given.
	2. C is the midpoint of $\overline{BE}$.	2. Given.
	3. $\overline{BC} \cong \overline{EC}$. (Side)	3. A midpoint divides a segment into two congruent segments.
	4. $\angle 1 \cong \angle 2$. (Angle)	4. Vertical angles are congruent.
	5. $\triangle ABC \cong \triangle DEC$.	5. ASA Postulate.

EXAMPLE 6.7

GIVEN $\overline{AB} \parallel \overline{CD}$,

$\overline{AB} \cong \overline{CD}$.

PROVE $\triangle DAB \cong \triangle BCD$.

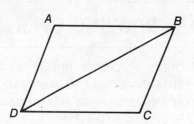

SOLUTION

PLAN **1.** Mark off the diagram.

2. Decide on the method to be used. In this example, SAS is the appropriate method.

3. Write up the formal proof.

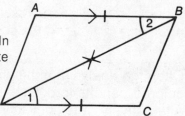

PROOF	Statements	Reasons
	1. $\overline{AB} \cong \overline{CD}$. (Side)	1. Given.
	2. $\overline{AB} \parallel \overline{CD}$.	2. Given.
	3. $\angle 1 \cong \angle 2$. (Angle)	3. If two lines are parallel, then their alternate interior angles are congruent.
	4. $\overline{BD} \cong \overline{BD}$. (Side)	4. Reflexive property of congruence.
	5. $\triangle DAB \cong \triangle BCD$.	5. SAS Postulate.

TO PROVE TWO TRIANGLES CONGRUENT

1. Mark off the diagram with the given.

2. Mark off any additional parts that may be deduced to be congruent such as vertical angles, or sides (or angles) shared by both triangles.

3. If appropriate, label angles referenced in the proof with numbers. This will make your job of writing up the proof easier.

4. Decide on which method of proving triangles congruent is to be applied.

5. Write up the formal proof. Next to each statement in which a required side or angle is established as being congruent, write "(Side)" or "(Angle)." When you look back at your proof, this will help you verify that you have satisfied the necessary conditions of the congruence postulate being used.

6.3 PROVING OVERLAPPING TRIANGLES CONGRUENT

Sometimes proofs appear to be more complicated than they actually are simply because the triangles are drawn so that one overlaps the other. For example, consider the following problem:

GIVEN $\overline{AB} \cong \overline{DC}$,

$\overline{AB} \perp \overline{BC}$, $\overline{DC} \perp \overline{BC}$.

PROVE $\triangle ABC \cong \triangle DCB$.

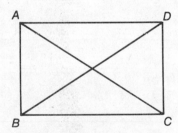

Often the analysis of the problem can be facilitated by using different colored pencils to outline the triangles to be proven congruent. Alternatively, it is sometimes helpful to redraw the diagram, "sliding" the triangles apart and then proceeding with marking the diagram:

PLAN Show $\triangle ABC \cong \triangle DCB$ by using SAS. Notice that $\overline{BC}$ is a side of both triangles. Overlapping triangles often share the same side or angle.

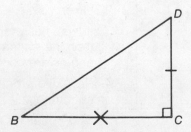
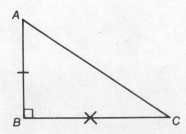

EXAMPLE 6.8

GIVEN $\overline{RA} \perp \overline{PQ}$, $\overline{QB} \perp \overline{PR}$,
$\overline{PA} \cong \overline{PB}$.

PROVE $\triangle PAR \cong \triangle PBQ$.

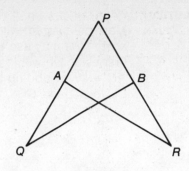

SOLUTION

PLAN 1. Separate the triangles and mark off corresponding congruent parts:

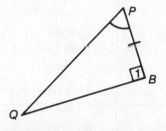 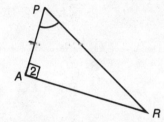

2. Observe that angle P is common to both triangles.

3. Prove triangles congruent by ASA.

PROOF

Statements	Reasons
1. $\angle P \cong \angle P$. (Angle)	1. Reflexive property of congruence.
2. $\overline{PA} \cong \overline{PB}$. (Side)	2. Given
3. $\overline{RA} \perp \overline{PQ}$ and $\overline{QB} \perp \overline{PR}$.	3. Given.
4. Angles 1 and 2 are right angles.	4. Perpendicular lines intersect to form right angles.
5. $\angle 1 \cong \angle 2$. (Angle)	5. All right angles are congruent.
6. $\triangle PAR \cong \triangle PBQ$.	6. ASA Postulate.

Sometimes the addition or subtraction properties must be applied in order to obtain a pair of congruent corresponding parts that will be needed in order to establish the congruence of a pair of triangles. Recall that we may only add or subtract the *measures* of segments or angles. In order to reduce the number of steps in a proof, we will freely convert from congruence to measure (in order to perform the arithmetic operation) and then convert back to congruence. This is illustrated in the following example (see steps 3 and 6).

EXAMPLE 6.9

GIVEN $\overline{AB} \cong \overline{DE}$, $\overline{AD} \cong \overline{FC}$,
$\overline{AB} \parallel \overline{DE}$.

PROVE $\triangle ABC \cong \triangle DEF$.

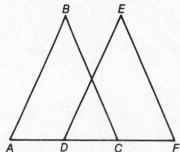

SOLUTION

PLAN Mark off diagram with the given.
NOTE: We may use SAS, *but* we must first establish within the proof that $\overline{AC} \cong \overline{DF}$ by adding the measure of $\overline{DC}$ to AD and CF.

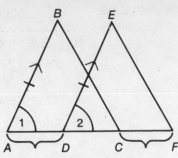

PROOF

Statements	Reasons
1. $\overline{AB} \cong \overline{DE}$. (Side)	1. Given.
2. $\angle 1 \cong \angle 2$. (Angle)	2. If two lines are parallel, then their corresponding angles are congruent.
3. $AD = FC$.	3. Given.
4. $DC = DC$.	4. Reflexive property of equality.
5. $AD + DC = FC + DC$.	5. Addition property of equality.
6. $\overline{AC} \cong \overline{FD}$. (Side)	6. Substitution.
7. $\triangle ABC \cong \triangle DEF$.	7. SAS Postulate.

EXAMPLE 6.10

GIVEN $\angle ABE \cong \angle CBD$,

$\angle BDE \cong \angle BED$,

$\overline{BD} \cong \overline{BE}$.

PROVE $\triangle DAB \cong \triangle ECB$.

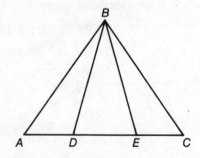

SOLUTION

PLAN Mark off the diagram. NOTE: We may use ASA, but we must establish within the proof that $\angle 3 \cong \angle 4$ by subtracting the measure of angle DBE from the measures of angles ABE and CBD.

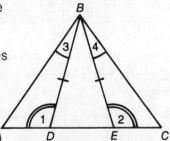

	Statements	Reasons
1.	$\angle BDE \cong \angle BED$.	1. Given.
2.	$\angle 1 \cong \angle 2$. (Angle)	2. Supplements of congruent angles are congruent.
3.	$\overline{BD} \cong \overline{BE}$. (Side)	3. Given.
4.	$m\angle ABE = m\angle CBD$.	4. Given.
5.	$m\angle DBE = m\angle DBE$.	5. Reflexive property of equality.
6.	$m\angle ABE - m\angle DBE = m\angle CBD - m\angle DBE$.	6. Subtraction property of equality.
7.	$\angle 3 \cong \angle 4$. (Angle)	7. Substitution.
8.	$\triangle DAB \cong \triangle ECB$.	8. ASA Postulate.

6.4 ADDITIONAL METHODS OF PROVING TRIANGLES CONGRUENT: AAS AND HY-LEG

Consider the following pair of triangles with the corresponding pairs of congruent parts already indicated:

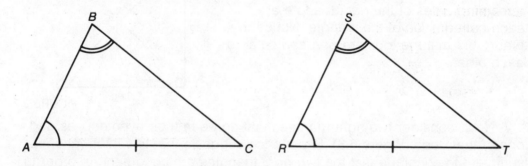

Are the triangles congruent? The triangles agree in two angles and the side *opposite* one of them (AAS), which means SSS, SAS, or ASA cannot be directly applied in establishing the congruence of the triangles. However, by Corollary 5.2.4, the third pair of angles of the triangles must be congruent. It follows that the triangles are congruent by application of the ASA Postulate. Thus, angle-angle-side $\cong$ angle-angle-side implies congruence of triangles.

> **ANGLE-ANGLE-SIDE (AAS) THEOREM**
>
> If the vertices of two triangles can be paired so that two angles and the side opposite one of them in one triangle are congruent to the corresponding parts of the second triangle, then the two triangles are congruent.

EXAMPLE 6.11

GIVEN $\overline{AB} \perp \overline{BD}$, $\overline{AC} \perp \overline{CD}$,

$\angle 1 \cong \angle 2$.

PROVE $\triangle ABD \cong \triangle ACD$.

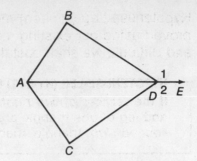

SOLUTION

OUTLINE OF PROOF Angles 3 and 4 are congruent since they are supplements of congruent angles. Angles 5 and 6 are congruent since all right angles are congruent. The triangles are, therefore, congruent by the AAS Theorem.

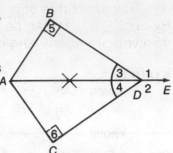

Some definitions related to a right triangle are necessary before we present the next (and final) method for proving triangles congruent.

In a right triangle, the side opposite the right angle is called the *hypotenuse* and the remaining sides of the right triangle are each called a *leg* of the triangle. Notice in Figure 6.7 that the legs are perpendicular to each other.

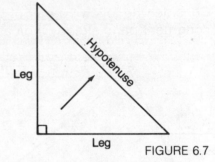

FIGURE 6.7

Next, consider two right triangles that agree in their hypotenuses and one of their legs (Figure 6.8). Experience suggests that this information is sufficient to conclude that the two *right* triangles are congruent. In order to make this assertion plausible, you may recall the Pythagorean Theorem from previous work in mathematics; a simple application of the Pythagorean Theorem implies that if two sides of a right triangle are congruent, the remaining sides must be congruent. Therefore, the triangles agree in all three sides and must be congruent by the SSS Postulate. Since the Pythagorean Theorem has yet to be formally presented, the preceding argument does *not* represent a valid proof. In fact, the so called

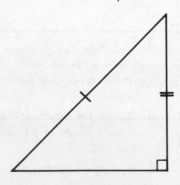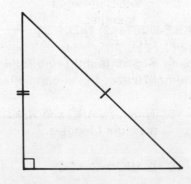

FIGURE 6.8

Hypotenuse-Leg (Hy-Leg) method of proving triangles congruent can be proved using our existing methods. However, since the proof is rather long and difficult, we shall postulate this method.

HYPOTENUSE-LEG (HY-LEG) POSTULATE

If the vertices of two *right* triangles can be paired so that the hypotenuse and leg of one triangle are congruent to the corresponding parts of the second *right* triangle, then the two *right* triangles are congruent.

This method is applicable only to pairs of *right* triangles. Before the Hy-Leg Postulate can be applied, it must be established that the triangles involved are right triangles.

EXAMPLE 6.12

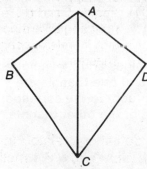

GIVEN $\overline{AB} \perp \overline{BC}$, $\overline{AD} \perp \overline{DC}$, $\overline{AB} \cong \overline{AD}$.

PROVE $\triangle ABC \cong \triangle ADC$.

SOLUTION

PLAN 1. Mark off diagram
2. Use the Hy-Leg Postulate.

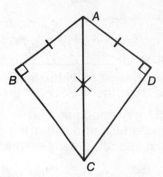

PROOF	Statements	Reasons
	1. $\overline{AC} \cong \overline{AC}$. (Hy)	1. Reflexive property of congruence.
	2. $\overline{AB} \cong \overline{AD}$. (Leg)	2. Given.
	NOTE Before we can invoke the Hy-Leg Postulate we must first establish that triangles *ABC* and *ADC* are right triangles.	
	3. $\overline{AB} \perp \overline{BC}$, $\overline{AD} \perp \overline{DC}$.	3. Given
	4. Angles *B* and *D* are right angles.	4. Perpendicular lines intersect to form right angles.
	5. Triangles *ABC* and *ADC* are right triangles.	5. A triangle which contains a right angle is a right triangle.
	6. $\triangle ABC \cong \triangle ADC$.	6. Hy-Leg Postulate.

■ You may conclude that two triangles are congruent if it can be shown that:

1. Three sides of one triangle are congruent to the corresponding parts of the other triangle.

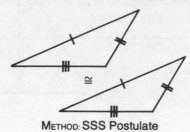

METHOD: SSS Postulate

3. Two angles and the included side of one triangle are congruent to the corresponding parts of the other triangle.

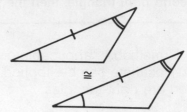

METHOD: ASA Postulate

2. Two sides and the included angle of one triangle are congruent to the corresponding parts of the other triangle.

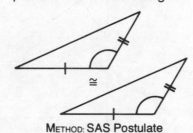

METHOD: SAS Postulate

4. Two angles and the side opposite one of them are congruent to the corresponding parts of the other triangle.

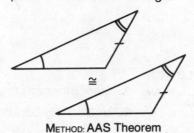

METHOD: AAS Theorem

■ You may conclude that two *right* triangles are congruent if the hypotenuse and either leg of one triangle are congruent to the corresponding parts of the other right triangle.

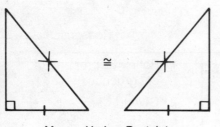 OR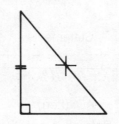

METHOD: Hy-Leg Postulate   METHOD: Hy-Leg Postulate

■ You may *not* conclude that two triangles are congruent when:

1. Two sides and an angle that is *not* included of one triangle are congruent to the corresponding parts of the other triangle.

SSA ≇ SSA

2. Three angles of one triangle are congruent to the corresponding parts of the other triangle.

AAA ≇ AAA

REVIEW EXERCISES FOR CHAPTER 6

1. **GIVEN** $\overline{BM} \perp \overline{AC}$, M is the midpoint of $\overline{AC}$.

    PROVE $\triangle ABM \cong \triangle CBM$.

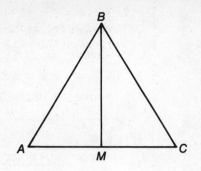

2. **GIVEN** $\overline{RT}$ bisects angles STW and SRW.

    PROVE $\triangle RST \cong \triangle RWT$.

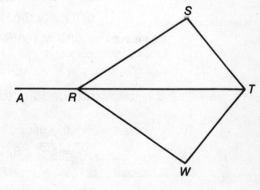

3. **GIVEN** $\overline{EF} \parallel \overline{AB}$, $\overline{ED} \parallel \overline{BC}$,
$\overline{AD} \cong \overline{FC}$.

    PROVE $\triangle ABC \cong \triangle FED$.

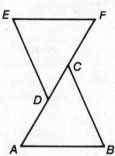

Use the following diagram to solve Exercises 4 and 5.

4. **GIVEN** $\angle R \cong \angle T$,
$\overline{SR} \cong \overline{ST}$.

    PROVE $\triangle SRH \cong \triangle STE$.

5. **GIVEN** $\overline{TE} \perp \overline{RS}$, $\overline{RH} \perp \overline{ST}$,
$\overline{EW} \cong \overline{HW}$.

    PROVE $\triangle EWR \cong \triangle HWT$.

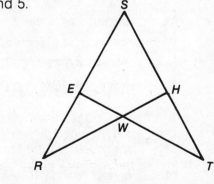

6. **GIVEN** $\overline{QU} \parallel \overline{DA}$, $\overline{QU} \cong \overline{DA}$.

    PROVE $\triangle UXQ \cong \triangle DXA$.

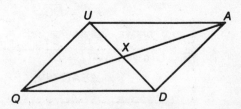

Use the following diagram to solve Exercises 7 and 8.

7. **GIVEN** $\overline{JK} \perp \overline{KT}$, $\overline{ET} \perp \overline{KT}$,
$\overline{KV} \cong \overline{TL}$, $\overline{JL} \cong \overline{EV}$.

PROVE $\triangle JKL \cong \triangle ETV$.

8. **GIVEN** $\overline{JK} \perp \overline{KT}$, $\overline{ET} \perp \overline{KT}$,
$\angle 1 \cong \angle 2$, $\overline{KL} \cong \overline{TV}$.

PROVE $\triangle JKL \cong \triangle ETV$.

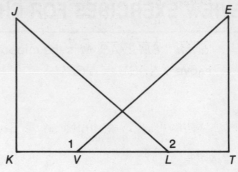

9. **GIVEN** $\overline{AR}$ bisects $\angle FRI$,
$\angle 1 \cong \angle 2$,
$\angle RFI \cong \angle RIF$.

PROVE $\triangle AFR \cong \triangle AIR$.

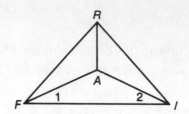

Use the following diagram to solve Exercises 10 and 11.

10. **GIVEN** S is the midpoint of $\overline{RT}$,
$\overline{SW} \cong \overline{SP}$, $\angle RSW \cong \angle TSP$.

PROVE $\triangle TSW \cong \triangle SRP$.

11. **GIVEN** $\overline{TW} \cong \overline{SP}$, $\overline{RP} \parallel \overline{SW}$,
$\overline{SP} \parallel \overline{TW}$.

PROVE $\triangle TSW \cong \triangle SRP$.

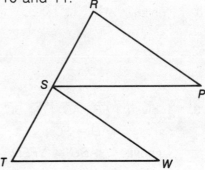

Use the following diagram to solve Exercises 12 and 13.

12. **GIVEN** $\overline{AB} \cong \overline{DE}$, M is the midpoint of $\overline{BE}$,
$\overline{AM} \cong \overline{KM}$, $\overline{DM} \cong \overline{KM}$.

PROVE $\triangle ABM \cong \triangle DEM$.

13. **GIVEN** $\overline{KM}$ is the $\perp$ bisector of $\overline{BE}$,
$\overline{KM}$ bisects $\angle AMD$,
$\overline{AB} \parallel \overline{KM} \parallel \overline{DE}$.

PROVE $\triangle ABM \cong \triangle DEM$.

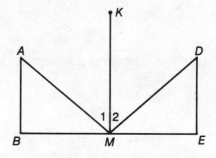

Use the following diagram to solve Exercises 14 and 15.

14. **GIVEN** $\overline{BD} \perp \overline{AC}$, $\overline{AF} \perp \overline{BC}$,
$\overline{FC} \cong \overline{DC}$.

PROVE $\triangle AFC \cong \triangle BDC$.

15. **GIVEN** $\overline{AD} \cong \overline{BF}$,
$\angle BAD \cong \angle ABF$.

PROVE $\triangle BAD \cong \triangle ABF$.

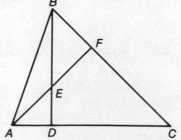

16. **GIVEN** $\overline{BF} \perp \overline{AC}$, $\overline{DE} \perp \overline{AC}$,
$\overline{AB} \cong \overline{DC}$, $\overline{AE} \cong \overline{CF}$.

PROVE $\triangle AFB \cong \triangle CED$.

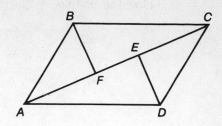

Use the following diagram to solve Exercises 17 and 18.

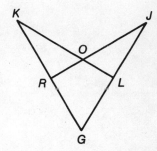

17. **GIVEN** $\overline{JR} \perp \overline{KG}$, $\overline{KL} \perp \overline{JG}$,
$\overline{KL} \cong \overline{JR}$.

PROVE $\triangle KLG \cong \triangle JRG$.

18. **GIVEN** $\overline{KG} \cong \overline{JG}$, $\overline{JR} \perp \overline{KG}$, $\overline{KL} \perp \overline{JG}$,
R is the midpoint of $\overline{KG}$,
L is the midpoint of $\overline{JG}$.

PROVE $\triangle KOR \cong \triangle JOL$.

19. **GIVEN** $\overline{AB} \perp \overline{BC}$, $\overline{DC} \perp \overline{BC}$,
$\overline{DB}$ bisects $\angle ABC$,
$\overline{AC}$ bisects $\angle DCB$,
$\overline{EB} \cong \overline{EC}$.

PROVE $\triangle BEA \cong \triangle CED$.

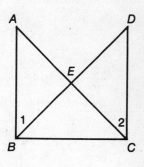

20. Prove if two triangles are congruent to the same triangle then they are
congruent to each other.

CHAPTER SEVEN
APPLYING CONGRUENT TRIANGLES

7.1 USING CONGRUENT TRIANGLES TO PROVE SEGMENTS AND ANGLES CONGRUENT

Based on the information marked off in Figure 7.1, can any conclusion be drawn concerning how the measures of angles *R* and *H* compare?

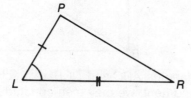 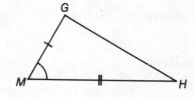

FIGURE 7.1

At first glance, there may seem to be insufficient information to draw any conclusion about these angles. However, by the SAS Postulate, △ *PLR* ≅ △ *GMH*. Do you recall the *definition* of congruent triangles? If two triangles have all *six* pairs of corresponding parts congruent, then the triangles are congruent. Since we have established that the two triangles under consideration are congruent, we may use the reverse of the definition of congruent triangles to conclude that *all* pairs of corresponding sides and *all* pairs of corresponding angles must be congruent. Since angles *R* and *H* are corresponding angles, they must be congruent.

This procedure represents an extremely useful method for proving segments or angles congruent, provided they are members of triangles which can be demonstrated to be congruent. *After* a pair of triangles are proven to be congruent, then *any* pair of corresponding parts may be concluded to be congruent, based on the principle that *corresponding parts of congruent triangles are congruent*, abbreviated CPCTC. The following series of examples will serve to further clarify this procedure.

EXAMPLE 7.1

GIVEN $\overline{AB} \cong \overline{AD}$,

$\overline{BC} \cong \overline{DC}$.

PROVE $\angle B \cong \angle D$.

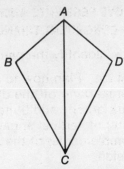

SOLUTION

PLAN Angles *B* and *D* are corresponding parts of triangles *ABC* and *ADC*, respectively. After proving these triangles congruent, we may then conclude that the desired pair of angles are congruent. Marking off the diagram suggests that the triangles can be proven congruent by the SSS Postulate.

PROOF

Statements	Reasons
1. $\overline{AB} \cong \overline{AD}$ and $\overline{BC} \cong \overline{DC}$.	1. Given
2. $\overline{AC} \cong \overline{AC}$.	2. Reflexive property of congruence.
3. $\triangle ABC \cong \triangle ADC$.	3. SSS Postulate.
4. $\angle B \cong \angle D$.	4. CPCTC.

EXAMPLE 7.2

GIVEN $\overline{AB} \cong \overline{DC}$,

$\overline{AB} \perp \overline{BC}$, $\overline{DC} \perp \overline{BC}$.

PROVE $\overline{AC} \cong \overline{DB}$.

SOLUTION

PLAN $\overline{AC}$ and $\overline{DB}$ are sides of triangles *ABC* and *DCB*, respectively. After proving these triangles congruent, we may then conclude that the desired pair of segments are congruent. Marking off the diagram suggests that the SAS method be applied.

PROOF

Statements	Reasons
1. $\overline{AB} \cong \overline{DC}$. (Side)	1. Given.
2. $\overline{AB} \perp \overline{BC}$ and $\overline{DC} \perp \overline{BC}$.	2. Given.
3. Angles *ABC* and *DCB* are right angles.	3. Perpendicular lines intersect to form right angles.
4. $\angle ABC \cong \angle DCB$. (Angle)	4. All right angles are congruent.
5. $\overline{BC} \cong \overline{BC}$. (Side)	5. Reflexive property of congruence.
6. $\triangle ABC \cong \triangle DCB$.	6. SAS Postulate.
7. $\overline{AC} \cong \overline{DB}$.	7. CPCTC.

As the diagrams and problems become more complicated, the need for a systematic procedure in proofs in which congruent triangles are used to prove segments and/or angles congruent becomes increasingly apparent. The following four-step approach is recommended.

┌──┐
│ **TO PROVE SEGMENTS AND/OR ANGLES CONGRUENT** ─────────────   │
│ **USING CONGRUENT TRIANGLES**   │
│   │
│ **1.** LOOK   Identify the pair of triangles which contain the desired parts. │
│   │
│ **2.** ANALYZE   Plan how to prove the selected pair of triangles congruent by │
│ first marking off the diagram with the given. In addition, be sure to │
│ mark off any additional pairs of parts which may be congruent as a │
│ result of vertical angles, perpendicular lines, parallel lines, supplements │
│ (complements) of the same (or congruent) angles, or a common angle │
│ or side.   │
│   │
│ **3.** DECIDE   Select the method of congruence to be used in establishing │
│ that the pair of triangles are congruent.   │
│   │
│ **4.** IMPLEMENT   Write up the formal proof.   │
└──┘

We illustrate this procedure in the next example.

EXAMPLE 7.3

GIVEN $\overline{MP} \cong \overline{ST}$,

$\overline{MP} \parallel \overline{ST}$,

$\overline{PL} \cong \overline{RT}$.

PROVE $\overline{RS} \parallel \overline{LM}$.

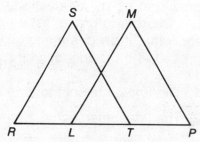

SOLUTION   In order to prove a pair of line segments parallel we must usually prove that an appropriate pair of angles are congruent.

1. Looking at the diagram, if $\overline{RS}$ is to be parallel to $\overline{LM}$ then it must be proven that $\angle SRT \cong \angle MLP$. This implies that the triangles which contain these angles must be proven congruent. Therefore, we must prove $\triangle RST \cong \triangle LMP$.

2. Mark off the diagram with the given.

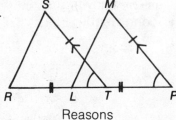

3. Use SAS Postulate.

4. The proof is given below.

PROOF

Statements	Reasons
1. $\overline{MP} \cong \overline{ST}$. (Side)	1. Given.
2. $\overline{MP} \parallel \overline{ST}$.	2. Given.
3. $\angle MPL \cong \angle STR$. (Angle)	3. If two lines are parallel, then their corresponding angles are congruent.
4. $\overline{PL} \cong \overline{RT}$. (Side)	4. Given.
5. $\triangle RST \cong \triangle LMP$.	5. SAS Postulate.
6. $\angle SRT \cong \angle MLP$.	6. CPCTC.
7. $\overline{RS} \parallel \overline{LM}$.	7. If two lines have their corresponding angles congruent, then they are parallel.

1. Notice that our approach is based on working backward beginning with the Prove. Once the solution path is clear, then the proof is written up, but working forward.

2. Although you are urged to follow the four-step approach that was presented in this section, these steps will often be consolidated into a "Plan" which will consist of key statements.

7.2 USING CONGRUENT TRIANGLES TO PROVE SPECIAL PROPERTIES OF LINES

In the previous section we illustrated how congruent triangles could be used to prove a pair of lines were parallel. Similarly, by first showing that an appropriate pair of segments or angles are congruent, a line may be demonstrated to be an angle or segment bisector or a pair of lines may be proved to be perpendicular.

In order to prove that a line bisects an angle or segment, we must show that it divides it into two congruent parts. A pair of lines may be proved to be perpendicular by showing either:

1. The lines intersect to form right angles, or

2. The lines intersect to form a pair of congruent adjacent angles.

For example, from the accompanying diagram, we may conclude that:

1. $\overline{BX}$ bisects $\overline{AC}$ by first proving that $\overline{AX} \cong \overline{XC}$.

2. $\overline{BX}$ bisects $\angle ABC$ by first proving that $\angle 1 \cong \angle 2$.

3. $\overline{BX} \perp \overline{AC}$ by first proving that $\angle 3 \cong \angle 4$ (that is, by showing a pair of adjacent angles are congruent).

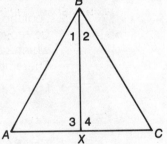

EXAMPLE 7.4

GIVEN $\angle 1 \cong \angle 2$,
$\overline{RM} \cong \overline{TM}$.

PROVE $\overline{SM}$ bisects $\angle RST$.

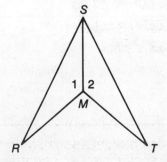

SOLUTION

PLAN Our goal is to prove $\angle RSM \cong \angle TSM$ by proving $\triangle RSM \cong \triangle TSM$. Marking off the diagram suggests that the SAS Postulate be applied.

PROOF	Statements	Reasons
	1. $\overline{RM} \cong \overline{TM}$. (Side)	1. Given.
	2. $\angle 1 \cong \angle 2$. (Angle)	2. Given.
	3. $\overline{SM} \cong \overline{SM}$. (Side)	3. Reflexive property of congruence.
	4. $\triangle RSM \cong \triangle TSM$.	4. SAS Postulate.
	5. $\angle RSM \cong \angle TSM$.	5. CPCTC.
	6. $\overline{SM}$ bisects $\angle RST$.	6. A segment which divides an angle into two congruent angles is an angle bisector. (NOTE: This is the *reverse* of the definition of angle bisector.)

EXAMPLE 7.5

GIVEN $\overline{LH} \cong \overline{LN}$,

$\overline{LB}$ bisects $\overline{HN}$.

PROVE $\overline{LB} \perp \overline{HN}$.

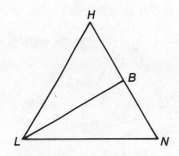

SOLUTION

PLAN Show $\angle LBH \cong \angle LBN$ by showing $\triangle LBH \cong \triangle LBN$ using SSS.

PROOF	Statements	Reasons
	1. $\overline{LH} \cong \overline{LN}$. (Side)	1. Given.
	2. $\overline{LB}$ bisects $\overline{HN}$.	2. Given.
	3. $\overline{BH} \cong \overline{BN}$. (Side)	3. A bisector divides a segment into two congruent segments.
	4. $\overline{LB} \cong \overline{LB}$. (Side)	4. Reflexive property of congruence.
	5. $\triangle LBH \cong \triangle LBN$.	5. SSS Postulate.
	6. $\angle LBH \cong \angle LBN$.	6. CPCTC.
	7. $\overline{LB} \perp \overline{HN}$.	7. If two lines intersect to form congruent adjacent angles, then the lines are perpendicular.

SUMMARY

- To prove a line bisects a segment (or an angle) show that it divides the segment (or angle) into two congruent segments (or angles).

- To prove a line is perpendicular to another line show that the lines meet to form right angles or, equivalently, that a pair of adjacent angles are congruent.

7.3 CLASSIFYING TRIANGLES AND SPECIAL SEGMENTS

In addition to classifying a triangle by the measures of its angles (acute, right, or obtuse), a triangle may be classified according to the number of its sides which are congruent. See Figure 7.2.

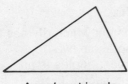

A *scalene* triangle has no congruent sides.

An *isosceles* triangle has at least two congruent sides.

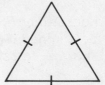

An *equilateral* triangle has three congruent sides.

FIGURE 7.2

Notice that an equilateral triangle is also isosceles. Some further definitions regarding the parts of an isosceles triangle are given in Figure 7.3.

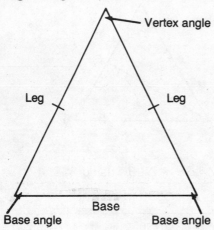

FIGURE 7.3 Legs, congruent sides; vertex angle, angle included between the legs; base, the side opposite the vertex angle; base angles, the angles which include the base and which lie opposite the legs.

DRAWING AUXILIARY LINES

Occasionally a proof will hinge on the drawing of an auxiliary line. For example, the proof of the theorem involving the sum of the measures of the angles of a triangle required that an auxiliary parallel line be drawn through a vertex of the triangle and parallel to the opposite side. Sometimes it may be necessary to draw other types of auxiliary lines. Given certain conditions, how many lines can be drawn which satisfy the stated conditions? One, more than one, or perhaps none. Some guidelines are needed.

We first take note of the *existence* of some special types of lines:

1. Every angle has a bisector.

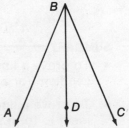

$\overrightarrow{BD}$ bisects angle *ABC*

2. Through a point not on a line segment, a line (or segment) may be drawn parallel to the segment, perpendicular to the segment, or to the midpoint of the segment (Figure 7.4).

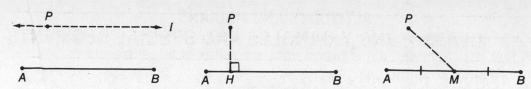

FIGURE 7.4   Each of these lines (or segments) is said to be *determined* since, based on the given conditions, *exactly one* such line (segment) can be drawn.

In describing what type of line is to be drawn, the line is said to be *under*determined if *too few* conditions are cited so that more than one line may be drawn to satisfy the stated condition(s).

For example, in Figure 7.5 more than one segment can be drawn through vertex *B* of triangle *ABC* so that it intersects side $\overline{AC}$.

A common error is to specify *too many* conditions so that the line to be drawn is *over*determined. When a line is *over*determined there does not necessarily exist a line which can be drawn which simultaneously meets all the given conditions. For example, in Figure 7.6 we try to draw a segment through vertex *B* of triangle *ABC* so that it is the perpendicular bisector of side $\overline{AC}$.

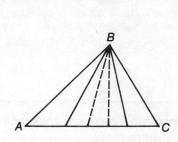

FIGURE 7.5   The line is *under*determined since more than one such line may be drawn.

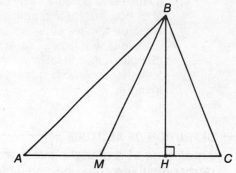

FIGURE 7.6   The line is *over*determined since we cannot be assured that the perpendicular dropped from point *B* will also intersect $\overline{AC}$ at its midpoint.

EXAMPLE 7.6   In the accompanying figure, classify the required line segment as underdetermined, determined, or overdetermined.

   a   Draw $\overline{BP}$.

   b   Through point *B* draw a line which bisects $\overline{AC}$.

   c   Through point *B* draw a line which intersects $\overline{AC}$.

   d   Draw a line through point *A* and parallel to $\overline{BC}$.

   e   Draw $\overline{BP}$ so that $\overline{BP}$ bisects angle *ABC*.

   f   Through point *B* draw a line which is perpendicular to $\overline{AC}$.

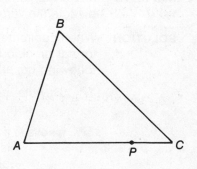

SOLUTION   **a** Determined   **b** Determined   **c** Underdetermined
               d Determined   **e** Overdetermined   **f** Determined

ALTITUDES AND MEDIANS

A line is *determined* if it is drawn from *any* vertex of a triangle either perpendicular to the opposite side *or* to the midpoint of the opposite side. These two segments are given special names: *altitude* (⊥) and *median*. See Figure 7.7.

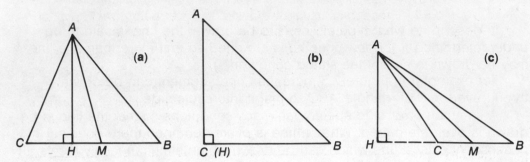

FIGURE 7.7   In each triangle an altitude ($\overline{AH}$) and a median ($\overline{AM}$) are drawn from vertex A to side BC. The formal definitions of altitude and median are also presented. (*a*) Every triangle has three altitudes and three medians. In this diagram, an altitude and a median may be drawn from vertex B to side $\overline{AC}$ and from vertex C to side $\overline{AB}$. (*b*) In a right triangle an altitude drawn to one of the legs will coincide with the other leg. In this diagram, the altitude $\overline{AH}$ drawn to leg $\overline{CB}$ coincides with leg $\overline{AC}$. (*c*) When drawing an altitude to a side of an obtuse triangle, it may be necessary to extend the side to which it is drawn. The altitudes drawn to a side which includes the obtuse angle will fall outside the triangle.

DEFINITION OF ALTITUDE

An *altitude* of a triangle is a segment drawn from *any* vertex of a triangle, perpendicular to the opposite side, extended, if necessary.

DEFINITION OF MEDIAN

A *median* of a triangle is a segment drawn from any *vertex* of a triangle to the midpoint of the opposite side.

EXAMPLE 7.7   Prove that the median drawn to the base of an isosceles triangle bisects the vertex angle.

SOLUTION   When confronted with a verbal statement of a problem requiring a formal proof, our first concern is to draw a suitable diagram. The following general approach is suggested:

1.   Rewrite, if necessary, the problem statement in "If . . . then . . ." form. For example,

   PROVE   If a median is drawn to the base of an isosceles triangle, then the vertex angle is bisected.

2.   Identify the hypothesis (given) which is contained in the "If" clause.

3.   Identify the conclusion (prove) which is found in the "then" clause.

4. Draw and label an appropriate diagram:

GIVEN $\overline{AB} \cong \overline{AC}$,

$\overline{AM}$ is a median to side $\overline{BC}$.

PROVE $\overline{AM}$ bisects ∢BAC.

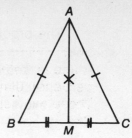

5. Proceed as usual by arriving at a plan before writing up the formal two-column proof.

PLAN $\overline{BM} \cong \overline{CM}$ since a median bisects the segment to which it is drawn. $\triangle AMB \cong \triangle AMC$ by the SSS Postulate. Angles 1 and 2 are congruent by CPCTC. $\overline{AM}$ bisects angle BAC (a segment which divides an angle into two congruent angles is an angle bisector). The formal two-column proof is left for you.

SUMMARY

DRAWING AUXILIARY SEGMENTS

Segments which do not appear in the diagram may be drawn provided too few conditions (*underdetermined* line) or too many conditions (*overdetermined* line) are not imposed. When exactly one line can be drawn which satisfies the given conditions, then the line is said to be *determined*.

SPECIAL SEGMENTS IN TRIANGLES

- An *altitude* of a triangle is perpendicular to the side to which it is drawn.

- A *median* of a triangle bisects the side to which it is drawn.

Altitudes and medians are determined segments. In every triangle, three altitudes and three medians can be drawn.

7.4 THE ISOSCELES TRIANGLE

Drawing several isosceles triangles (Figure 7.8) leads one to suspect that there is a relationship between the measures of the angles which lie opposite the congruent sides. This observation is stated in the following theorem:

THEOREM 7.1 BASE ANGLES THEOREM

If two sides of a triangle are congruent, then the angles opposite those sides are congruent.

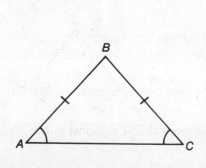
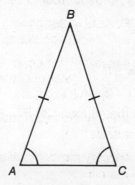

FIGURE 7.8

The proof is instructive since it illustrates the drawing of an auxiliary line.

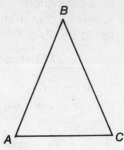

GIVEN $\overline{AB} \cong \overline{CB}$.

PROVE $\angle A \cong \angle C$.

PLAN   Based upon past experience we look to prove angle *A* congruent to angle *C* by proving that they are corresponding angles of congruent triangles. In order to form two triangles which can be proven congruent, we may draw the bisector of angle *ABC* which will intersect the base at some point, say point *R*. The resulting triangles may be proven congruent by the SAS Postulate. (NOTE: It is also possible to prove this theorem by drawing an altitude or median to side $\overline{AC}$.)

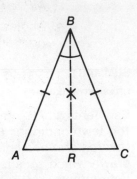

PROOF

Statements	Reasons
1. $\overline{AB} \cong \overline{CB}$. (Side)	1. Given.
2. Draw the bisector of angle *ABC*, naming the point it intersects $\overline{AC}$, point *R*.	2. An angle has exactly one bisector.
3. $\angle ABR \cong \angle CBR$. (Angle)	3. An angle bisector divides an angle into two congruent angles.
4. $\overline{BR} \cong \overline{BR}$. (Side)	4. Reflexive property of congruence.
5. $\triangle ABR \cong \triangle CBR$.	5. SAS Postulate.
6. $\angle A \cong \angle C$.	6. CPCTC.

EXAMPLE 7.8   The measure of the vertex angle of an isosceles triangle is three times as great as the measure of a base angle. Find the measure of a base angle of the triangle.

SOLUTION
$$3x + x + x = 180$$
$$5x = 180$$
$$x = 36$$

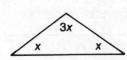

EXAMPLE 7.9

GIVEN $\overline{SR} \cong \overline{ST}$,
$\overline{MP} \perp \overline{RS}$, $\overline{MQ} \perp \overline{ST}$,
M is the midpoint of $\overline{RT}$.

PROVE $\overline{MP} \cong \overline{MQ}$.

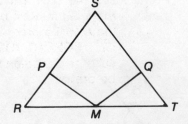

SOLUTION

PLAN By application of Theorem 7.1, $\angle R \cong \angle T$. Marking off the diagram suggests that triangles *MPR* and *MQT* may be proved congruent by using the AAS Theorem.

PROOF

Statements	Reasons
1. $\overline{SR} \cong \overline{ST}$.	1. Given.
2. $\angle R \cong \angle T$. (Angle)	2. If two sides of a triangle are congruent, then the angles opposite those sides are congruent.
3. $\overline{MP} \perp \overline{RS}$, $\overline{MQ} \perp \overline{ST}$.	3. Given.
4. Angles *MPR* and *MQT* are right angles.	4. Perpendicular lines intersect to form right angles.
5. $\angle MPR \cong \angle MQT$. (Angle)	5. All right angles are congruent.
6. *M* is the midpoint of $\overline{RT}$.	6. Given.
7. $\overline{RM} \cong \overline{TM}$. (Side)	7. A midpoint divides a segment into two congruent segments.
8. $\triangle MPR \cong \triangle MQT$.	8. AAS Theorem.
9. $\overline{MP} \cong \overline{MQ}$.	9. CPCTC.

PROVING A TRIANGLE IS ISOSCELES

The converse of the Base Angles Theorem is presented as theorem 7.2.

> **THEOREM 7.2 CONVERSE OF THE BASE ANGLES THEOREM**
> If two angles of a triangle are congruent, then the sides opposite are congruent.

OUTLINE OF PROOF

GIVEN $\angle A \cong \angle C$.
PROVE $\overline{AB} \cong \overline{CB}$.

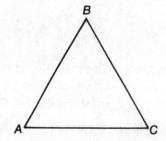

PLAN Draw the bisector of angle *B*, intersecting side $\overline{AC}$ at *R*. The resulting pair of triangles may be proven congruent by the AAS Theorem. It follows that $\overline{AB} \cong \overline{CB}$ by CPCTC. (Other auxiliary lines, such as a median or altitude, may also be drawn.)

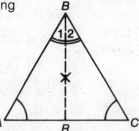

Theorem 7.2 is particularly useful in proving that a triangle is isosceles. A triangle may be proven to be isosceles by showing that either:

1. A pair of sides are congruent, or

2. A pair of angles are congruent (since by Theorem 7.2 the sides opposite must be congruent).

EXAMPLE 7.10

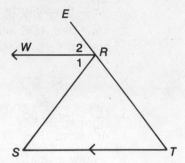

GIVEN $\overline{WR} \parallel \overline{ST}$,

$\overline{WR}$ bisects ∡SRE.

PROVE △SRT is isosceles.

SOLUTION

PLAN Show the base angles are congruent to each other by showing that each is congruent to one of the pairs of angles formed by the angle bisector. Since the base angles are congruent to congruent angles, they are congruent to each other and the triangle is isosceles.

PROOF

Statements	Reasons
1. $\overline{WR}$ bisects ∡SRE.	1. Given.
2. ∡1 ≅ ∡2.	2. A bisector divides an angle into two congruent angles.
3. $\overline{WR} \parallel \overline{ST}$.	3. Given.
4. ∡1 ≅ ∡S.	4. If two lines are parallel, then their alternate interior angles are congruent.
5. ∡2 ≅ ∡T.	5. If two lines are parallel, then their corresponding angles are congruent.
6. ∡S ≅ ∡T.	6. Transitive property of congruence.
7. △SRT is isosceles.	7. A triangle that has a pair of congruent angles is isosceles.

EXAMPLE 7.11 Prove that if two altitudes of a triangle are congruent, then the triangle is isosceles.

SOLUTION

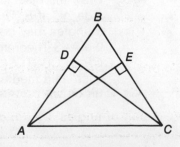

GIVEN $\overline{CD}$ is the altitude to $\overline{AB}$,

$\overline{AE}$ is the altitude to $\overline{BC}$,

$\overline{CD} ≅ \overline{AE}$.

PROVE △ABC is isosceles.

PLAN Our goal is to show ∡BAC = ∡BCA by proving △ADC ≅ △CEA. Marking off the diagram suggests the Hy-Leg method be used:

$$\overline{AC} \cong \overline{AC} \text{ (Hy)} \qquad \text{and} \qquad \overline{CD} \cong \overline{AE} \text{ (Leg)}$$

PROOF

Statements	Reasons
1. $\overline{CD}$ is the altitude to $\overline{AB}$, $\overline{AE}$ is the altitude to $\overline{BC}$.	1. Given.
2. Triangles *ADC* and *CEA* are right triangles.	2. A triangle which contains a right angle is a right triangle. (*Note*: This step consolidates several obvious steps.)
3. $\overline{CD} \cong \overline{AE}$ (Leg)	3. Given.
4. $\overline{AC} \cong \overline{AC}$ (Hy)	4. Reflexive property of congruence.
5. △ADC ≅ △CEA	5. Hy-Leg Postulate.
6. ∡BAC ≅ ∡BCA	6. CPCTC.
7. △ABC is isosceles.	7. A triangle that has a pair of congruent angles is isosceles.

SUMMARY

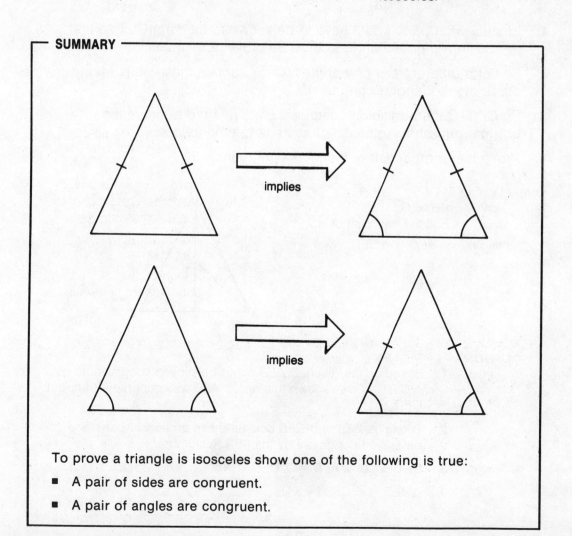

To prove a triangle is isosceles show one of the following is true:

- A pair of sides are congruent.
- A pair of angles are congruent.

7.5 DOUBLE CONGRUENCE PROOFS

In some problems it may appear that there is insufficient information provided in the given in order to be able to prove a pair of triangles congruent. Upon closer examination, however, it may be possible to

1. Prove a *different* pair of triangles which are contained in the diagram congruent, and then to

2. Obtain additional pairs of congruent parts (by CPCTC) which can then be used to

3. Prove the original pair of triangles congruent.

Thus, these problems are characterized by proving one pair of triangles congruent in order to obtain congruent corresponding parts which can then be used in proving the desired pair of triangles congruent. These problems tend to be difficult and will require a certain amount of trial-and-error work on your part. The following general approach will prove helpful when confronted with a problem in which one pair of congruent triangles must be used to prove a second pair of triangles congruent.

1. Identify what parts would have to be known to be congruent so that the desired pair of triangles could be proven congruent.

2. Select a different pair of triangles which contain these parts and prove these triangles congruent.

3. By CPCTC the additional congruent parts needed to prove the original pair of triangles congruent has been obtained.

4. Prove the original pair of triangles congruent.

EXAMPLE 7.12

GIVEN $\overline{AB} \cong \overline{CB}$,
E is the midpoint of $\overline{AC}$.

PROVE $\triangle AED \cong \triangle CED$.

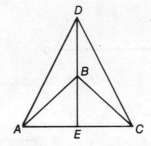

SOLUTION

PLAN 1. Based on the given, $\triangle AED$ could be proven congruent to $\triangle CED$ if it was known that angle AED was congruent to angle CED.

2. Triangles AEB and CEB contain these angles as parts and can be proved congruent by the SSS Postulate.

3. By CPCTC, $\angle AED = \angle CED$.

4. $\triangle AED \cong \triangle CED$ by SAS.

PROOF	Statements	Reasons

Part I. To Prove △*AEB* ≅ △*CEB*:

1. $\overline{AB} \cong \overline{CB}$. (Side)	1. Given.
2. *E* is the midpoint of $\overline{AC}$.	2. Given.
3. $\overline{AE} \cong \overline{CE}$. (Side)	3. A midpoint divides a segment into two congruent segments.
4. $\overline{BE} \cong \overline{BE}$.	4. Reflexive property of congruence.
5. △*AEB* ≅ △*CEB*.	5. SSS Postulate.

Part II. To Prove △*AED* ≅ △*CED*:

6. ∡*AED* ≅ ∡*CED*. (Angle)	6. CPCTC.
7. $\overline{DE} \cong \overline{DE}$. (Side)	7. Reflexive property of congruence.
8. △*AED* ≅ △*CED*.	8. SAS Postulate.

EXAMPLE 7.13

GIVEN $\overline{BC} \cong \overline{AD}$,
$\overline{BC} \parallel \overline{AD}$,
$\overline{AR} \cong \overline{CS}$.

PROVE $\overline{BR} \cong \overline{DS}$.

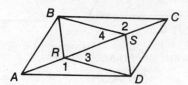

SOLUTION

PLAN **1.** The desired pair of segments can be proven congruent if it can be proven that △*BRS* ≅ △*DSR*. (*Note:* Although the desired pair of segments are also contained in triangles *ARB* and *CSD*, efforts to prove these triangles congruent would prove fruitless.)

2. By first proving △*ARD* ≅ △*CSB* we may obtain the congruent parts necessary to prove the desired pair of triangles congruent. Triangles *ARD* and *CSB* are congruent by SAS:

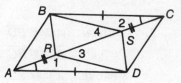

3. By CPCTC, $\overline{RD} \cong \overline{SB}$ and ∡1 ≅ ∡2. Since supplements of congruent angles are congruent, angles 3 and 4 are congruent.

4. $\overline{RS} \cong \overline{RS}$ so △*BRS* ≅ △*DSR* by SAS, and $\overline{BR} \cong \overline{DS}$ by CPCTC.

The formal proof is left for you.

REVIEW EXERCISES FOR CHAPTER 7

1. Find the value of *x*:

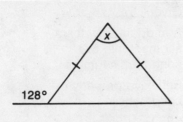

a

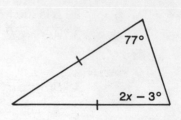

b

c

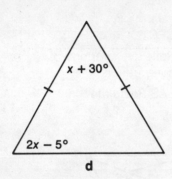

d

e

$\overline{AD}$ and $\overline{CD}$ are angle bisectors.

f

2. GIVEN $\overline{FH}$ bisects $\angle GHJ$, $\overline{GH} \cong \overline{JH}$.

PROVE $\angle 1 \cong \angle 2$.

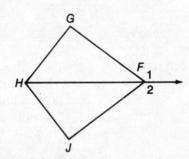

3. GIVEN $\overline{AC} \cong \overline{BD}$, $\overline{AB} \perp \overline{BC}$, $\overline{DC} \perp \overline{BC}$.

PROVE $\angle 1 \cong \angle 2$.

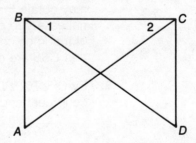

For Exercises 4 and 5, use the following diagram.

4. GIVEN $\angle J \cong \angle K$, $\overline{PJ} \cong \overline{PK}$.

PROVE $\overline{JY} \cong \overline{KX}$.

5. GIVEN $\overline{PX} \cong \overline{PY}$, $\overline{XJ} \cong \overline{YK}$.

PROVE $\overline{KX} \cong \overline{JY}$.

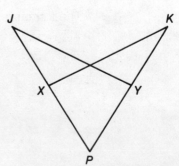

6. **GIVEN** $\overline{UT} \parallel \overline{DW}$, $\overline{UT} \cong \overline{DW}$,
$\overline{QW} \cong \overline{AT}$.

PROVE $\overline{UQ} \parallel \overline{AD}$.

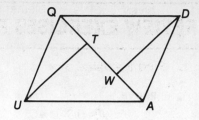

7. **GIVEN** $\angle S \cong \angle H$,
$\overline{SR} \perp \overline{RW}$, $\overline{HW} \perp \overline{RW}$,
$\overline{ST} \cong \overline{HT}$.

PROVE T is the midpoint of $\overline{RW}$.

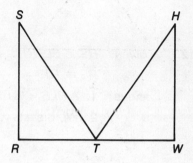

8. **GIVEN** $\overline{AB} \cong \overline{CB}$, $\overline{AD} \cong \overline{CD}$.

PROVE $\overline{DB}$ bisects $\angle ADC$.

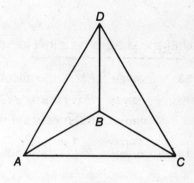

9. **GIVEN** $\overline{BA} \perp \overline{MA}$, $\overline{CD} \perp \overline{MD}$.
Point M is the midpoint of $\overline{BC}$.

PROVE $\overline{BC}$ bisects $\overline{AD}$.

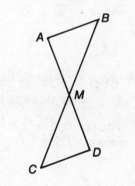

10. **GIVEN** $\angle 1 \cong \angle 2$,
$\overline{HK}$ bisects $\angle RHN$,
$\overline{HR} \cong \overline{HN}$.

PROVE $\overline{HK} \perp \overline{RN}$.

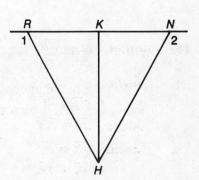

11. **GIVEN** ∡1 is supplementary to ∡2,
$\overline{RC} \cong \overline{AT}$,
$\overline{SR} \cong \overline{AB}$,
$\overline{ST} \cong \overline{BC}$.

PROVE **1.** $\overline{ST} \perp \overline{TR}$.
2. $\overline{BC} \perp \overline{AC}$.

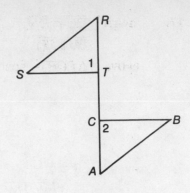

12. **GIVEN** $\overline{RS} \perp \overline{SL}$, $\overline{RT} \perp \overline{LT}$,
$\overline{RS} \cong \overline{RT}$.

PROVE **1.** $\triangle RLS \cong \triangle RLT$.
2. $\overline{WL}$ bisects ∡SWT.

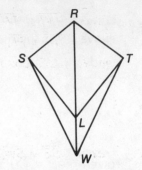

Use the diagram below to solve Exercises 13 and 14.

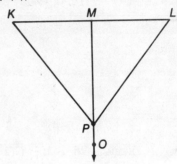

13. **GIVEN** $\overline{PM}$ is the altitude to $\overline{KL}$,
$\overline{PK} \cong \overline{PL}$.

PROVE Point M is the midpoint of $\overline{KL}$.

14. **GIVEN** $\overline{PM}$ is the median to $\overline{KL}$,
$\overline{KP} \cong \overline{LP}$.

PROVE $\overline{PM} \perp \overline{KL}$.

15. **GIVEN** $\overline{PS}$ and $\overline{LT}$ are altitudes
to sides $\overline{LM}$ and $\overline{PM}$, respectively.
∡LPT ≅ ∡SLP.

PROVE $\overline{PS} \cong \overline{LT}$.

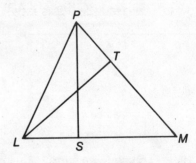

16. **GIVEN** $\overline{XL} \cong \overline{XP}$,
$\overline{XL} \perp \overline{TR}$, $\overline{XP} \perp \overline{TS}$,
X is the midpoint of $\overline{RS}$.

PROVE $\triangle RTS$ is isosceles.

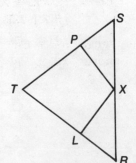

17. GIVEN $\overline{QL} \cong \overline{QM}$,
$\overline{LM} \parallel \overline{PR}$.

PROVE $\triangle PQR$ is isosceles.

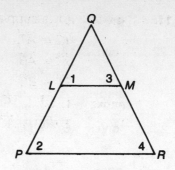

18. GIVEN $\overline{LW} \perp \overline{TW}$, $\overline{FX} \perp \overline{XP}$,
$\overline{TF} \cong \overline{PL}$, $\overline{WL} \cong \overline{XF}$.

PROVE $\triangle FML$ is isosceles.

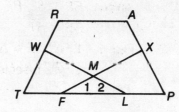

19. GIVEN $\overline{SE} \cong \overline{SW}$,
$\angle 1 \cong \angle 2$,
$\overline{EL} \cong \overline{WB}$.

PROVE $\overline{KL} \cong \overline{AB}$.

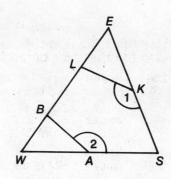

20. GIVEN $\overline{OV} \cong \overline{LV}$,
$\overline{KO} \cong \overline{ZL}$.

PROVE $\triangle KVZ$ is isosceles.

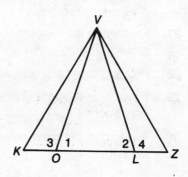

21. GIVEN $\angle JHL \cong \angle JLH$,
$\overline{BH} \perp \overline{HJ}$, $\overline{KL} \perp \overline{LJ}$.

PROVE $\overline{JB} \cong \overline{JK}$.

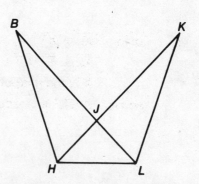

Use the following diagram to solve Exercises 22 and 23.

22. **GIVEN** $\overline{AB} \cong \overline{AD}$,

   $\overline{EA}$ bisects $\angle DAB$.

   PROVE $\triangle BCE \cong \triangle DCE$.

23. **GIVEN** $\overline{BE} \cong \overline{DE}$,

   $\overline{BC} \cong \overline{DC}$.

   PROVE $\overline{CA}$ bisects $\angle DAB$.

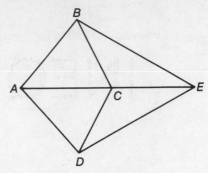

24. **GIVEN** $\overline{AB} \parallel \overline{CD}$,

   $\overline{AB} \cong \overline{CD}$,

   $\overline{AL} \cong \overline{CM}$.

   PROVE $\angle CBL = \angle ADM$.

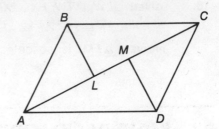

Use the following diagram to solve Exercises 25 and 26.

25. **GIVEN** $\angle FAC \cong \angle FCA$,

   $\overline{FD} \perp \overline{AB}, \overline{FE} \perp \overline{BC}$.

   PROVE $\overline{BF}$ bisects $\angle DBE$.

26. **GIVEN** $\overline{BD} \cong \overline{BE}$,

   $\overline{FD} \cong \overline{FE}$.

   PROVE $\triangle AFC$ is isosceles.

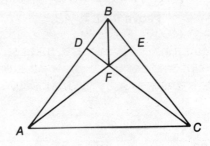

27. Prove that if two triangles are congruent, then the altitudes drawn to a pair of corresponding sides are congruent.

28. Prove that an equilateral triangle is equiangular.

29. Prove that the altitudes drawn to the legs of an isosceles triangle are congruent.

30. Prove that the medians drawn to the legs of an isosceles triangle are congruent.

CHAPTER EIGHT
INEQUALITIES AND REASONING

8.1 SOME BASIC PROPERTIES OF INEQUALITIES

In everyday life we are constantly making comparisons. Which of two people earns more money? Which of two athletes can run faster? Which of two students has the higher grade average? Comparisons also play an important role in mathematics. It is often essential that we know how two numbers compare in magnitude. In comparing two quantities, say a and b, there are exactly three possibilities:

TABLE 8.1

CONDITION	NOTATION	EXAMPLE
a is less than b.	$a < b$	$3 < 5$
a is equal to b.	$a = b$	$4 = 4$
a is greater than b.	$a > b$	$7 > 2$

The *direction* or *sense* of an inequality refers to whether the inequality symbol is pointing to the right ($>$) or to the left ($<$). For example,

1. $a > b$ and $c > d$ are two inequality expressions that have the same direction or *sense* since they are *both* greater than.

2. By changing $a > b$ to $a < b$ (or vice versa), we have reversed the direction or *sense* of the inequality.

We shall assume a number of properties related to inequalities are true.

TABLE 8.2

PROPERTY	FORMAL STATEMENT	EXAMPLE
Addition	If $a < b$ then $a + c < b + c$.	$\begin{array}{r} 3 < 5 \\ +4 = 4 \\ \hline 7 < 9 \end{array}$
	If $a < b$ and $c < d$, then $a + c < b + d$.	$\begin{array}{r} 3 < 5 \\ +6 < 10 \\ \hline 9 < 15 \end{array}$
Subtraction	If $a < b$ then $a - c < b - c$.	$\begin{array}{r} 8 < 13 \\ -6 = 6 \\ \hline 2 < 7 \end{array}$
	If $a < b$ and $c < d$, then $a - d < b - c$.	$\begin{array}{r} 8 < 13 \\ 1 < 4 \\ \hline (8-4) < (13-1) \end{array}$
Multiplication	If $a < b$ and $c > 0$, then $ac < bc$.	$\begin{array}{r} 5 < 8 \\ 2(5) < 2(8) \\ \text{or} \quad 10 < 16 \end{array}$
Transitive	If $a < b$ and $b < c$, then $a < c$.	$\begin{array}{r} 4 < 7 \\ \text{and} \quad 7 < 10, \\ \text{then} \quad 4 < 10. \end{array}$

Some comments regarding the previous set of inequality properties should be made.

1. Although each property was expressed in terms of the less than relation ($<$), the properties clearly hold for the greater than (and equal) relation.

2. As a corollary to the multiplication property, we note that "halves of unequals are unequal in the same sense." For example, if $a < b$, then $a/2 < b/2$.

3. A quantity may be substituted for its equal in an equation as well as in an inequality expression. For example, if $a + b < c$, and $b = d$, then $a + d < c$.

4. If $a < b$ and $c > d$, then no conclusion can be stated regarding how the values of $a + c$ and $b + d$ compare.

5. $a < b$ and $b > a$ are equivalent expressions.

Turning our attention to inequalities in geometry, we begin by making two obvious remarks.

1. If point X is *between* points A and B, then $AX < AB$ and $XB < AB$.

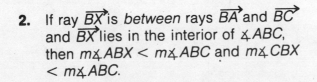

2. If ray $\overrightarrow{BX}$ is *between* rays $\overrightarrow{BA}$ and $\overrightarrow{BC}$ and $\overrightarrow{BX}$ lies in the interior of $\angle ABC$, then $m\angle ABX < m\angle ABC$ and $m\angle CBX < m\angle ABC$.

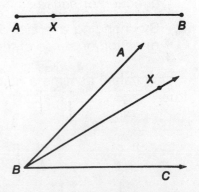

In both situations the conclusions drawn ultimately rest on the "betweenness" property of points and rays. We shall agree that when it becomes necessary to use these facts as statements in a formal proof, the reason we shall offer is simply Definition of Betweenness of Points (or Rays).

8.2 INEQUALITY RELATIONSHIPS IN A TRIANGLE

We have previously encountered an example of a geometric inequality relationship when discussing the exterior angle of a triangle theorem (see Theorem 5.4). The lengths of the sides of a triangle must conform to an inequality relationship based on the fact that the shortest distance between two points is the length of the segment joining them. See Figure 8.1.

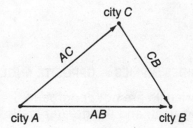

FIGURE 8.1

The shortest trip from city A to city B is the direct route represented by segment AB. This direct route must be less than any indirect route such as traveling from city A to city C and then from city C to city B. That is,

$$AB < AC + CB$$

TRIANGLE INEQUALITY POSTULATE

The length of *each* side of a triangle must be less than the sum of the lengths of the remaining two sides.

This postulate provides us with a convenient method for determining whether a set of three numbers can represent the lengths of the sides of a triangle. All we need do is test that *each* number is less than the sum of the other two. This is illustrated in the following example.

EXAMPLE 8.1 Which of the following sets of numbers cannot represent the sides of a triangle?

a 9, 40, 41   **b** 7, 7, 3   **c** 4, 5, 1   **d** 6, 6, 6

SOLUTION Choice **c**: $4 < 5 + 1$ and $1 < 4 + 5$, but 5 is *not* less than $4 + 1$.

COMPARING ANGLES OF A TRIANGLE

The base angles theorem tells us that if two sides of a triangle are congruent, then the angles opposite these sides are congruent (See Figure 8.2a.). What conclusion can be drawn if the two sides of the triangle are *not* congruent? In Figure 8.2b, clearly the angles cannot be congruent. Furthermore, it appears that the greater angle lies opposite the longer side.

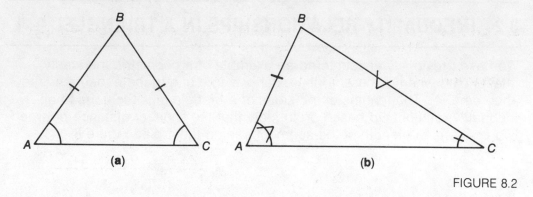

(a)   **(b)**

FIGURE 8.2

THEOREM 8.1 ≢ **SIDES IMPLIES** ≢ **OPPOSITE ANGLES**

If two sides of a triangle are not congruent, then the angles opposite these sides are not congruent, and the greater angle is opposite the longer side.

OUTLINE OF PROOF

GIVEN $BC > BA$.

PROVE $m\angle BAC > m\angle C$.

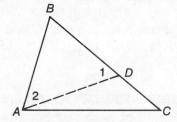

PLAN   Since $BC > BA$, there exists a point D on $\overline{BC}$ such that $AB = DB$.

1 Draw $\overline{AD}$. (See diagram.)

2 $m\angle 1 > m\angle C$ (Exterior angle of a triangle theorem, Theorem 5.4.)

3 $m\angle 1 = m\angle 2$ (Base angles theorem.)

4 $m\angle 2 > m\angle C$ (Substitution.)

5 $m\angle BAC > m\angle 2$ (Definition of betweenness of rays; see diagram.)

6 $m\angle BAC > m\angle C$ (Transitive property.)

The converse of Theorem 8.1 is also true.

THEOREM 8.2 ≢ **ANGLES IMPLIES** ≢ **OPPOSITE SIDES**

If two angles of a triangle are not congruent, then the sides opposite these angles are not congruent, and the longer side is opposite the greater angle.

EXAMPLE 8.2   In △ABC, AB = 3, BC = 5, and AC = 7. What is the largest angle of the triangle? the smallest angle of the triangle?

SOLUTION   ∡B is the largest angle.
∡C is the smallest angle.

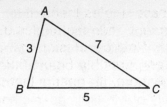

EXAMPLE 8.3   The measure of the vertex angle S of isosceles triangle RST is 80. What is the longest side of the triangle?

SOLUTION $\overline{RT}$ is the longest side.

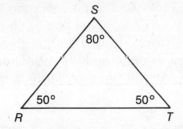

EXAMPLE 8.4

GIVEN $\overline{BD}$ bisects ∡ABC.

PROVE   AB > AD.

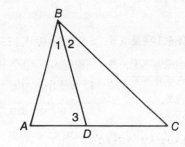

SOLUTION

PLAN   By Theorem 8.2 in order to prove AB > AD we must first establish that m∡3 (the angle opposite $\overline{AB}$) is greater than m∡1 (the angle opposite $\overline{AD}$).

PROOF

Statements	Reasons
1. m∡3 > m∡2.	1. The measure of an exterior angle of a triangle is greater than the measure of either nonadjacent interior angle.
2. $\overline{BD}$ bisects ∡ABC.	2. Given.
3. m∡1 = m∡2.	3. A bisector divides an angle into two angles having the same measure.
4. m∡3 > m∡1.	4. Substitution property of inequalities.
5. AB > AD.	5. If two angles of a triangle are not equal in measure, then the sides opposite are not equal and the longer side is opposite the greater angle.

8.3   THE INDIRECT METHOD OF PROOF

If Sherlock Holmes had a list of three suspects, he might arrive at the identity of the criminal *indirectly* by proving that two of the suspects could not have possibly been at the scene of the crime, thereby indirectly establishing the guilt of the only remaining suspect. Needless to say, the validity of Mr. Holmes' method is based on the assumption that he has correctly surmised that there were exactly three suspects.

The indirect method of proof in mathematics is based on a similar strategy. To prove an assertion *indirectly*, we account for all possibilities (suspects). We then investigate each of the possibilities (except the one to be proved) and show that each of these lead to a contradiction or an inconsistency based on the given information. The result is that the remaining possibility (the desired conclusion) must be true. Several examples will help to clarify this procedure.

EXAMPLE 8.5   Use an indirect method of proof to show that a triangle has at most one obtuse angle.

SOLUTION

INDIRECT
PROOF
- There are exactly two possibilities:

    1. the triangle has *at most* one obtuse angle.

    2. the triangle has *more than* one obtuse angle.

- Since our goal is to establish that the first possibility is true, we investigate the second possibility.

- Assume the second possibility is true. Suppose the triangle has two obtuse angles. Since the sum of these two obtuse angles exceeds 180, this situation is impossible.

- The second possibility is false. By elimination, the first possibility (which is the only remaining possibility) must be true.

TO PROVE AN ASSERTION USING THE INDIRECT METHOD OF PROOF

1. Examine the prove and list the different possibilities. Often this will simply be a matter of listing the prove and its *negation*. The negation of the statement "$\overline{AB}$ is congruent to $\overline{CD}$" is "$\overline{AB}$ is *not* congruent to $\overline{CD}$." The negation of the statement "line *l* is *not* parallel to line *m*" is "line *l* *is* parallel to line *m*."

2. Assume that each possibility is true (except the desired one).

3. By reasoning directly, show that each possibility ultimately leads to a contradiction of the *given* or of some established geometric principle. After each contradiction is illustrated, state that the corresponding assumption must be false.

4. State the desired conclusion must be true by elimination since all other possibilities have been accounted for and have been demonstrated to be false.

EXAMPLE 8.6

GIVEN   ∡1 is not congruent to ∡2.

PROVE   Line *l* is not parallel to line *m*.

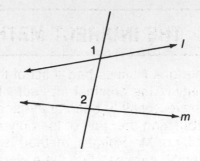

SOLUTION   Use an indirect method of proof:

1. There are two possibilities: the original statement and its negation: line *l* is parallel to line *m*.

2. Assume the negation is true; that is, line *l* *is* parallel to line *m*.

3. If the lines are parallel, then corresponding angles are congruent which means that ∡1 ≅ ∡2. But this contradicts the given.

4. Since its negation is false, the statement "line *l* is not parallel to line *m*" must be true since it is the only remaining possibility.

EXAMPLE 8.7

GIVEN $\overline{AB} \cong \overline{DB}$.

PROVE $\overline{AB}$ is not ≅ to $\overline{BC}$.

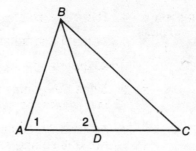

SOLUTION

INDIRECT PROOF   Assume the negation of the desired conclusion is true. That is, assume $\overline{AB}$ *is* ≅ to $\overline{BC}$. If $\overline{AB} \cong \overline{BC}$, then ∡1 ≅ ∡C. From the given, ∡1 ≅ ∡2. Hence, ∡2 ≅ ∡C. But this is impossible since the measure of an exterior angle of a triangle must be greater than the measure of either nonadjacent interior angle. Hence, our assumption is false and the only remaining possibility, that $\overline{AB}$ is not ≅ to $\overline{BC}$, must be true.

EXAMPLE 8.8

GIVEN $\overline{TW} \perp \overline{RS}$,
∡1 is not ≅ to ∡2.

PROVE $\overline{TW}$ is not the median to side $\overline{RS}$.

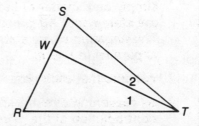

SOLUTION

INDIRECT PROOF   Assume the negation of the desired conclusion is true: assume $\overline{TW}$ is the median to side $\overline{RS}$. Then $\overline{WR} \cong \overline{WS}$ and △TWS is congruent to △TWR by SAS. By CPCTC, ∡1 ≅ ∡2. But this contradicts the given. Hence, the only remaining possibility, that $\overline{TW}$ is *not* the median to side $\overline{RS}$, must be true.

Proofs which rely on the indirect method of proof may also be organized in our familiar two-column format. The two-column format of the proof for Example 8.8 is shown.

PROOF	Statements	Reasons
	1. ∡1 is *not* ≅ to ∡2.	1. Given.
	2. Either $\overline{TW}$ is *not* the median to side $\overline{RS}$ or $\overline{TW}$ *is* the median to side $\overline{RS}$. Assume $\overline{TW}$ *is* the median to side $\overline{RS}$.	2. A statement is either true or false.
	3. $\overline{WR} \cong \overline{WS}$. (Side)	3. A median divides a side into two congruent segments.
	4. $\overline{TW} \perp \overline{RS}$.	4. Given.
	5. Angles *TWS* and *TWR* are right angles.	5. Perpendicular lines meet to form right angles.
	6. ∡*TWS* ≅ ∡*TWR*. (Angle)	6. All right angles are congruent.
	7. $\overline{TW} \cong \overline{TW}$. (Side)	7. Reflexive property of congruence.
	8. △*TWS* ≅ △*TWR*.	8. SAS Postulate.
	9. ∡1 ≅ ∡2.	9. CPCTC.
	10. $\overline{TW}$ is *not* the median to side $\overline{RS}$.	10. Statement 9 contradicts statement 1. The assumption made in statement 2 must therefore be false. By elimination, statement 10 is established.

8.4 LOGIC AND REASONING

We have seen that statements of the form, "If *statement 1,* then *statement 2,*" are commonly encountered in geometry. In order to simplify our notation, we shall use the letter *p* to represent *statement 1* (the if clause) and the letter *q* to represent *statement 2* (the then clause):

General Form of a Conditional Statement: If *p*, then *q*.

In addition to forming the converse of a statement, two other types of statements may be formed which can give us clues when exploring the validity of new geometric relationships. The Table 8.3 lists the variety of different combinations of statements that can be formed from a statement of the If/then form.

TABLE 8.3

FORM OF THE STATEMENT	TYPE OF STATEMENT
If *p*, then *q*.	Original statement
If *q*, then *p*.	Converse
If *not p*, then *not q*.	Inverse
If *not q*, then *not p*.	Contrapositive

Note that the *inverse* of a statement is formed by taking the negation of both *p* (the hypothesis) and *q* (the conclusion) of the original statement. Similarly, the *contrapositive* is formed by negating both parts of the converse of the original statement.

EXAMPLE 8.9 Form the inverse, converse, and contrapositive of the following theorem and state whether the resulting statement is true.

SOLUTION If two angles are right angles, then they are congruent.

TYPE	EXAMPLE	TRUE/FALSE
Inverse	If two angles are not right angles, then they are not congruent.	False
Converse	If two angles are congruent, then they are right angles.	False
Contrapositive	If two angles are not congruent, then they are not right angles.	True

In general, *the contrapositive of a true statement will always be true.* The converse and inverse of a true statement may or may not be true.

A statement and its contrapositive are said to be *logically equivalent* since if the statement is true, then its contrapositive must be true; if the statement is false, then its contrapositive is false. As a point of interest we note that the indirect method of proof that was illustrated in the previous section is based on the relationship between the statement to be proven and the statement's contrapositive. For example, let us reconsider Example 8.6:

GIVEN (IF) ∢1 is not congruent to ∢2.
PROVE (THEN) Line *l* is not parallel to line *m*.

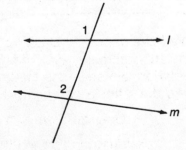

In this proof we are asked to prove that if angle 1 is not congruent to angle 2, then line *l* is not parallel to line *m*. Using the indirect method of proof, we assumed that if line *l* is parallel to line *m*, then angle 1 is congruent to angle 2, which contradicted the given and thus completed the proof. Notice that the indirect proof used the contrapositive of the statement to be proven:

Original Statement of the Proof	Contrapositive
If two corresponding angles are not congruent, then the lines are not parallel.	If two lines are parallel, then the corresponding angles are congruent.

In other words, a perfectly valid proof for Example 8.6 would proceed as follows:

ALTERNATIVE PROOF   The statement to be proved is of the form "if a pair of corresponding angles are not congruent, then the lines are not parallel." The statement's contrapositive is, "if two lines are parallel, then corresponding angles are congruent." The contrapositive is true. Since the contrapositive and statement are logically equivalent, the original statement must also be true.

Nevertheless, the indirect method of proof illustrated in Section 8.3 is a powerful tool. Although some proofs can be approached using either a direct or indirect method, the indirect proof is often easier. Other proofs can only be performed using an indirect approach (see Example 8.10).

Another interesting application of the indirect method of proof is in providing an alternative means for developing the properties of parallel lines.

EXAMPLE 8.10   Prove Postulate 4.1: If two lines are parallel, then their alternate interior angles are congruent.

GIVEN $l \parallel m$.

PROVE $\angle 1 \cong \angle 2$.

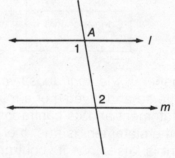

SOLUTION   Using an indirect method of proof, we assume that angle 1 is *not* congruent to angle 2. It is therefore possible to construct at point A a line k such that angle 3 is congruent to angle 2:

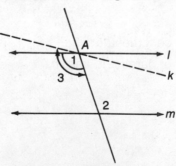

Since $\angle 3 \cong \angle 2$, line k is parallel to line m since if alternate interior angles are congruent, the lines are parallel. Through point A, line l has been drawn parallel to line m (given) and line k has been drawn parallel to line m. This contradicts the Parallel Postulate which says exactly one such line can be drawn. Hence our assumption that angles 1 and 2 are not congruent leads to a contradiction which implies that angles 1 and 2 must be congruent.

REVIEW EXERCISES FOR CHAPTER 8

Form the converse, inverse, and contrapositive of the statements in Exercises 1 to 3. Determine whether the resulting statement is true or false.

1. If I live in Los Angeles, then I live in the United States.

2. If today is Monday, then tomorrow is Tuesday.

3. If two angles of a triangle are congruent, then the sides opposite these angles are congruent.

4. In $\triangle ABC$, $BC > AB$ and $AC < AB$. Which is the longest side of the triangle?

5. In $\triangle RST$, $m\angle R < m\angle T$ and $m\angle S > m\angle T$. Which is the largest angle of the triangle?

6.   **GIVEN** $\overline{AB} \cong \overline{BD}$,
                   $m\angle 5 > m\angle 6$.

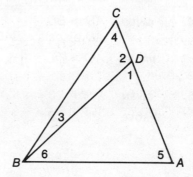

State whether each of the following inequality relationships are true or false.

  a $m\angle 1 > m\angle 3$.

  b $m\angle 5 > m\angle 1$.

  c $m\angle 3 < m\angle ABC$.

  d $m\angle 2 > m\angle 1$.

  e $BC > BA$.

7. In $\triangle ABC$, $AB > AC$, and $BC > AC$. Name the smallest angle of $\triangle ABC$.

8. In $\triangle RST$, $ST > RT$, and $RT > RS$.

  a   If one of the angles of the triangle is obtuse, which angle of the triangle must it be?

  b   If the measure of one of the angles of the triangle is 60, which angle of the triangle must it be?

9. Determine whether each of the following sets of numbers can represent the lengths of the sides of a triangle.

  a   8, 17, 15   **b** $\frac{1}{2}, \frac{1}{3}, \frac{1}{6}$

  c   1, 1, 3   **d**   6, 6, 7

10.   **GIVEN** $\overline{AB} \cong \overline{CB}$.

       PROVE $AB > BD$.

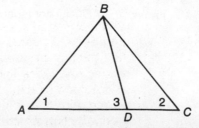

11. GIVEN $m \angle 1 = m \angle 2$.

PROVE $AD > ED$.

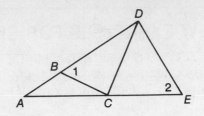

Use the following diagram to solve Exercises 12 to 14.

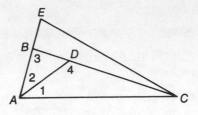

12. GIVEN Triangles AEC and ABC.

PROVE $m \angle 4 > m \angle AEC$.

13. GIVEN $AC > BC$.

PROVE $AD > BD$.

14. GIVEN $AD > BD$,
AD bisects $\angle BAC$.

PROVE $AC > DC$.

15. GIVEN $\angle 1 \cong \angle 3$.

PROVE $\angle 1 \not\cong \angle 2$.

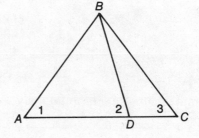

16. GIVEN $\angle 1 \cong \angle 2$.

PROVE $\overline{AB} \not\cong \overline{BC}$.

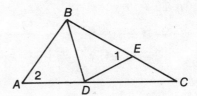

17. GIVEN $RS = TS$.

PROVE $RW \neq WL$.

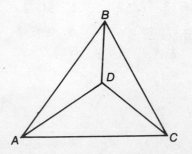

18. GIVEN $\triangle ABC$ is *not* isosceles,
$\angle ADB \cong \angle CDB$.

PROVE $\triangle ADC$ is *not* isosceles.

19. **GIVEN** $AC > AB$,
$\overline{DE} \cong \overline{CE}$.

PROVE $\overline{AB}$ is *not* parallel to $\overline{DE}$.

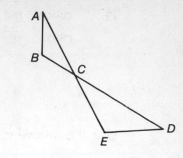

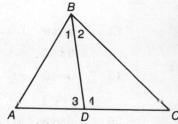

20. **GIVEN** $\triangle ABC$ is scalene,
$\overline{BD}$ bisects $\angle ABC$.

PROVE $\overline{BD}$ is *not* $\perp \overline{AC}$.

21. Prove that the length of the line segment drawn from any vertex of an equilateral triangle to a point on the opposite side is less than the length of any side of the triangle.

22. Prove that if the vertex angle of an isosceles triangle is obtuse, then the base is longer than either leg.

23. Prove the shortest distance from a point to a line is the length of the perpendicular segment from the point to the line.

CHAPTER NINE
SPECIAL QUADRILATERALS

9.1 INTRODUCTION

In this chapter we shall extend our knowledge of geometry by exploring the properties of some special types of *quadrilaterals*. Quadrilaterals have four sides and four interior angles. When referring to pairs of angles or to pairs of sides of a quadrilateral, the terms *opposite* and *consecutive* are used.

A pair of *consecutive sides* of a quadrilateral intersect at a vertex of the quadrilateral. *Opposite sides* do *not* intersect. See Figure 9.1.

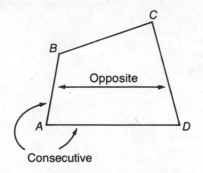 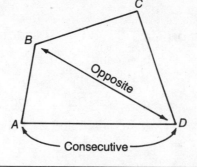

FIGURE 9.1

PAIRS OF CONSECUTIVE SIDES	PAIRS OF OPPOSITE SIDES
$\overline{AD}$ and $\overline{AB}$	$\overline{AD}$ and $\overline{BC}$
$\overline{AB}$ and $\overline{BC}$	$\overline{AB}$ and $\overline{DC}$
$\overline{BC}$ and $\overline{CD}$	
$\overline{CD}$ and $\overline{DA}$	

Pairs of *opposite angles* are located at alternate vertices of a quadrilateral. *Consecutive angle* pairs are found at adjacent vertices. See Figure 9.1.

PAIRS OF CONSECUTIVE ANGLES	PAIRS OF OPPOSITE ANGLES
∡A and ∡D	∡A and ∡C
∡A and ∡B	∡D and ∡B
∡B and ∡C	
∡C and ∡D	

A *diagonal* is a line segment which joins opposite (alternate) vertices. See Figure 9.2.

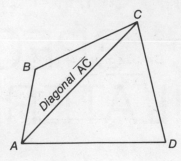

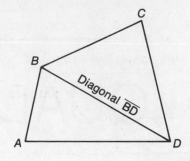

FIGURE 9.2

EXAMPLE 9.1   In the accompanying diagram, name all pairs of:

    a Opposite sides

    b Consecutive sides

    c Opposite angles

    d Consecutive angles

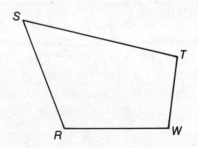

SOLUTION   **a** $\overline{RS}$ and $\overline{WT}$; $\overline{ST}$ and $\overline{RW}$

    b $\overline{RS}$ and $\overline{ST}$; $\overline{ST}$ and $\overline{TW}$; $\overline{TW}$ and $\overline{WR}$; $\overline{WR}$ and $\overline{RS}$

    c Angles R and T; angles W and S

    d Angles R and S; S and T; T and W; W and R

CLASSIFYING QUADRILATERALS

It may happen that a quadrilateral's sides, angles, or diagonals may exhibit some special properties. For example, a quadrilateral may have one or both pairs of opposite sides parallel. If you consider the relationships between the sides and angle pairs of quadrilaterals to be genetic traits, then a family tree of quadrilaterals may be developed with a special name given to each type of quadrilateral which displays a special trait. This is illustrated in Figure 9.3.

    The family tree of quadrilaterals shows that a quadrilateral has two major types of descendants. One, called a *trapezoid,* has exactly one pair of parallel sides. The other special type of quadrilateral has two pairs of parallel sides and is called a *parallelogram.*

> **DEFINITION OF A PARALLELOGRAM**
> A *parallelogram* is a quadrilateral having two pairs of parallel sides.
>   NOTATION:□*ABCD* is read as ''parallelogram *ABCD*.'' The letters *A, B, C,* and *D* represent consecutive vertices of the quadrilateral and the symbol which precedes these letters is a miniature parallelogram.

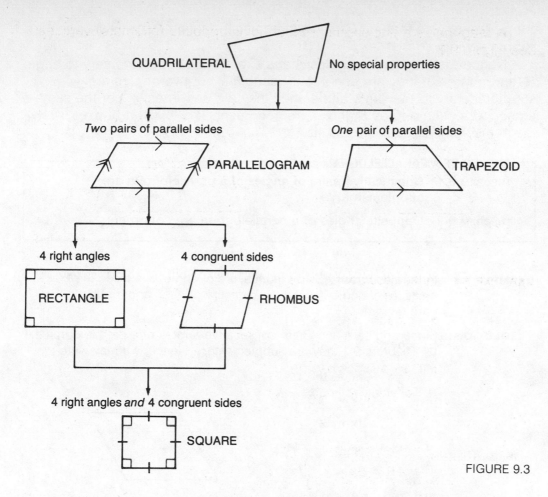

FIGURE 9.3

Of the special quadrilaterals which are listed in Figure 9.3, only the trapezoid is *not* a parallelogram. A rectangle, rhombus, and square are special types of parallelograms. The remainder of this chapter will be devoted to developing the special properties of each of these figures.

9.2 PROPERTIES OF A PARALLELOGRAM

ANGLES OF A PARALLELOGRAM

Consider parallelogram *ABCD* (Figure 9.4) in which the measure of angle *A* is 70. What are the measures of angles *B*, *C*, and *D*?

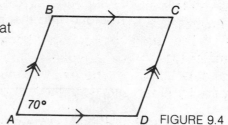

FIGURE 9.4

In parallelogram *ABCD*, $\overline{AB}$ may be considered to be a transversal, intersecting parallel line segments $\overline{AD}$ and $\overline{BC}$. Since interior angles on the same side of a transversal intersecting parallel lines are supplementary, $m\angle B = 110$ (180 − 70 = 110). Similarly, angles *B* and *C* are supplementary so that $m\angle C = 70$. In addition, angles A and D are supplementary so that $m\angle D = 110$.

Notice also that since angles A and C are each supplementary to angle B (and angle D), they are congruent (recall that, "if two angles are supplementary to the same angle, then they are congruent"). For the same reason, opposite angles B and D are congruent. The following two theorems summarize these results.

ANGLES OF A PARALLELOGRAM

THEOREM 9.1   Consecutive pairs of angles of a parallelogram are supplementary.

THEOREM 9.2   Opposite angles of a parallelogram are congruent.

EXAMPLE 9.2   In parallelogram $ABCD$ the measure of angle B is twice the measure of angle A. Find the measure of each angle of the parallelogram.

SOLUTION   Since angles A and B are consecutive angles of a parallelogram, by Theorem 9.1 they are supplementary. Hence, we may write

$$x + 2x = 180$$
$$3x = 180$$
$$x = 60$$
$$m\angle A = x = 60$$
$$m\angle B = 2x = 120$$
$$m\angle C = m\angle A = 60$$
$$m\angle D = m\angle B = 120$$

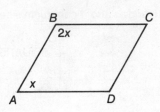

DIAGONALS AND SIDES OF A PARALLELOGRAM

Draw either diagonal of the parallelogram in Figure 9.5. In each case two triangles are formed. Is there a relationship between the triangles formed by a diagonal? Consider diagonal $\overline{BD}$ in Figure 9.5a. Notice that $\overline{BC}$, $\overline{BD}$, and $\overline{DA}$ form a Z. Since $\overline{BC}$ is parallel to $\overline{AD}$, $\angle 1 \cong \angle 2$. Similarly, $\overline{AB}$, $\overline{BD}$, and $\overline{DC}$ form a Z. Since, $\overline{AB} \parallel \overline{DC}$, $\angle 3 \cong \angle 4$. In addition, $\overline{BD} \cong \overline{BD}$, from which it follows that $\triangle BAD \cong \triangle DCB$ by ASA. Using the same approach it can be easily shown that in Figure 9.5b $\overline{AC}$ divides parallelogram $ABCD$ into two triangles such that $\triangle ABC \cong \triangle CDA$. These observations are summarized in the following theorem.

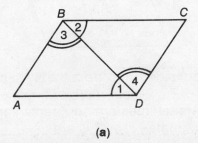

(a)

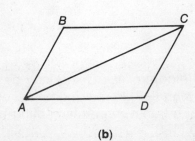

(b)

FIGURE 9.5

In a parallelogram either diagonal separates the parallelogram into two congruent triangles.

Theorem 9.3 allows us to draw some conclusions regarding the parts of the parallelogram. Applying the CPCTC principle, we may conclude that $\overline{AD} \cong \overline{BC}$ and $\overline{AB} \cong \overline{DC}$. This result is stated formally in Theorem 9.4.

THEOREM 9.4

Opposite sides of a parallelogram are congruent.

EXAMPLE 9.3

GIVEN $\square ABCD$, diagonals $\overline{AC}$ and $\overline{BD}$ intersect at point E.

PROVE **a** $\overline{AE} \cong \overline{EC}$

b $\overline{BE} \cong \overline{ED}$

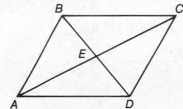

SOLUTION

PLAN Prove $\triangle BEC \cong \triangle DEA$ by ASA.

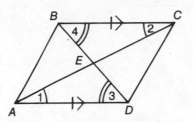

PROOF

Statements	Reasons
1. $\square ABCD$.	1. Given.
2. $\overline{BC} \parallel \overline{AD}$.	2. Opposite sides of a parallelogram are parallel.
3. $\angle 1 \cong \angle 2$. (Angle)	3. If two lines are parallel, then their alternate interior angles are congruent.
4. $\overline{AD} \cong \overline{BC}$. (Side)	4. Opposite sides of a parallelogram are congruent.
5. $\angle 3 \cong \angle 4$. (Angle)	5. Same as reason 3.
6. $\triangle BEC \cong \triangle DEA$.	6. ASA Postulate.
7. $\overline{AE} \cong \overline{EC}$ and $\overline{BE} \cong \overline{ED}$.	7. CPCTC.

Example 9.3 establishes the following theorem:

THEOREM 9.5

The diagonals of a parallelogram bisect each other.

SUMMARY OF THE PROPERTIES OF A PARALLELOGRAM

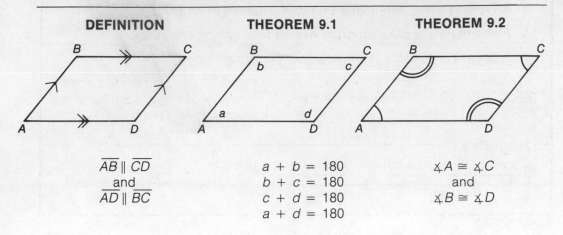

DEFINITION	THEOREM 9.1	THEOREM 9.2
$\overline{AB} \parallel \overline{CD}$ and $\overline{AD} \parallel \overline{BC}$	$a + b = 180$ $b + c = 180$ $c + d = 180$ $a + d = 180$	$\angle A \cong \angle C$ and $\angle B \cong \angle D$

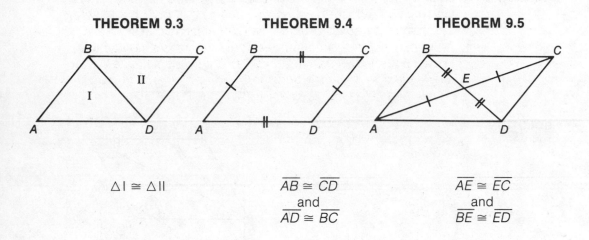

THEOREM 9.3	THEOREM 9.4	THEOREM 9.5
$\triangle I \cong \triangle II$	$\overline{AB} \cong \overline{CD}$ and $\overline{AD} \cong \overline{BC}$	$\overline{AE} \cong \overline{EC}$ and $\overline{BE} \cong \overline{ED}$

IN A PARALLELOGRAM

1 Opposite sides are parallel.

2 Opposite sides are congruent.

3 Opposite angles are congruent.

4 Consecutive angles are supplementary.

5 Diagonals bisect each other.

9.3 PROPERTIES OF SPECIAL PARALLELOGRAMS

An equiangular parallelogram is called a *rectangle*. An *equilateral* parallelogram is known as a *rhombus*. A *square* is a parallelogram which is both equiangular and equilateral. It will be convenient to use the following definitions in our work with these figures.

DEFINITIONS OF SPECIAL PARALLELOGRAMS

1 A *rectangle* is a parallelogram having four right angles.

2 A *rhombus* is a parallelogram having four congruent sides.

3 A *square* is a rectangle having four congruent sides.

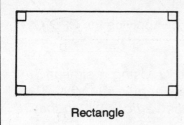 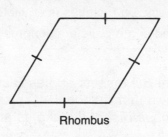 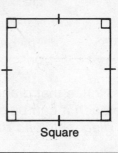

Rectangle   Rhombus   Square

REMARKS   **1**   Recall that a good definition does not include any unnecessary information. Although not as convenient, it is possible to define each of these figures so that their definitions contain less information. For example, a rectangle can be defined as a parallelogram having *a* right angle. The properties of a parallelogram can then be used to *prove* that each of the remaining angles of a rectangle must be a right angle.

2   Although a square has been defined in terms of a *rectangle,* it could be defined as a *rhombus* which contains four right angles.

EXAMPLE 9.4

   GIVEN   Rectangle *ABCD.*

   PROVE $\overline{AC} \cong \overline{DB}.$

SOLUTION

OUTLINE OF   Prove $\triangle BAD \cong \triangle CDA$ by SAS.

   PROOF $\overline{AB} \cong \overline{CD}$ since opposite sides of a rectangle are congruent.

Angles *BAD* and *CDA* are congruent since they are right angles.

$\overline{AD} \cong \overline{AD}.$

Since the triangles are congruent, $\overline{AC} \cong \overline{DB}$ by CPCTC.

Example 9.4 establishes the following theorem:

THEOREM 9.6

The diagonals of a rectangle are congruent.

EXAMPLE 9.5

   GIVEN   Rhombus *ABCD.*

   PROVE   **a** $\angle 1 \cong \angle 2.$

            b $\angle 3 \cong \angle 4.$

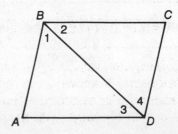

SOLUTION

OUTLINE OF PROOF

Prove $\triangle BAD \cong \triangle BCD$ by SAS.

$\overline{AB} \cong \overline{CB}$ since a rhombus is equilateral.

Angles A and C are congruent since opposite angles of a rhombus are congruent.

$\overline{AD} \cong \overline{CD}$ since a rhombus is equilateral.

The desired angle pairs are congruent by applying the CPCTC principle.

Notice that diagonal $\overline{BD}$ bisects angles B and D. Using a similar approach it can be shown that diagonal $\overline{AC}$ bisects angles A and C. Thus, the *diagonals of a rhombus bisect the angles at the vertices which they join.*

EXAMPLE 9.6   Prove the diagonals of a rhombus are perpendicular to each other.

SOLUTION

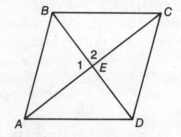

GIVEN   Rhombus $ABCD$, diagonals $\overline{AC}$ and $\overline{BD}$ intersect at E.

PROVE $\overline{AC} \perp \overline{BD}$.

OUTLINE OF PROOF

Prove $\triangle AEB \cong \triangle CEB$ (other triangle pairs may be selected) by SAS.

$\overline{AB} \cong \overline{CB}$ since a rhombus is equilateral.

Angles ABE and CBE are congruent by Theorem 9.7 since a diagonal of a rhombus bisects the angle formed at each vertex.

$\overline{BE} \cong \overline{BE}$. By CPCTC, angles 1 and 2 are congruent.

$\overline{AC} \perp \overline{BD}$ since the lines intersect to form a congruent pair of adjacent angles.

DIAGONALS OF A RHOMBUS

THEOREM 9.7   The diagonals of a rhombus bisect the angles at the vertices which they join (see Example 9.5).

THEOREM 9.8   The diagonals of a rhombus are perpendicular to each other (see Example 9.6).

Since a rhombus includes all the properties of a parallelogram, the diagonals of a rhombus bisect each other. Each diagonal of a rhombus is the *perpendicular bisector* of the other diagonal.

EXAMPLE 9.7   Given $ABCD$ in the accompanying figure is a rhombus and $m\angle 1 = 40$. Find the measure of each of the following angles.

a $m\angle 2$ **b** $m\angle 3$ **c** $m\angle ADC$

SOLUTION **a** Triangle ABC is isosceles since $\overline{AB} \cong \overline{BC}$. Hence, the base angles of the triangle must be congruent.

$$m\angle 1 = m\angle 2 = 40$$

b In triangle AEB, angle AEB is a right angle since the diagonals of a rhombus are perpendicular to each other. Since the sum of the measures of the angles of a triangle is 180, the measure of angle 3 must be 50.

c Since the diagonals of a rhombus bisect the angles of the rhombus, if $m\angle 3 = 50$, then $m\angle ABC = 100$. Since opposite angles of a rhombus are equal in measure, $m\angle ADC$ must also equal 100.

Keep in mind that a rhombus is *not* necessarily a rectangle since it may or may not contain four right angles. A rectangle is *not* necessarily a rhombus since it may or may not contain four congruent sides. A square combines the properties of a rectangle with the properties of a rhombus. The diagonals of a square are congruent, bisect its opposite angles, and intersect at right angles.

EXAMPLE 9.8

GIVEN Square $ABCD$,
$\overline{AE} \cong \overline{DF}$.

PROVE $\overline{AF} \cong \overline{BE}$.

SOLUTION

PLAN Prove $\triangle BAE \cong \triangle ADF$ by SAS.

PROOF

Statements	Reasons
1. $ABCD$ is a square.	1. Given.
2. $\overline{BA} \cong \overline{DA}$. (Side)	2. A square is equilateral.
3. Angles A and D are right angles.	3. A square contains four right angles.
4. $\angle A \cong \angle D$. (Angle)	4. All right angles are congruent.
5. $\overline{AE} \cong \overline{DF}$. (Side)	5. Given.
6. $\triangle BAE \cong \triangle ADF$.	6. SAS Postulate.
7. $\overline{AF} \cong \overline{BE}$.	7. CPCTC.

SUMMARY OF PROPERTIES OF THE RECTANGLE, RHOMBUS, AND SQUARE

PROPERTY	RECTANGLE	RHOMBUS	SQUARE
1. All the properties of a parallelogram?	Yes	Yes	Yes
2. Equiangular (4 right angles)?	Yes	No	Yes
3. Equilateral (4 congruent sides)?	No	Yes	Yes
4. Diagonals congruent?	Yes	No	Yes
5. Diagonals bisect opposite angles?	No	Yes	Yes
6. Diagonals perpendicular?	No	Yes	Yes

9.4 PROVING A QUADRILATERAL IS A PARALLELOGRAM

The previous sections were devoted to developing the properties of quadrilaterals that were known to be parallelograms. We now consider the other side of the coin. How can we prove that a quadrilateral is a parallelogram? What is the minimum information required in order to be able to conclude that a quadrilateral is a parallelogram?

Using the reverse of the definition of a parallelogram, we know that if *both pairs of sides of a quadrilateral are parallel, then the quadrilateral is a parallelogram.* We may use this fact in establishing alternative methods for proving that a quadrilateral is a parallelogram. For example, try drawing a quadrilateral in which the same pair of sides is both congruent and parallel. Imposing this condition forces the remaining pair of sides to be parallel. We now formally state and prove this result.

THEOREM 9.9

If a quadrilateral has a pair of sides that are both parallel and congruent, then the quadrilateral is a parallelogram.

PROOF

GIVEN   Quadrilateral *ABCD*,

       $\overline{AD} \parallel \overline{BC}$, $\overline{AD} \cong \overline{BC}$.

PROVE   Quadrilateral *ABCD* is a parallelogram.

PLAN   Show $\overline{AB} \parallel \overline{CD}$.

OUTLINE OF PROOF   Draw diagonal $\overline{AC}$. Prove $\triangle ABC \cong \triangle CDA$ by SAS:

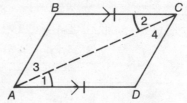

     $\overline{BC} \cong \overline{AD}$ (Side)

     $\angle 1 \cong \angle 2$ (Angle)

     $\overline{AC} \cong \overline{AC}$ (Side)

Angles 3 and 4 are congruent by CPCTC. This implies that $\overline{AB} \parallel \overline{CD}$ (since alternate interior angles are congruent, the lines are parallel). Since both pairs of sides of quadrilateral *ABCD* are parallel, *ABCD* is a parallelogram.

Are there any additional methods for proving a quadrilateral is a parallelogram? The converses of the theorems which state the properties of quadrilaterals which are parallelograms may offer some clues. Recall that these theorems take the general form

> If a quadrilateral is a parallelogram, then *a certain property is true.*

The converse of this statement takes the form

> If a quadrilateral has *a certain property*, then it is a parallelogram.

Keep in mind that the converse of a theorem is *not* necessarily true. We therefore need to investigate whether the converse holds for each special property of parallelograms. Table 9.1 summarizes the results of these investigations.

TABLE 9.1

IF A QUADRILATERAL HAS . . .	THEN IS IT A PARALLELOGRAM (?)
Congruent opposite sides 	Yes, since $\triangle ABC \cong \triangle CDA$ by SSS. By CPCTC, angles 1 and 2 are congruent which implies that $AB \parallel CD$. Similarly, angles 3 and 4 are congruent making $AD \parallel BC$. Since both pairs of sides are parallel, $ABCD$ is a parallelogram.
Congruent opposite angles 	Yes. Since the sum of the interior angles of a quadrilateral is 360, $$x + y + x + y = 360$$ $$2x + 2y = 360$$ $$2(x + y) = 360$$ $$x + y = 180$$ If interior angles on the same side of a transversal are supplementary, the lines are parallel. It follows that $AB \parallel DC$ and $AD \parallel BC$. $ABCD$ is therefore a parallelogram.
Diagonals which bisect each other 	Yes. $\triangle AED \cong \triangle BEC$ by SAS. By CPCTC, $AD \cong BC$ and $\angle 1 \cong \angle 2$. It follows that $AD \parallel BC$. Since AD is both congruent and parallel to BC, $ABCD$ is a parallelogram (see Theorem 9.9).

We may now summarize three additional methods for proving that a quadrilateral is a parallelogram.

> **THEOREM 9.10**
> **MORE WAYS OF PROVING A QUADRILATERAL IS A PARALLELOGRAM**
>
> A quadrilateral is a parallelogram if any one of the following are true:
>
> 1   Opposite sides are congruent.
> 2   Opposite angles are congruent.
> 3   Diagonals bisect each other.

EXAMPLE 9.9 Draw a diagram to help prove or disprove that a quadrilateral is a parallelogram if:

    a One pair of opposite sides are congruent.

    b One pair of sides are parallel.

    c Two pairs of sides are congruent.

SOLUTION **a** Not necessarily a parallelogram:

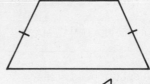

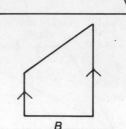

    b Not necessarily a parallelogram:

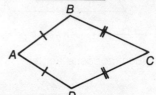

    c Not necessarily a parallelogram:

       NOTE: This type of figure is referred to as a *kite*.

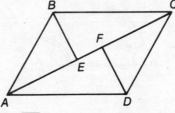

EXAMPLE 9.10

GIVEN $\overline{BE} \perp \overline{AC}$, $\overline{DF} \perp \overline{AC}$,
    $\overline{BE} \cong \overline{DF}$,
    $\angle EBC \cong \angle FDA$.

PROVE *ABCD* is a parallelogram.

SOLUTION

PLAN Prove $\overline{BC}$ is parallel and congruent to $\overline{AD}$ by first proving that $\triangle BEC \cong \triangle DFA$ by ASA.

PROOF

Statements	Reasons
1. $\angle EBC \cong \angle FDA$. (Angle)	1. Given.
2. $\overline{BE} \cong \overline{DF}$. (Side)	2. Given.
3. $\overline{BE} \perp \overline{AC}$ and $\overline{DF} \perp \overline{AC}$.	3. Given.
4. $\angle BEC \cong \angle DFA$. (Angle)	4. Perpendicular lines intersect to form right angles. All right angles are congruent. (*Note*: We have consolidated steps.)
5. $\triangle BEC \cong \triangle DFA$.	5. ASA Postulate.
6. $\overline{AD} \cong \overline{BC}$ and $\angle BCE \cong \angle DAF$.	6. CPCTC.
7. $\overline{AD} \parallel \overline{BC}$.	7. If alternate interior angles are congruent, then the lines are parallel.
8. Quadrilateral *ABCD* is a parallelogram.	8. If a quadrilateral has a pair of sides that are both parallel and congruent, then the quadrilateral is a parallelogram (Theorem 9.10).

EXAMPLE 9.11

GIVEN $\Box ABCD$,
$\overline{AE}$ bisects $\angle BAD$,
$\overline{CF}$ bisects $\angle BCD$.

PROVE Quadrilateral $AECF$ is a parallelogram.

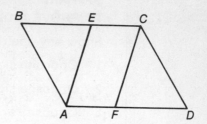

SOLUTION

PLAN Prove $\triangle ABE \cong \triangle CDF$ by ASA. $AE = CF$ and $BE = DF$ by CPCTE. $EC = AF$ by subtraction. Hence, $AECF$ is a parallelogram since opposite sides have the same length.

PROOF

Statements	Reasons
1. $ABCD$ is a parallelogram.	1. Given.
2. $m\angle BAD = m\angle DCB$.	2. Opposite angles of a parallelogram are equal in measure.
3. $\overline{AE}$ bisects $\angle BAD$. $\overline{CF}$ bisects $\angle BCD$.	3. Given.
4. $\angle BAE \cong \angle DCF$. (Angle)	4. Halves of equals are equal (and therefore congruent).
5. $\overline{AB} \cong \overline{DC}$. (Side)	5. Opposite sides of a parallelogram are congruent.
6. $\angle B \cong \angle D$. (Angle)	6. Same as reason 2.
7. $\triangle ABE \cong \triangle CDF$.	7. ASA Postulate.
8. $AE = FC$.	8. Corresponding sides of congruent triangles are equal in length.
To show $EC = AF$ use subtraction:	
9. $BE = DF$.	9. Same as reason 8.
10. $BC = AD$.	10. Opposite sides of a parallelogram are equal in length.
11. $EC = AF$.	11. Subtraction property of equality.
12. $AECF$ is a parallelogram.	12. If the opposite sides of a quadrilateral are equal in length (that is, congruent), then the quadrilateral is a parallelogram.

REMARK Note that throughout this proof, statements and corresponding reasons have been expressed in terms of equality of measures rather than congruence. This is necessary since arithmetic operations (taking halves of equals and subtraction) were performed on these quantities.

┌─ **SUMMARY** ──────────────────────────────
│ **TO PROVE A QUADRILATERAL IS A PARALLELOGRAM**
│
│ Show any one of the following is true:
│
│ **1** Opposite sides are parallel.
│
│ **2** Opposite sides are congruent.
│
│ **3** Opposite angles are congruent.
│
│ **4** Diagonals bisect each other.
│
│ **5** A pair of sides is both parallel and congruent.
│
│ **6** Consecutive angle pairs are supplementary.
└──

In the above summary, method 1 was derived from taking the reverse of the definition of a parallelogram. Methods 2, 3, and 4 represent the *converses* of theorems that specified the properties of a parallelogram. Method 5 was established by the proof of Theorem 9.9. The proof of method 6 is based on the fact that lines which form supplementary consecutive angle pairs are parallel.

Using a similar approach we can develop methods for proving that a quadrilateral (or parallelogram) is a rectangle, rhombus, or square:

1 To prove a quadrilateral is a *rectangle,* show that it is a *parallelogram* having one of the following properties:

   a Contains at least one right angle.

   b Diagonals are congruent.

2 To prove a quadrilateral is a *rhombus,* show that it is a *parallelogram* having one of the following properties:

   a Contains at least one pair of congruent adjacent sides.

   b Diagonals intersect at right angles.

   c Diagonals bisect the vertex angles.

3 To prove that a quadrilateral is a *square,* show that it is

   a A *rectangle* having an adjacent pair of congruent sides; or

   b A *rhombus* having at least one right angle.

9.5 APPLICATIONS OF PARALLELOGRAMS

ABCD in Figure 9.6*a* is a rectangle. If $BD = 10$, what is the length of $\overline{AM}$? The diagonals of a rectangle are congruent making $AC = 10$. Since the diagonals bisect each other, $AM = 5$ (and $MC = 5$). We can redraw the diagram as Figure 9.6*b* by deleting $\overline{BC}$, $\overline{MC}$, and $\overline{CD}$. Since point *M* is the midpoint of $\overline{BD}$, $\overline{AM}$ is the median to hypotenuse $\overline{BD}$ of right triangle *BAD*. This suggests the following theorem.

┌─ **THEOREM 9.11** ──────────────────────────
│ The length of the median drawn to the hypotenuse of a right triangle is one-half the length of the hypotenuse.
└──

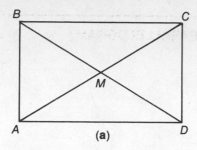 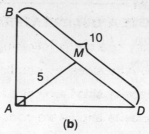

FIGURE 9.6

The theorem that we will now establish states that if the midpoints of *any* two sides of a triangle are connected by a line segment, then this segment must be parallel to the remaining side of the triangle. Furthermore, its length must be exactly one-half the length of the remaining side of the triangle. This theorem may be stated formally as follows.

THEOREM 9.12   MIDPOINTS OF A TRIANGLE THEOREM

The line segment joining the midpoints of two sides of a triangle is parallel to the third side and is one-half its length.

GIVEN D and E are midpoints of sides $\overline{AB}$ and $\overline{CB}$, respectively.

PROVE **a** $\overline{DE} \parallel \overline{AC}$.

b $DE = \frac{1}{2}AC$.

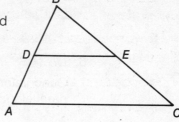

OUTLINE OF PROOF Extend $\overline{DE}$ so that $\overline{DE} \cong \overline{EF}$, and then draw $\overline{CF}$.

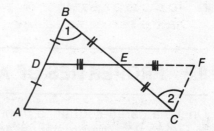

Prove $ADFC$ is a parallelogram by showing that $\overline{CF}$ is parallel and congruent to $\overline{AD}$ as follows:

1 $\triangle DBE \cong \triangle FCE$ (by SAS).

2 By CPCTC, angles 1 and 2 are congruent which implies $\overline{FC} \parallel \overline{AD}$. Also, $\overline{CF} \cong \overline{DB} \cong \overline{AD}$.

Since $ADFC$ ix a parallelogram, $\overline{DE} \parallel \overline{AC}$. Also, $DE = \frac{1}{2}DF = \frac{1}{2}AC$.

EXAMPLE 9.12 In triangle RST, A is the midpoint of $\overline{RS}$ and B is the midpoint of $\overline{RT}$.

a If $ST = 18$, find AB.

b If $AB = 7$, find ST.

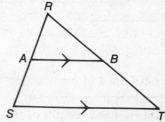

SOLUTION **a** $AB = \frac{1}{2}ST = \frac{1}{2}(18) = 9$.

b $ST = 2(AB) = 2(7) = 14$.

EXAMPLE 9.13

GIVEN Points Q, R, and S are midpoints.

PROVE $PQRS$ is a parallelogram.

SOLUTION

PLAN Show that $\overline{QR}$ is parallel and congruent to $\overline{PS}$.

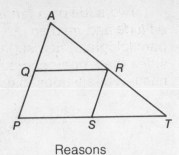

PROOF

Statements	Reasons
1. Points Q, R, and S are midpoints.	1. Given.
2. $\overline{QR} \parallel \overline{PT}$.	2. The line segment joining the midpoints of two sides of a triangle is parallel to the third side and is one-half its length.
3. $\overline{QR} \parallel \overline{PS}$.	3. Segments of parallel lines are parallel.
4. $QR = \frac{1}{2}PT$.	4. Same as reason 2.
5. $PS = \frac{1}{2}PT$.	5. A midpoint divides a segment into two congruent segments.
6. $\overline{QR} \cong \overline{PS}$.	6. Transitive property.
7. Quadrilateral $PQRS$ is a parallelogram.	7. If a quadrilateral has a pair of sides that are both parallel and congruent, then the quadrilateral is a parallelogram.

9.6 PROPERTIES OF A TRAPEZOID

Unlike a parallelogram, a trapezoid has exactly one pair of parallel sides. In Figure 9.7, quadrilaterals $ABCD$ and $RSTW$ are examples of trapezoids. Quadrilateral $JKLM$ is *not* a trapezoid since it does not include one pair of parallel sides.

> **DEFINITION OF A TRAPEZOID**
>
> A *trapezoid* is a quadrilateral which has exactly *one* pair of parallel sides. The parallel sides are called the *bases* of the trapezoid. The nonparallel sides are referred to as the *legs* of the trapezoid.

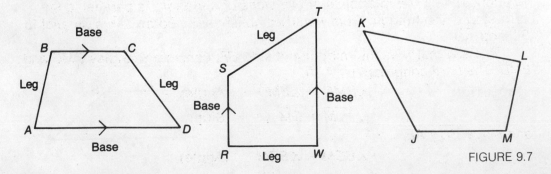

FIGURE 9.7

Two additional terms that are used in connection with trapezoids are *altitude* and *median*. An *altitude* of a trapezoid (or for that matter, a parallelogram) is a segment drawn from any point on one of the parallel sides (base) perpendicular to the opposite side (other base). An infinite number of altitudes may be drawn in a trapezoid. See Figure 9.8.

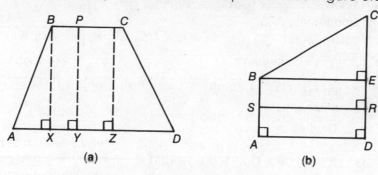

FIGURE 9.8 (a) $\overline{BX}$, $\overline{PY}$, and $\overline{CZ}$ are examples of *altitudes*. (b) $\overline{AD}$, $\overline{SR}$, and $\overline{BE}$ are examples of *altitudes*.

A *median* of a trapezoid is the segment which joins the midpoints of the nonparallel sides (legs). A trapezoid has exactly one median. See Figure 9.9.

As Figure 9.9 seems to indicate, the median of a trapezoid appears to be parallel to the bases. In addition, there is a relationship between the length of the median and the lengths of the bases of the trapezoid. Our strategy in developing the properties of the median of a trapezoid is to apply Theorem 9.12. To accomplish this, an auxiliary segment must be drawn so that the median of the trapezoid becomes the segment which joins the midpoints of a newly formed triangle.

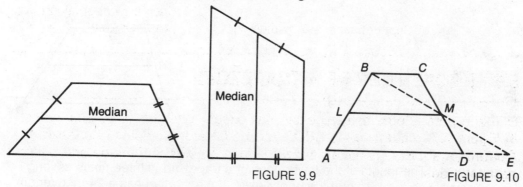

FIGURE 9.9   FIGURE 9.10

In trapezoid *ABCD* of Figure 9.10, median $\overline{LM}$ has been drawn. By drawing line segment $\overline{BM}$ and extending it so that it meets the extension of line segment $\overline{AD}$, triangle *ABE* is formed. Point *L* is the midpoint of side $\overline{AB}$ of triangle *ABE*. If we can show that point *M* is the midpoint of side $\overline{BE}$ of triangle *ABE*, then we can apply Theorem 9.12 which tells us that the line segment joining the midpoints of two sides of a triangle is parallel to the third side of the triangle. This would establish that median $\overline{LM}$ is parallel to $\overline{BC}$ and $\overline{AD}$.

To show that *M* is the midpoint of side $\overline{BE}$, consider triangles *BMC* and *EMD*. They are congruent by ASA,

$$\angle BMC \cong \angle EMD \qquad \text{(Angle)}$$

$$\overline{CM} \cong \overline{DM} \qquad \text{(Side)}$$

and since $\overline{BC} \parallel \overline{AE}$,

$$\angle BCM \cong \angle EDM \qquad \text{(Angle)}$$

By CPCTC, $\overline{BM} \cong \overline{EM}$. Hence, Theorem 9.12 applies.

Theorem 9.12 also informs us that the length of $\overline{LM}$ (the segment joining the midpoints of two sides of triangle ABE) must be one-half the length of the third side. Thus,

$$LM = \tfrac{1}{2}(AE) = \tfrac{1}{2}(AD + DE)$$

Since we have established that $\triangle BMC \cong \triangle EMD$, $\overline{DE} \cong \overline{BC}$. Applying the substitution principle,

$$LM = \tfrac{1}{2}(AD + BC) = \tfrac{1}{2}(\text{sum of the lengths of the bases})$$

The result of our investigation is stated formally in Theorem 9.13.

THEOREM 9.13   MEDIAN OF A TRAPEZOID THEOREM

The median of a trapezoid is parallel to the bases and has a length equal to one-half the sum of the lengths of the bases.

EXAMPLE 9.14   In trapezoid $ABCD$ $\overline{BC} \parallel \overline{AD}$ and $\overline{RS}$ is the median.

    a If $AD = 13$ and $BC = 7$, find RS.

    b If $BC = 6$ and $RS = 11$, find AD.

SOLUTION   **a** $RS = \tfrac{1}{2}(13 + 7) = \tfrac{1}{2}(20) = 10$

    b If median $= \tfrac{1}{2}(\text{sum of bases})$, then

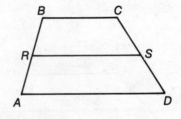

$$\text{Sum of bases} = 2 \times \text{median}$$
$$AD + 6 = 2 \times 11$$
$$AD + 6 = 22$$
$$AD = 22 - 6$$
$$AD = 16$$

EXAMPLE 9.15   The length of the lower base of a trapezoid is three times as long as the length of the upper base. If the median has a 24-inch length, find the lengths of the bases.

SOLUTION

Let a = length of upper base.
Then $3a$ = length of lower base.

Since, sum of bases $= 2 \times$ median,

$$a + 3a = 2 \times 24$$
$$4a = 48$$
$$a = \frac{48}{4}$$

$$\boxed{a = 12}$$

and

$$\boxed{3a = 36}$$

SUMMARY
SPECIAL MEDIAN RELATIONSHIPS

1   The length of the median drawn to the hypotenuse of a right triangle is one-half the length of the hypotenuse.

2   The length of the median of a trapezoid may be found by taking the average of the lengths of the two bases:

    Median = ½(upper base length + lower base length)

3   The median of a trapezoid is parallel to both bases.

THE ISOSCELES TRAPEZOID

If the legs of a trapezoid are congruent, then the trapezoid is called an *isosceles trapezoid*. See Figure 9.11. An *isosceles trapezoid* features some special properties which are not found in all trapezoids.

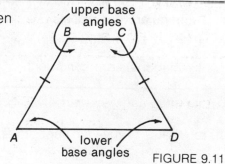

FIGURE 9.11

THEOREM 9.14

The lower base angles of an isosceles trapezoid are congruent.

OUTLINE OF PROOF

> **GIVEN**   Isosceles trapezoid *ABCD*.
>
> **PROVE** $\angle A \cong \angle D$.

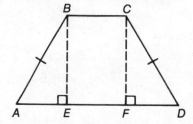

Draw altitudes $\overleftrightarrow{BE}$ and $\overline{CF}$, thereby forming triangles *ABE* and *DCF*. Prove these triangles congruent by applying the Hy-Leg method:

$$\overline{AB} \cong \overline{DC} \quad \text{(Hypotenuse)}$$

Since parallel lines are everywhere equidistant,

$$\overline{BE} \cong \overline{CF} \quad \text{(Leg)}$$

By CPCTC, $\angle A \cong \angle D$.

As a corollary to Theorem 9.14, we note that since supplements of congruent angles are congruent, the upper base angles are congruent. We now turn our attention to the diagonals of an isosceles trapezoid.

THEOREM 9.15

The diagonals of an isosceles trapezoid are congruent.

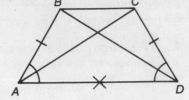

GIVEN Isosceles trapezoid *ABCD*,
diagonals $\overline{AC}$ and $\overline{DB}$.

PROVE $\overline{AC} \cong \overline{DB}$.

Diagonals $\overline{AC}$ and $\overline{DB}$ form a pair of overlapping triangles, $\triangle ACD$ and $\triangle DBA$. Prove these triangles congruent by using SAS. The included angles, angles *BAD* and *CDA*, are congruent as a result of Theorem 9.14. By CPCTC, $\overline{AC} \cong \overline{PB}$.

PROPERTIES OF AN ISOSCELES TRAPEZOID

1 All the properties of a trapezoid.

2 Legs are congruent.

3 Base angles are congruent.

4 Diagonals are congruent.

EXAMPLE 9.16 Prove the opposite angles of an isosceles trapezoid are supplementary.

SOLUTION

GIVEN Isosceles trapezoid *ABCD*.

PROVE **a** Angles *A* and *C* are supplementary.

b Angles *D* and *B* are supplementary.

PROOF

Statements	Reasons
1. *ABCD* is an isosceles trapezoid.	1. Given.
2. $\overline{AD} \parallel \overline{BC}$.	2. The bases of a trapezoid are parallel.
3. $m\angle D + m\angle C = 180$	3. If two lines are parallel, then interior angles on the same side of the transversal are supplementary.
4. $m\angle A = m\angle D$	4. Base angles of an isosceles trapezoid are equal in measure.
5. $m\angle A + m\angle C = 180$	5. Substitution.
6. Angles *A* and *C* are supplementary.	6. If the sum of the measures of two angles is 180, then the angles are supplementary.

In a similar fashion, it can be shown easily that angles *D* and *B* are supplementary.

For the sake of completeness we state without proof Theorem 9.16 which offers methods of proving that a trapezoid is an isosceles trapezoid.

THEOREM 9.16 WAYS OF PROVING A TRAPEZOID IS ISOSCELES

A trapezoid is an isosceles trapezoid if any one of the following is true:

1 Legs are congruent.

2 Base angles are congruent.

3 Diagonals are congruent.

REMARK Part 1 results from considering the reverse of the definition of an isosceles trapezoid. Parts 2 and 3 are derived from taking the converses of Theorems 9.14 and 9.15.

SUMMARY OF PROPERTIES OF SPECIAL QUADRILATERALS

SPECIAL QUAD-RILATERAL	OPPOSITE SIDES		ANGLES		DIAGONALS BISECT		DIAGONALS	
	$\parallel$	$\cong$	OPPO-SITE	CONSEC-UTIVE	EACH OTHER	ANGLES	$\cong$	$\perp$
Parallelogram	yes	yes	Con-gruent	Supple-mentary	yes	NA*	NA	NA
Rectangle	yes	yes	4 right angles		yes	NA	yes	NA
Rhombus	yes	all sides $\cong$	Con-gruent	Supple-mentary	yes	yes	NA	yes
Square	yes	all sides $\cong$	4 right angles		yes	yes	yes	yes
Trapezoid	1 pair (bases)			Upper and lower base angles are supple-mentary			NA	
Isosceles Trapezoid	1 pair (bases)	1 pair (legs)	supple-mentary	Base angles $\cong$			yes	

* NA = *Not Always.*

REVIEW EXERCISES FOR CHAPTER 9

1. In rhombus *RSTW*, diagonal $\overline{RT}$ is drawn. If $m\angle RST = 108$, find $m\angle SRT$.

2. In parallelogram *MATH* the measure of angle *T* exceeds the measure of angle *H* by 30. Find the measure of each angle of the parallelogram.

3. In parallelogram *TRIG*, $m\angle R = 2x + 19$ and $m\angle G = 4x - 17$. Find the measure of each angle of the parallelogram.

4. The length of the median drawn to the hypotenuse of a right triangle is represented by the expression $3x - 7$ while the hypotenuse is represented by $5x - 4$. Find the length of the median.

5. In triangle RST, E is the midpoint of $\overline{RS}$ and F is the midpoint of $\overline{ST}$. If $EF = 5y - 1$ and $RT = 7y + 10$, find the length of $\overline{EF}$ and $\overline{RT}$.

6. In trapezoid $BYTE$, $\overline{BE} \parallel \overline{YT}$ and median $\overline{LM}$ is drawn.

    a $LM = 35$. If the length of $\overline{BE}$ exceeds the length of $\overline{YT}$ by 13, find the lengths of the bases.

    b If $YT = x + 9$, $LM = x + 15$, and $BE = 2x - 5$, find the lengths of $\overline{YT}$, $\overline{LM}$, and $\overline{BE}$.

7. In parallelogram $RSTW$ diagonals $\overline{RT}$ and $\overline{SW}$ intersect at point A. If $SA = x - 13$ and $AW = 2x - 37$, find SW.

8. The lengths of the sides of a triangle are 9, 40, and 41. Find the perimeter of the triangle formed by joining the midpoints of the sides.

9.   **GIVEN** $\square ABCD$, $\overline{AE} \cong \overline{CF}$.

       PROVE $\angle ABE \cong \angle CDF$.

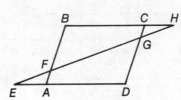

10.   **GIVEN** $\square ABCD$, $\overline{EF} \cong \overline{HG}$.

       PROVE $\overline{AF} \cong \overline{CG}$.

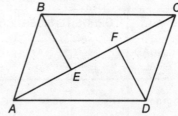

11.   **GIVEN** $\square ABCD$, B is the midpoint of $\overline{AE}$.

       PROVE $\overline{EF} \cong \overline{FD}$.

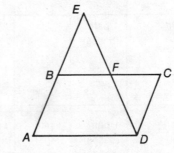

12.   **GIVEN** Rectangle $ABCD$, M is the midpoint of $\overline{BC}$.

       PROVE $\triangle AMD$ is isosceles.

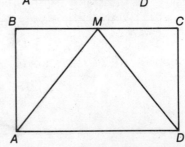

13.   **GIVEN** Rhombus $ABCD$.

       PROVE $\triangle ASC$ is isosceles.

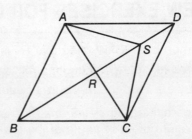

14. **GIVEN** Rectangle $ABCD$,
$\overline{BE} \cong \overline{CE}$.

PROVE $\overline{AF} \cong \overline{DG}$.

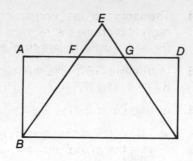

15. **GIVEN** $ABCD$ is a parallelogram,
$AD > DC$.

PROVE $m\angle BAC > m\angle DAC$.

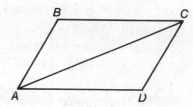

16. **GIVEN** $ABCD$ is a parallelogram,
$\overline{BR}$ bisects $\angle ABC$,
$\overline{DS}$ bisects $\angle CDA$.

PROVE $BRDS$ is a parallelogram.

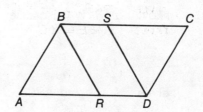

Use the following diagram to solve Exercises 17 and 18.

17. **GIVEN** $\square BMDL$,
$\overline{AL} \cong \overline{CM}$.

PROVE $ABCD$ is a parallelogram.

18. **GIVEN** $\square ABCD$,
$\angle ABL \cong \angle CDM$.

PROVE $BLDM$ is a parallelogram.

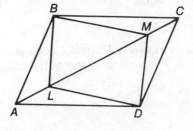

19. **GIVEN** $ABCD$ is a rhombus,
$\overline{BL} \cong \overline{CM}$, $\overline{AL} \cong \overline{BM}$.

PROVE $ABCD$ is a square.

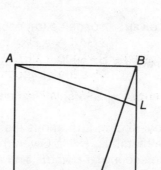

20. **GIVEN** $\square ABCD$,
$m\angle 2 > m\angle 1$.

PROVE $\square ABCD$ is *not* a rectangle.

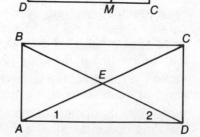

21. **GIVEN** $\overline{DE} \cong \overline{DF}$,
points D, E, and F are the midpoints
of $\overline{AC}$, $\overline{AB}$, and $\overline{BC}$, respectively.

PROVE $\triangle ABC$ is isosceles.

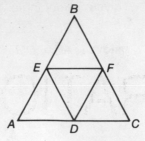

22. **GIVEN** $\square RSTW$;
in $\triangle WST$, B and C are midpoints.

PROVE $WACT$ is a parallelogram.

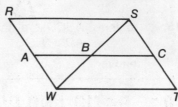

23. **GIVEN** Isosceles trapezoid $RSTW$.

PROVE $\triangle RPW$ is isosceles.

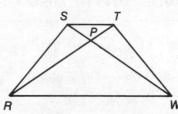

24. **GIVEN** Trapezoid $ABCD$,
$\overline{EF} \cong \overline{EG}$,
$\overline{AF} \cong \overline{DG}$,
$\overline{BG} \cong \overline{CF}$.

PROVE Trapezoid $ABCD$ is isosceles.

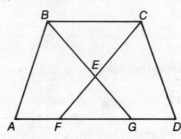

25. **GIVEN** Trapezoid $ABCD$ with median $\overline{LM}$,
P is the midpoint of $\overline{AD}$,
$\overline{LP} \cong \overline{MP}$.

PROVE Trapezoid $ABCD$ is isosceles.

26. **GIVEN** Isosceles trapezoid $ABCD$,
$\angle BAK \cong \angle BKA$.

PROVE $BKDC$ is a parallelogram.

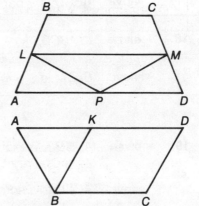

27. Prove that in a rhombus the longer diagonal lies opposite the larger angle of the rhombus. (*Hint:* Given rhombus $ABCD$ with diagonals $\overline{AC}$ and $\overline{DB}$ intersecting at point E, and assume $m\angle CDA > m\angle BAD$. Prove $AC > BD$. Work with $\triangle AED$ and first establish that $AE > DE$.)

28. Prove parallel lines are everywhere equidistant. (*Hint:* Given any pair of parallel lines, select two distinct points on one of the lines and from each point draw a segment that is perpendicular to the other parallel line. Show that the lengths of these perpendicular segments are equal.)

29. Prove the quadrilateral formed by joining consecutively the midpoints of the sides of a parallelogram is a parallelogram.

30. Prove if the midpoints of the sides of a rectangle are joined consecutively, the resulting quadrilateral is a rhombus.

CHAPTER TEN

RATIO, PROPORTION, AND SIMILARITY

10.1 RATIO AND PROPORTION

Allan is 30 years old and Bob is 10 years old. How do their ages compare? Allan is 20 years older than Bob. It is sometimes desirable to compare two numbers by determining how many times larger (or smaller) one number is compared to a second number. This can be accomplished by dividing the first number by the second number:

$$\frac{\text{Allan's age}}{\text{Bob's age}} = \frac{30}{10} = \frac{3}{1} \quad \text{or} \quad 3$$

Allan is three times as old as Bob. The result of dividing two numbers is called a *ratio*.

> **DEFINITION OF RATIO**
>
> The *ratio* of two numbers a and b ($b \neq 0$) is the quotient of the numbers. The numbers a and b are referred to as the *terms* of the ratio.

REMARKS **1** The ratio of two numbers, a and b, may be written in a variety of ways. For example,

$$\frac{a}{b} \quad a \div b \quad a \text{ to } b \quad a:b$$

We will normally express a ratio using either the first or last of these forms.

2 In writing the ratio of two numbers it is customary to express the ratio (fraction) in *simplest* form. For example, the ratio of 50 to 100 would be expressed as follows:

$$\frac{50}{100} = \frac{1}{2} \quad \text{or} \quad 1:2 \text{ (read as ``1 is to 2'')}$$

In forming the ratio of two numbers, each number may be expressed in different units of measurement. For example, if a person travels 120 miles in a car in 3 hours, then the ratio of the distance traveled to the time traveled is

$$\frac{120 \text{ miles}}{3 \text{ hours}} = 40 \frac{\text{miles}}{\text{hour}}$$

157

The value 40 miles/hour corresponds to the average speed during the trip. When forming a ratio, in order to determine how many times larger or smaller one value is compared to another, both quantities must be expressed in the same units of measurement. For example, if the length of $\overline{AB}$ is 2 feet and the length of $\overline{XY}$ is 16 inches, then in order to determine how many times larger $\overline{AB}$ is compared to $\overline{XY}$ we must convert one of the units of measurement into the other. In this example it is convenient to express feet in terms of inches. Since 2 feet is equivalent to 24 inches, we may write

$$\frac{AB}{XY} = \frac{24}{16} = \frac{3}{2}$$

The ratio of the length of $\overline{AB}$ to the length of $\overline{XY}$ is 3:2. Since the decimal representation of $\frac{3}{2}$ is 1.5, we may say that the length of $\overline{AB}$ is 1.5 times the length of $\overline{XY}$.

EXAMPLE 10.1   If the measure of angle A is 60 and angle B is a right angle, find the ratio of the measure of angle A to the measure of angle B.

SOLUTION $$\frac{m\angle A}{m\angle B} = \frac{60}{90} = \frac{2}{3} \quad \text{or} \quad 2:3$$

EXAMPLE 10.2   Find each of the following ratios using the figure provided:

  a $AB:XY$ **b** $BC:YZ$ **c** $XZ:AC$

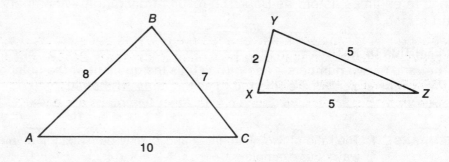

SOLUTION   **a** $\frac{8}{2} = 4:1$ **b** $\frac{7}{5} = 7:5$ **c** $\frac{5}{10} = 1:2$

EXAMPLE 10.3   The measures of a pair of consecutive angles of a parallelogram are in the ratio of 1:8. Find the measure of the smaller of these angles.

SOLUTION   Let x = the measure of the smaller of the angles.
Then $8x$ = the measure of the larger of the angles.

Since consecutive angles of a parallelogram are supplementary,

$$x + 8x = 180$$
$$9x = 180$$
$$\frac{9x}{9} = \frac{180}{9}$$
$$\boxed{x = 20}$$

The measure of the smaller angle is 20.

EXAMPLE 10.4 In an isosceles triangle, the ratio of the measure of the vertex angle to the measure of a base angle is 2:5. Find the measure of each angle of the triangle.

SOLUTION

Let $2x$ = the measure of the vertex angle.
Then $5x$ = the measure of one base angle
and $5x$ = the measure of the other base angle.

Since the sum of the measures of the angles of a triangle is 180,

$$2x + 5x + 5x = 180$$
$$12x = 180$$
$$x = \tfrac{180}{12}$$
$$\boxed{x = 15}$$
$$2x = 2(15) = 30$$
$$5x = 5(15) = 75$$
$$5x = 5(15) = 75$$

The measures of the three angles of the triangle are 30, 75, and 75.

PROPORTIONS

The ratio $\tfrac{24}{16}$ may be simplified and written as $\tfrac{3}{2}$. That is,

$$\frac{24}{16} = \frac{3}{2}$$

An equation which states that two ratios are equal is called a *proportion*. The previous proportion may also be written in the form, $24:16 = 3:2$.

DEFINITION OF A PROPORTION

A *proportion* is an equation which states that two ratios are equal:

$$\frac{a}{b} = \frac{c}{d} \quad \text{or} \quad a:b = c:d \quad \text{(provided } b \neq 0 \text{ and } d \neq 0)$$

Each term of a proportion is given a special name according to its position in the proportion.

First proportional ———⎤   ⎡——— Third proportional

$$\frac{a}{b} = \frac{c}{d}$$

Second proportional ———⎦   ⎣——— Fourth proportional

The pair of terms which form the first and fourth proportionals are referred to as the *extremes* of a proportion; the second and third proportionals of a proportion are called the *means* of a proportion.

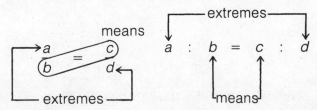

┌─ **DEFINITION OF THE TERMS OF A PROPORTION** ─────────

In the proportion $\dfrac{a}{b} = \dfrac{c}{d}$, a is called the *first* proportional, b is called the
second proportional, c is called the *third* proportional, and d is called the
fourth proportional. The pair of terms a and d are referred to as the
extremes of the proportion; the pair of terms b and c are referred to as the
means of the proportion.
└──────────────────────────────────────

Returning to the proportion $\frac{24}{16} = \frac{3}{2}$, 24 and 2 (the outermost terms) are
the extremes. The two innermost terms, 16 and 3, are the means. Notice
that the cross-products of a proportion are equal:

$$\frac{24}{16} \diagdown\!\!\!\!\diagup \frac{3}{2}$$

$$16 \times 3 = 24 \times 2$$

$$48 = 48$$

┌─ **THEOREM 10.1   EQUAL CROSS-PRODUCTS THEOREM** ─────────

In a proportion the product of the means is equal to the product of the
extremes. (Forming this product is sometimes referred to as cross-
multiplying.)
└──────────────────────────────────────

EXAMPLE 10.5   Find the first proportional if the remaining terms of a proportion are
3, 14, and 21.

SOLUTION   Let x = the first proportional. Then

$$\frac{x}{3} = \frac{14}{21}$$

By Theorem 10.1,

$$21x = 42$$

$$\frac{21x}{21} = \frac{42}{21}$$

$$\boxed{x = 2}$$

The first proportional is 2.

REMARK   Although it is not essential, it is sometimes more convenient to
simplify the initial proportion *before* applying Theorem 10.1.

$$\frac{x}{3} = \frac{14}{21}$$

$$\frac{x}{3} = \frac{2}{3}$$

$$3x = 6$$

$$\boxed{x = 2}$$

EXAMPLE 10.6   The first term of a proportion is 2 and the second and third terms are both 8. Find the fourth proportional.

SOLUTION   Let x = the fourth proportional. Then,

$$\frac{2}{8} = \frac{8}{x}$$

Simplify *before* applying Theorem 10.1:

$$\frac{1}{4} = \frac{8}{x}$$

Cross-multiply:

$$\boxed{x = 32}$$

The fourth proportional is 32.

In Example 10.6, the means were equal since the second and third terms of the proportion were both equal to 8. Whenever the means of a proportion are identical, then the value which appears in the means is referred to as the *mean proportional* (or *geometric mean*) between the first and fourth terms of the proportion. In Example 10.6, 8 is said to be the mean proportional between 2 and 32.

DEFINITION OF MEAN PROPORTIONAL

If the second and third terms of a proportion are the same, then either term is referred to as the mean proportional or geometric mean between the first and fourth terms of the proportion:

 $\dfrac{a}{m} = \dfrac{m}{d}$ mean proportional   or $a : m = m : d$
↑——mean——↑
proportional

EXAMPLE 10.7   Find the mean proportional between each pair of extremes.
a 3 and 27   **b** 5 and 7

SOLUTION   **a** Let m = the mean proportional between 3 and 27. Then

$$\frac{3}{m} = \frac{m}{27}$$

$$m^2 = 3(27)$$

$$m^2 = 81$$

$$m = \sqrt{81}$$

$$\boxed{m = 9}$$

b Let m = the mean proportional between 5 and 7.

$$\frac{5}{m} = \frac{m}{7}$$

$$m^2 = 35$$

$$\boxed{m = \sqrt{35}}$$

REMARK   The mean proportional may be an irrational number. (An irrational number is a number that does *not* have an exact integer or decimal representation.)

It is sometimes useful to be able to determine whether a pair of ratios are in proportion. Two ratios are in proportion if *the product of the means of the resulting proportion is equal to the product of the extremes*. Are the ratios $\frac{2}{5}$ and $\frac{12}{30}$ in proportion? We write a tentative proportion and then determine whether the cross-products are equal:

$$\frac{2}{5} \overset{?}{=} \frac{12}{30}$$

$$5 \times 12 \overset{?}{=} 30 \times 2$$

$$60 = 60$$

Therefore, $\frac{2}{5}$ and $\frac{12}{30}$ are in proportion.

Now repeat this procedure, this time investigating whether the ratios $\frac{8}{12}$ and $\frac{6}{8}$ are in proportion.

$$\frac{8}{12} \overset{?}{=} \frac{6}{8}$$

$$12 \times 6 \overset{?}{=} 8 \times 8$$

$$72 \neq 64$$

This implies that $\frac{8}{12}$ and $\frac{6}{8}$ are *not* in proportion.

Some algebraic properties of proportions are worth noting.

PROPERTY 1 If the numerators and denominators of a proportion are switched, then an equivalent proportion results.

If $\dfrac{a}{b} = \dfrac{c}{d}$, then $\dfrac{b}{a} = \dfrac{d}{c}$

(provided a, b, c, and d are nonzero numbers).

PROPERTY 2 If either pair of opposite terms of a proportion are interchanged, then an equivalent proportion results.

a If $\dfrac{a}{b} = \dfrac{c}{d}$, then $\dfrac{d}{b} = \dfrac{c}{a}$.

b If $\dfrac{a}{b} = \dfrac{c}{d}$, then $\dfrac{a}{c} = \dfrac{b}{d}$.

PROPERTY 3 If the denominator is added or subtracted from the numerator on each side of the proportion, then an equivalent proportion results.

a If $\dfrac{a}{b} = \dfrac{c}{d}$, then $\dfrac{a+b}{b} = \dfrac{c+d}{d}$.

b If $\dfrac{a}{b} = \dfrac{c}{d}$, then $\dfrac{a-b}{b} = \dfrac{c-d}{d}$.

PROPERTY 4 If the product of two nonzero numbers equals the product of another pair of nonzero numbers, then a proportion may be formed by making the factors of one product the extremes, and making the factors of the other product the means. That is, if $R \times S = T \times W$, then we may

a make R and S the *extremes*: $\dfrac{R}{T} = \dfrac{W}{S}$

or

b make R and S the *means*: $\dfrac{T}{R} = \dfrac{S}{W}$.

10.2 PROPORTIONS IN A TRIANGLE

In the previous chapter we saw that a line passing through the midpoints of two sides of a triangle was parallel to the third side (and one-half of its length). Suppose we draw a line parallel to a side of a triangle so that it intersects the other two sides, but not necessarily at their midpoints. Many such lines can be drawn, as shown in Figure 10.1.

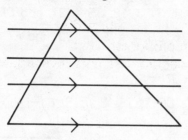

FIGURE 10.1

Consider one of these lines and the segments that it forms on the sides of the triangle, as shown in Figure 10.2.

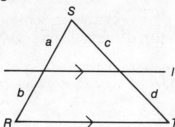

FIGURE 10.2

It will be convenient to postulate that line l divides $\overline{RS}$ and $\overline{ST}$ such that lengths of corresponding segments on each side have the same ratio:

$$\frac{a}{b} = \frac{c}{d} \qquad or \qquad \frac{a}{RS} = \frac{c}{ST} \qquad or \qquad \frac{b}{RS} = \frac{d}{ST}$$

If any of the above ratios hold, then the line segments are said to be *divided proportionally*. Notice that these ratios have the form

$$\frac{\text{Upper segment of side } \overline{RS}}{\text{Lower segment of side } \overline{RS}} = \frac{\text{upper segment of side } \overline{ST}}{\text{lower segment of side } \overline{ST}}$$

or

$$\frac{\text{Upper segment of side } \overline{RS}}{\text{Whole side } (\overline{RS})} = \frac{\text{upper segment of side } \overline{ST}}{\text{whole side } (\overline{ST})}$$

or

$$\frac{\text{Lower segment of side } \overline{RS}}{\text{Whole side } (\overline{RS})} = \frac{\text{lower segment of side } \overline{ST}}{\text{whole side } (\overline{ST})}$$

Keep in mind that the algebraic properties of proportions allow these three proportions to be expressed in equivalent forms. For example, the numerator and denominator of each fraction may be interchanged (that is, each ratio may be *inverted*).

┌─ **POSTULATE 10.1** ──┐
│ A line parallel to one side of a triangle and intersecting the other two │
│ sides *divides these sides proportionally.* │
└──┘

EXAMPLE 10.8 In triangle RST, line segment $\overline{EF}$ is parallel to side $\overline{RT}$, intersecting side $\overline{RS}$ at point E and side $\overline{TS}$ at point F.

a If $SE = 8$, $ER = 6$, $FT = 15$, find SF.

b If $SF = 4$, $ST = 12$, $SR = 27$, find SE.

c If $SE = 6$, $ER = 4$, $ST = 20$, find FT.

SOLUTION **a** $\dfrac{SE}{ER} = \dfrac{SF}{FT}$

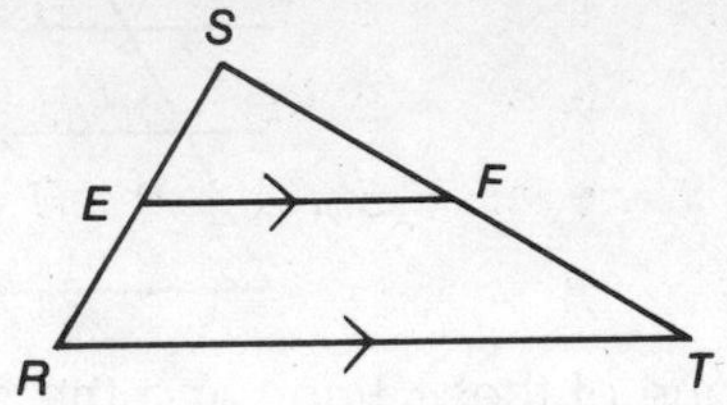

$\dfrac{8}{6} = \dfrac{SF}{15}$

$\dfrac{4}{3} = \dfrac{SF}{15}$

$3 \cdot SF = 60$

$SF = \dfrac{60}{3}$

$\boxed{SF = 20}$

b $\dfrac{SE}{SR} = \dfrac{SF}{ST}$

$\dfrac{SE}{27} = \dfrac{4}{12}$

$\dfrac{SE}{27} = \dfrac{1}{3}$

$3 \cdot SE = 27$

$SE = \dfrac{27}{3}$

$\boxed{SE = 9}$

c $\dfrac{ER}{RS} = \dfrac{FT}{ST}$

$\dfrac{4}{4 + 6} = \dfrac{FT}{20}$

$\dfrac{4}{10} = \dfrac{FT}{20}$

$\dfrac{2}{5} = \dfrac{FT}{20}$

$5 \cdot FT = 40$

$FT = \dfrac{40}{5}$

$\boxed{FT = 8}$

164 RATIO, PROPORTION, AND SIMILARITY

The converse of Postulate 10.1 is also true. If a line is drawn so that the ratio of the segment lengths it cuts off on one side of a triangle is equal to the ratio of the segment lengths it cuts off on a second side of a triangle, then the line must be parallel to the third side of the triangle.

POSTULATE 10.2

A line that divides two sides of a triangle *proportionally* is parallel to the third side of the triangle.

EXAMPLE 10.9 Determine if $\overline{AB} \parallel \overline{KJ}$ if

a $KA = 2$, $AL = 5$, $JB = 6$, and $BL = 15$.

b $AL = 3$, $KL = 8$, $JB = 10$, and $JL = 16$.

c $AL = 5$, $KA = 9$, $LB = 10$, and $JB = 15$.

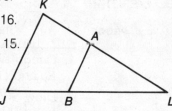

SOLUTION In each instance, we must write a tentative proportion and determine whether the proportion is true. If it is true, then $\overline{AB}$ is parallel to $\overline{KJ}$. The proportion written must conform to the meaning of divided proportionally.

a Based upon the information provided, we use the proportion $\dfrac{KA}{AL} = \dfrac{JB}{BL}$ (equivalent proportions may also be formed). We determine whether this proportion is true using the numbers provided:

$$\frac{2}{5} \stackrel{?}{=} \frac{6}{15}$$

$$2 \times 15 = 5 \times 6$$

$$30 = 30$$

Therefore, $\overline{AB} \parallel \overline{KJ}$.

b Use the proportion $\dfrac{AL}{KL} = \dfrac{BL}{JL}$ (other proportions can also be used). Note that $BL = 16 - 10 = 6$.

$$\frac{3}{8} \stackrel{?}{=} \frac{6}{16}$$

$\overline{AB}$ is parallel to $\overline{KJ}$ since $3 \times 16 = 8 \times 6$.

c The numbers provided suggest the proportion $\dfrac{AL}{KA} = \dfrac{LB}{JB}$ be used.

$$\frac{5}{9} \stackrel{?}{=} \frac{10}{15}$$

$\overline{AB}$ is *not* parallel to $\overline{KJ}$ since $5 \times 15 \neq 9 \times 10$.

EXAMPLE 10.10

GIVEN   Quadrilateral *RSTW* with $\overline{KJ}$ and $\overline{LM}$ drawn,

$$\frac{RK}{KS} = \frac{RJ}{JW},$$

$$\frac{TL}{LS} = \frac{TM}{MW}.$$

PROVE $\overline{KJ} \parallel \overline{LM}$.

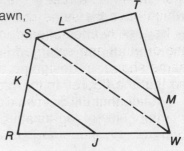

SOLUTION

PLAN   Draw $\overline{SW}$. Applying Postulate 10.2, $\overline{KJ} \parallel \overline{SW}$ and $\overline{LM} \parallel \overline{SW}$. Hence, $\overline{KJ} \parallel \overline{LM}$.

PROOF

Statements	Reasons
1. Quadrilateral *RSTW* with $\overline{KJ}$ and $\overline{LM}$ drawn.	1. Given.
2. Draw $\overline{SW}$. In △*SRW*:	2. Two points determine a line.
3. $\dfrac{RK}{KS} = \dfrac{RJ}{JW}$.	3. Given.
4. $\overline{KJ} \parallel \overline{SW}$.	4. A line that divides two sides of a triangle proportionally is parallel to the third side of the triangle.
In △*WTS*:	
5. $\dfrac{TL}{LS} = \dfrac{TM}{MW}$.	5. Given.
6. $\overline{LM} \parallel \overline{SW}$.	6. Same as reason 4.
7. $\overline{KJ} \parallel \overline{LM}$.	7. Two lines parallel to the same line are parallel to each other.

10.3   DEFINING SIMILAR POLYGONS

Compare the three triangles in Figure 10.3. Triangles I and III have exactly the same size and shape since they agree in three pairs of angles *and* in three pairs of sides. Triangle I is *congruent* to triangle III.

Triangles II and III have three pairs of congruent corresponding angles, but *each* side of triangle II is twice the length of the corresponding side of

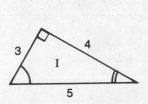
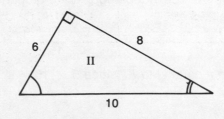
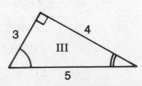

FIGURE 10.3

triangle III. As a result, triangles II and III have the same shape. Polygons which have the same shape are said to be *similar*. The concept of similarity is frequently encountered in everyday life. When a photograph is enlarged, the original and enlarged figures are *similar* since they have exactly the same shape. In designing a blueprint, everything must be drawn to scale so that the figures in the blueprint are *similar* to the actual figures.

Mathematically speaking, two polygons have the same shape when corresponding angles are congruent *and* the ratios of the lengths of all pairs of corresponding sides are equal.

DEFINITION OF SIMILAR POLYGONS

Two polygons are *similar* if their vertices can be paired so that corresponding angles are congruent *and* the lengths of corresponding sides are in proportion.

REMARKS
1. The definition assumes that there exists a one-to-one correspondence between the vertices of the polygons. This means that each vertex of the first polygon is matched with exactly one vertex of the second polygon and vice versa.

2. *Congruent* polygons have the same size *and* shape, while *similar* polygons have only the same shape. The symbol for the same shape is ~. The expression $\triangle ABC \sim \triangle RST$ is read, "triangle ABC is similar to triangle RST."

3. If it is known that two polygons are similar, then it may be concluded that each pair of corresponding angles are congruent *and* the ratios of the lengths of each pair of corresponding sides are equal.

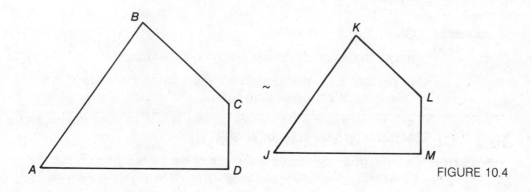

FIGURE 10.4

In Figure 10.4, if quadrilateral $ABCD$ is similar to quadrilateral $JKLM$, then the following relationships must hold:

CORRESPONDING ANGLES ARE CONGRUENT	LENGTHS OF CORRESPONDING SIDES ARE IN PROPORTION
$\angle A \cong \angle J$	$\dfrac{AB}{JK} = \dfrac{BC}{KL} = \dfrac{CD}{LM} = \dfrac{AD}{JM}$
$\angle B \cong \angle K$	
$\angle C \cong \angle L$	
$\angle D \cong \angle M$	

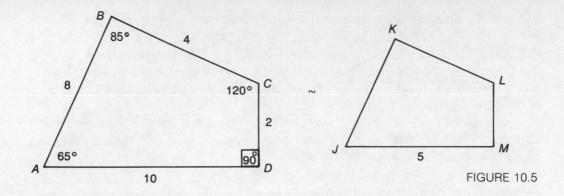

FIGURE 10.5

For example, let's find the measures of the parts of quadrilateral *JKLM* in Figure 10.5. Since the quadrilaterals are given to be similar,

$$m\angle J = m\angle A = 65$$

$$m\angle K = m\angle B = 85$$

$$m\angle L = m\angle C = 120$$

$$m\angle M = m\angle D = 90$$

The ratio of the lengths of each pair of corresponding sides must be equal to

$$\frac{JM}{AD} = \frac{5}{10} = \frac{1}{2}$$

The ratio of the length of any side of quadrilateral *JKLM* to the length of the corresponding side of quadrilateral *ABCD* must also be 1:2. Hence, *JK* = 4, *KL* = 2, and *LM* = 1.

EXAMPLE Given △*HLX* ~ △*WKN*.
10.11
  a List the pairs of corresponding congruent angles.

  b Write an extended proportion which forms the ratios of the lengths of corresponding sides.

  c If *HL* = 8, *LX* = 14, *HX* = 18, and *WN* = 27, find the lengths of the remaining sides of △*WKN*.

SOLUTION   **a** The order in which the vertices of the triangles are written is significant—it defines the pairs of corresponding vertices:

$$H \leftrightarrow W \qquad L \leftrightarrow K \qquad X \leftrightarrow N$$

Hence, ∠*H* ≅ ∠*W*, ∠*L* ≅ ∠*K*, and ∠*X* ≅ ∠*N*.

  b Corresponding sides connect corresponding vertices:

CORRESPONDENCE	CORRESPONDING SIDES
△*H L X* ~ △*W K N*	$\overline{HL} \leftrightarrow \overline{WK}$
△*H L X* ~ △*W K N*	$\overline{LX} \leftrightarrow \overline{KN}$
△*H L X* ~ △*W K N*	$\overline{HX} \leftrightarrow \overline{WN}$

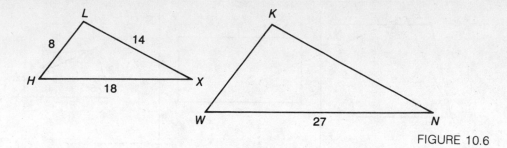

FIGURE 10.6

The resulting proportion may be written as follows:

$$\frac{\text{Side of } \triangle HLX}{\text{Corresponding side of } \triangle WKN} = \frac{HL}{WK} = \frac{LX}{KN} = \frac{HX}{WN}$$

After you have determined the pairs of corresponding vertices, you will probably find it easier to determine the corresponding sides of two similar figures by drawing a diagram and then applying the principle that *corresponding sides lie opposite corresponding angles* (vertices). See Figure 10.6. $\overline{HX}$ and $\overline{WN}$ lie opposite corresponding angles L and K; $\overline{HX}$ and $\overline{WN}$ therefore represent a pair of corresponding sides.

c Since $\overline{HX}$ and $\overline{WN}$ are corresponding sides, the lengths of each pair of corresponding sides must be the same as the ratio of HX to WN:

$$\frac{HX}{WN} = \frac{18}{27} = \frac{2}{3}$$

The length of any side of triangle HLX to the length of the corresponding side of triangle WKN must be in the ratio of $2:3$. Using this fact, the lengths of the remaining sides of $\triangle WKN$ may be found.

To find WK,

$$\frac{2}{3} = \frac{HL}{WK}$$

$$\frac{2}{3} = \frac{8}{WK}$$

$$2 \cdot WK = 24$$

$$WK = \frac{24}{2}$$

$$\boxed{WK = 12}$$

To find KN,

$$\frac{2}{3} = \frac{LX}{KN}$$

$$\frac{2}{3} = \frac{14}{KN}$$

$$2 \cdot KN = 42$$

$$KN = \frac{42}{2}$$

$$\boxed{KN = 21}$$

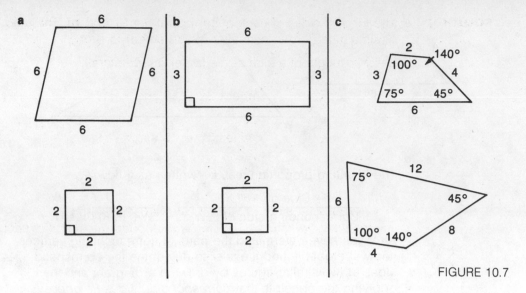

FIGURE 10.7

EXAMPLE 10.12 Determine whether each of the pairs of polygons in Figure 10.7 are similar.

SOLUTION **a** No. The ratios of the lengths of corresponding sides are equal (3:1), but corresponding angles are *not* congruent.

b No. Corresponding angles are congruent (each figure contains four right angles), but the ratios of the lengths of corresponding sides are *not* all equal.

c Yes. Corresponding angles are congruent and the ratios of the lengths of each pair of corresponding sides is the same (1:2).

If two polygons are similar then the ratio of the lengths of any pair of corresponding sides is called the *ratio of similitude*. If the lengths of a pair of corresponding sides of two similar polygons are 3 and 12, then the ratio of similitude is $\frac{3}{12}$ or 1:4.

Are congruent polygons similar? Yes, since they satisfy the two conditions of similarity; all corresponding pairs of angles are congruent and the ratios of the lengths of all pairs of corresponding sides are the same (1:1). Are similar polygons also congruent? Generally speaking, no! Similar polygons are congruent only if their ratio of similitude is 1:1.

EXAMPLE 10.13 Two quadrilaterals are similar and have a ratio of similitude of 1:3. If the lengths of the sides of the smaller quadrilateral are 2, 5, 8, and 12, find the lengths of the sides of the larger quadrilateral.

SOLUTION Since the ratio of similitude is 1:3, the length of each side of the larger quadrilateral is three times as great as the corresponding side in the smaller quadrilateral. The lengths of the sides of the larger quadrilateral are 6, 15, 24, and 36.

EXAMPLE 10.14 Two quadrilaterals are similar. The length of the sides of the smaller quadrilateral are 4, 6, 12, and 18. The length of the longest side of the larger quadrilateral is 27. Determine each of the following:

a The ratio of similitude.

b The lengths of the remaining sides of the larger quadrilateral.

c The ratio of the perimeters of the two quadrilaterals.

SOLUTION **a** The longest sides of each quadrilateral are 27 and 18. The ratio of similitude (larger to smaller quadrilateral) is $\frac{27}{18}$ or $\frac{3}{2}$.

b Let x = length of a side of the larger quadrilateral.

$$\frac{3}{2} = \frac{x}{4} \quad \text{implies } 2x = 12 \text{ and } x = 6$$

$$\frac{3}{2} = \frac{x}{6} \quad \text{implies } 2x = 18 \text{ and } x = 9$$

$$\frac{3}{2} = \frac{x}{12} \quad \text{implies } 2x = 36 \text{ and } x = 18$$

The sides of the larger quadrilateral are 6, 9, 18, and 27. Alternatively, we could have simply multiplied the length of each side of the smaller quadrilateral by the ratio of similitude ($\frac{3}{2}$) in order to obtain the length of the corresponding side in the larger quadrilateral.

c $\dfrac{\text{Perimeter of larger quadrilateral}}{\text{Perimeter of smaller quadrilateral}} = \dfrac{6 + 9 + 18 + 27}{4 + 6 + 12 + 18}$

$$= \frac{60}{40} = \frac{3}{2}$$

Notice that in part **c** of Example 10.14 the ratio of the perimeters of the quadrilaterals is the same as the ratio of the lengths of a pair of corresponding sides.

THEOREM 10.2   RATIO OF PERIMETERS THEOREM

The perimeters of a pair of similar polygons have the same ratio as the lengths of any pair of corresponding sides.

10.4   PROVING TRIANGLES SIMILAR

In order to prove triangles *congruent* we did not have to show that three pairs of angles were congruent *and* three pairs of sides were congruent. Instead, we were able to develop shortcut methods which depended on showing that a particular set of *three* parts of one triangle were congruent to the corresponding parts of the second triangle (e.g., SSS, ASA, SAS, and AAS).

The definition of similarity requires that to show two triangles are similar, three pairs of corresponding angles must be proven congruent *and* that the lengths of three pairs of corresponding sides have the same ratio. A formidable task! In a manner analogous to our work with congruent triangles, we will present several shortcut methods which can be used to prove a pair of *triangles* similar.

Let us try the following experiment. Suppose we wished to draw a triangle that has at least two angles which have the same measure as two

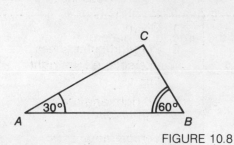

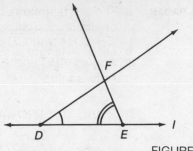

FIGURE 10.8   FIGURE 10.9

angles of triangle *ABC* (Figure 10.8). We may begin by drawing a line *l* and choosing *any* two distinct points on line *l*, say points *D* and *E* (Figure 10.9). At points *D* and *E* we may use a protractor to draw angles having the same measures as angles *A* and *B*. Call the point at which the two rays having *D* and *E* as endpoints meet, point *F*.

How do the measures of angles *C* and *F* compare? They must be congruent since if two angles of a triangle are congruent to two angles of another triangle, then the third pair of angles must be congruent (since the sum must equal 180 in each case). Another interesting thing happens. If you would actually measure the sides of each triangle and then compare the ratios $\frac{AB}{DE}$, $\frac{BC}{EF}$, and $\frac{AC}{DF}$, you would find that they are equal! Hence, $\triangle ABC$ must be similar to $\triangle DEF$ since the requirements of the definition of similarity are satisfied. Recall that we began by making two angles of triangle *DEF* congruent to two angles of triangle *ABC*. This forced the third pair of angles to be congruent and the lengths of the corresponding sides to be in proportion.

THEOREM 10.3   THE ANGLE-ANGLE THEOREM OF SIMILARITY

If two angles of one triangle are congruent to two angles of another triangle, then the triangles are similar. (It will be convenient to refer to this result as the AA Theorem.)

This theorem provides a simple method for establishing that two triangles are similar. All we need do is show that *two* angles of one triangle are congruent to *two* angles of another triangle. The proof of this theorem will be taken up in the exercise section (see Exercise 16).

EXAMPLE 10.15
    GIVEN $\overline{CB} \perp \overline{BA}$,
               $\overline{CD} \perp \overline{DE}$.
    PROVE $\triangle ABC \sim \triangle EDC$.

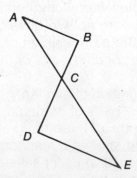

    SOLUTION
     PLAN   Use the AA Theorem. The two triangles include right and vertical angles which yield two pairs of congruent angles.

PROOF	Statements	Reasons
	1. $\overline{CB} \perp \overline{BA}$ and $\overline{CD} \perp \overline{DE}$.	1. Given.
	2. Angles ABC and EDC are right angles.	2. Perpendicular lines intersect to form right angles.
	3. $\angle ABC \cong \angle EDC$. (Angle)	3. All right angles are congruent.
	4. $\angle ACB \cong \angle ECD$. (Angle)	4. Vertical angles are congruent.
	5. $\triangle ABC \sim \triangle EDC$.	5. AA Theorem.

SOME ADDITIONAL METHODS FOR PROVING TRIANGLES SIMILAR

Two less frequently used methods for proving triangles similar will be stated without their proofs being given. Each method involves showing that a special relationship exists between *three* parts of one triangle and the corresponding parts of a second triangle.

Suppose we do not have any information regarding the measures of the angles of a pair of triangles. If we do know that their sides are in proportion, then it can be proven that the angles of the triangle are congruent which implies that the pair of triangles are similar.

THEOREM 10.4   SIDE-SIDE-SIDE THEOREM OF SIMILARITY

If the vertices of two triangles can be paired so that the lengths of corresponding sides are in proportion, then the triangles are similar. (Abbreviation: SSS Theorem of Similarity.)

Theorem 10.4 can be used to determine whether the pair of triangles in Figure 10.10 are similar. Since

$$\frac{AB}{JK} = \frac{BC}{KL} = \frac{AC}{JL} = \frac{1}{3}$$

the *three* pairs of corresponding sides have their lengths in proportion which, according to Theorem 10.4, makes triangle ABC similar to triangle JKL.

Theorem 10.4 is analogous to the SSS Theorem of congruence. The next method for proving triangles similar is analogous to the SAS method of proving triangles congruent.

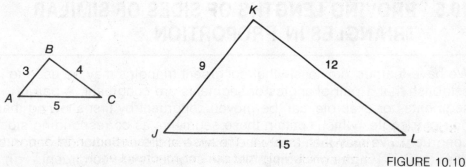

FIGURE 10.10

┌─ THEOREM 10.5 SIDE-ANGLE-SIDE THEOREM OF SIMILARITY ─┐

If the vertices of two triangles can be paired so that a pair of corresponding angles are congruent *and* the sides which include this angle are in proportion, then the two triangles are similar. (Abbreviation: SAS Theorem of Similarity.)

└──────────────────────┘

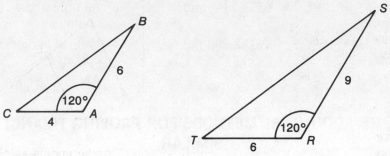

FIGURE 10.11

Theorem 10.5 can be used to determine whether the pair of triangles in Figure 10.11 are similar.

$$\angle A \cong \angle R$$

In addition,

$$\frac{AB}{RS} = \frac{AC}{RT} = \frac{2}{3}$$

Triangle *ABC* is therefore similar to triangle *RST*.

┌─ SUMMARY ─────────────────────────────┐

TO PROVE THAT A PAIR OF TRIANGLES ARE SIMILAR, PROVE ONE OF THE FOLLOWING:

1 Two pair of corresponding angles are congruent. (AA Theorem)

2 Three pairs of corresponding sides are in proportion. (SSS Theorem of Similarity)

3 A pair of corresponding angles is congruent *and* the sides which include this angle are in proportion. (SAS Theorem of Similarity)

NOTE: In the overwhelming majority of proofs which involve similar triangles, the AA Theorem will be applicable.

└──────────────────────┘

10.5 PROVING LENGTHS OF SIDES OF SIMILAR TRIANGLES IN PROPORTION

We have learned previously that congruent triangles may be used to establish that a pair of angles or segments are congruent. A pair of segments, for example, can be proven congruent by first showing that a pair of triangles which contain these segments as corresponding sides are congruent. We may then apply the reverse of the definition of congruence (that is, CPCTC) in concluding that the segments are congruent.

In like fashion, we can establish that the lengths of four segments are in proportion by first showing that a pair of triangles which contain these segments as corresponding sides are similar. We may then apply the reverse of the definition of similarity and conclude that the lengths of the segments are in proportion, using as a reason: *the lengths of corresponding sides of similar triangles are in proportion.*

EXAMPLE 10.16

GIVEN $\overline{AB} \parallel \overline{DE}$.

PROVE $\dfrac{EC}{BC} = \dfrac{ED}{AB}$.

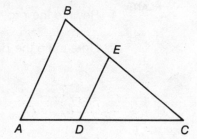

SOLUTION

PLAN **1** Select the triangles which contain these segments as sides. Read across the proportion:

$$\triangle ECD$$

$$\frac{EC}{BC} = \frac{ED}{AB}$$

$$\triangle BCA$$

2 Mark off the diagram with the given and all pairs of corresponding congruent angle pairs.

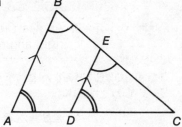

3 Write up the proof.

PROOF

Statements	Reasons
1. $\overline{AB} \parallel \overline{DE}$.	1. Given.
2. $\angle CED \cong \angle CBA$. (Angle) $\angle CDE \cong \angle CAB$. (Angle)	2. If two lines are parallel, then their corresponding angles are congruent.
3. $\triangle ECD \sim \triangle BCA$.	3. AA Theorem.
4. $\dfrac{EC}{BC} = \dfrac{ED}{AB}$	4. The lengths of corresponding sides of similar triangles are in proportion.

REMARKS **1** The statement "the lengths of corresponding sides of similar triangles are in proportion" is analogous to the statement "corresponding sides of *congruent* triangles are congruent" which was abbreviated as CPCTC. We will *not* introduce an abbreviation for the expression which appears as reason 4 in the previous proof.

2 Statement 3 of the proof establishes that a *line intersecting two sides of a triangle and parallel to the third side, forms two similar triangles.*

EXAMPLE 10.17

GIVEN $\overline{AC} \perp \overline{CB}, \overline{ED} \perp \overline{AB}$.

PROVE $EB:AB = ED:AC$.

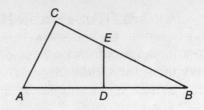

SOLUTION

PLAN

1 Rewrite the prove in fractional form: $\dfrac{EB}{AB} = \dfrac{ED}{AC}$.

2 Determine the pair of triangles which must be proven similar.

$$\triangle EBD$$
$$\downarrow \qquad \downarrow$$
$$\frac{EB}{AB} = \frac{ED}{AC}$$
$$\uparrow \qquad \uparrow$$
$$\triangle ABC$$

3 Mark off the diagram with the given and all pairs of corresponding congruent angle pairs.

4 Write up the proof.

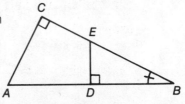

PROOF

Statements	Reasons
1. $\overline{AC} \perp \overline{CB}, \overline{ED} \perp \overline{AB}$.	1. Given.
2. Angles C and EDB are right angles.	2. Perpendicular lines intersect to form right angles.
3. $\angle C \cong \angle EDB$. (Angle)	3. All right angles are congruent.
4. $\angle B \cong \angle B$. (Angle)	4. Reflexive property of congruence.
5. $\triangle EBD \sim \triangle ABC$.	5. AA Theorem.
6. $EB:AB = ED:AC$	6. The lengths of corresponding sides of similar triangles are in proportion.

SUMMARY

To prove that the lengths of segments are in proportion:

1 Use the proportion provided in the prove to help identify the triangles which contain the desired segments as sides.

2 Mark off the diagram with the given as well as with any additional information which may be deduced (for example, vertical angles, right angles, congruent alternate interior angles, and so on).

3 Determine the similarity method of proof to be used (AA Theorem, SSS Theorem, SAS Theorem). In most cases, look to apply the AA Theorem.

4 Prove the triangles similar.

5 Write the desired proportion using as a reason, the lengths of corresponding sides of similar triangles are in proportion.

ALTITUDES AND MEDIANS OF SIMILAR TRIANGLES

The key concept to be developed here is that in similar triangles the lengths of altitudes or medians drawn to corresponding sides have the same ratio as the lengths of any pair of corresponding sides.

THEOREM 10.6   RATIO OF ALTITUDES IN SIMILAR TRIANGLES THEOREM

If a pair of triangles are similar, then the lengths of a pair of corresponding altitudes have the same ratio as the lengths of any pair of corresponding sides.

OUTLINE OF PROOF

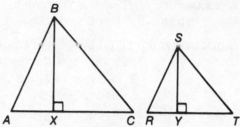

GIVEN $\triangle ABC \sim \triangle RST$, $\overline{BX}$ and $\overline{SY}$ are altitudes.

PROVE $\dfrac{BX}{SY} = \dfrac{BC}{ST} = \dfrac{AC}{RT} = \dfrac{AB}{RS}$.

PLAN   Show $\triangle BXC \sim \triangle SYT$.

$\angle C \cong \angle T$ since $\triangle ABC \sim \triangle RST$. Angles BXC and SYT are congruent since they are right angles. By the AA Theorem, $\triangle BXC \sim \triangle SYT$. Hence,

$$\frac{BX}{SY} = \frac{BC}{ST}$$

But, since the original pair of triangles are similar,

$$\frac{BC}{ST} = \frac{AC}{RT} = \frac{AB}{RS}$$

By the transitive property,

$$\frac{BX}{SY} = \frac{BC}{ST} = \frac{AC}{RT} = \frac{AB}{RS}$$

A theorem similar to Theorem 10.6 may be stated for corresponding medians of similar triangles.

THEOREM 10.7   RATIO OF MEDIANS IN SIMILAR TRIANGLES THEOREM

If a pair of triangles are similar, then the lengths of a pair of corresponding medians have the same ratio as the lengths of any pair of corresponding sides.

OUTLINE OF PROOF

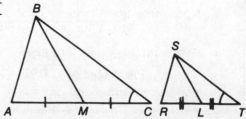

GIVEN $\triangle ABC \sim \triangle RST$, $\overline{BM}$ and $\overline{SL}$ are medians.

PROVE $\dfrac{BM}{SL} = \dfrac{BC}{ST} = \dfrac{AC}{RT} = \dfrac{AB}{RS}$.

PLAN Show $\triangle BMC \sim \triangle SLT$ by SAS Similarity Theorem.

Since $\triangle ABC \sim \triangle RST$, $\angle C \cong \angle T$ and $\dfrac{AC}{RT} = \dfrac{BC}{ST}$. Points M and L bisect segments $\overline{AC}$ and $\overline{RT}$, respectively. Hence

$$\frac{MC}{LT} = \frac{\frac{1}{2}AC}{\frac{1}{2}RT} = \frac{AC}{RT} = \frac{BC}{ST}$$

This implies that $\triangle BMC \sim \triangle SLT$ by the SAS Theorem of Similarity. The remainder of the proof immediately follows from the fact that lengths of corresponding sides of similar triangles are in proportion.

EXAMPLE 10.18 $\triangle RST \sim \triangle KLM$. Altitude $\overline{SA}$ exceeds the length of altitude $\overline{LB}$ by 5. If $RT = 9$ and $KM = 6$, find the length of each altitude.

SOLUTION Let x = the length of altitude $\overline{LB}$.
Then $x + 5$ = the length of altitude $\overline{SA}$.

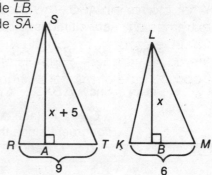

Applying Theorem 10.6,

$$\frac{x + 5}{x} = \frac{9}{6}$$

$$9x = 6(x + 5)$$

$$9x = 6x + 30$$

$$3x = 30$$

$$x = 10$$

Altitude $LB = x = 10$

Altitude $SA = x + 5 = 15$

SUMMARY

The lengths of corresponding *altitudes*, the lengths of corresponding *medians*, and the *perimeters* of similar triangles have the same ratio as the lengths of any pair of corresponding sides:

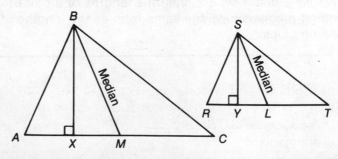

$$\frac{\text{Perimeter of } \triangle ABC}{\text{Perimeter of } \triangle RST} = \frac{BX}{SY} = \frac{BM}{SL} = \frac{AB}{RS} = \frac{BC}{ST} = \frac{AC}{RT}$$

10.6 PROVING PRODUCTS OF SEGMENT LENGTHS EQUAL

We know that if $\dfrac{A}{B} = \dfrac{C}{D}$, then $A \times D = B \times C$. The reason is that in a proportion the product of the means equals the product of the extremes. Instead of generating a product from a proportion, it is sometimes necessary to be able to take a product and determine the related proportion that would yield that product.

Suppose the lengths of four segments are related such that

$$KM \times LB = LM \times KD$$

What proportion gives this result when the products of its means and extremes are set equal to each other? A true proportion may be derived from the product by making a pair of terms appearing on the same side of the equal sign the extremes (say, KM and LB). The pair of terms on the opposite side of the equal sign then becomes the means (LM and KD):

$$\frac{KM}{LM} = \frac{KD}{LB}$$

NOTE: An equivalent proportion results if KM and LB were made the means rather than the extremes.

The ideas presented here are essential in problems in which similar triangles are used to prove products of segment lengths equal. As an illustration, let's look at the following problem.

GIVEN $\square ABCD$.

PROVE $KM \times LB = LM \times KD$.

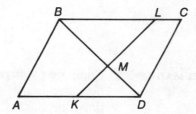

We analyze the solution to this type of problem by *working backwards*, beginning with the Prove:

1 Express the product as an equivalent proportion (this was accomplished in our preliminary discussion):

$$\frac{KM}{LM} = \frac{KD}{LB}$$

2 From the proportion (and in some problems, in conjunction with the given), determine the pair of triangles to be proven similar.

$$\triangle KMD$$

$$\frac{KM}{LM} = \frac{KD}{LB}$$

$$\triangle LMB$$

3 Mark off the diagram with the given and decide how to show the triangles similar.
STRATEGY: Use AA Theorem.

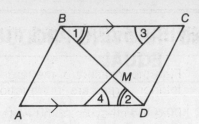

4 Write up the formal two-column proof. The steps in the proof should reflect the logic of this analysis, proceeding from step 4 back to step 1 (proving the triangles similar, forming the appropriate proportion, and, lastly, writing the product):

PROOF	Statements	Reasons
	1. $\square ABCD$.	1. Given.
	2. $\overline{AD} \parallel \overline{BC}$.	2. Opposite sides of a parallelogram are parallel.
	3. $\angle 1 \cong \angle 2$, $\angle 3 \cong \angle 4$.	3. If two lines are parallel, then their alternate interior angles are congruent.
	4. $\triangle KMD \sim \triangle LMB$.	4. AA Theorem.
	5. $\dfrac{KM}{LM} = \dfrac{KD}{LB}$.	5. The lengths of corresponding sides of similar triangles are in proportion.
	6. $KM \times LB = LM \times KD$	6. In a proportion, the product of the means equals the product of the extremes.

REMARKS
1 Compared with our previous work with similar triangles, the only new step is the last statement/reason of the proof. In order to prove products of segment lengths equal, we must first prove that a related proportion is true; in order to prove the proportion is true, we must first establish that the triangles which contain these segments as sides are similar.

2 In our analysis of the product $KM \times LB = LM \times KD$, suppose you formed the proportion,

$$\frac{KM}{KD} = \frac{LM}{LB}$$

Reading across the top (K-M-L), we do not find a set of letters which correspond to the vertices of a triangle. When this happens, switch the terms in either the means or the extremes of the proportion:

$$\frac{KM}{LM} \nearrow \frac{KD}{LB}$$

Reading across the top (K-M-D) now gives us the vertices of one of the desired triangles.

REVIEW EXERCISES FOR CHAPTER 10

1. Find the measure of the largest angle of a triangle if the measures of its interior angles are in the ratio $3:5:7$.

2. Find the measure of the vertex angle of an isosceles triangle if the measures of the vertex angle and a base angle have the ratio $4:3$.

3. The measures of a pair of consecutive angles of a parallelogram have the ratio $5:7$. Find the measure of each angle of the parallelogram.

4. Solve for x.

   a $\dfrac{2}{6} = \dfrac{8}{x}$ **b** $\dfrac{2}{x} = \dfrac{x}{50}$ **c** $\dfrac{2x - 5}{3} = \dfrac{9}{4}$

   d $\dfrac{4}{x + 3} = \dfrac{1}{x - 3}$ **e** $\dfrac{3}{x} = \dfrac{x - 4}{7}$

5. Find the mean proportional between each pair of extremes.

   a 4 and 16   **b** $3e$ and $12e^3$

   c $\frac{1}{2}$ and $\frac{1}{8}$ **d** 6 and 9

6. Determine whether the following pairs of ratios are in proportion:

   a 1/2 and 9/18   **b** 12/20 and 3/5

   c 4/9 and 12/36   **d** 15/25 and 20/12

7. In $\triangle BAG$, $\overline{LM}$ intersects side $\overline{AB}$ at L and side $\overline{AG}$ at M such that $\overline{LM}$ is parallel to $\overline{BG}$. Write at least 3 different true proportions. (Do *not* generate equivalent proportions by inverting the numerators and denominators of each ratio.)

8. For each of the following segment lengths, determine whether $\overline{TP} \parallel \overline{BC}$.

   a $AT = 5$, $TB = 15$, $AP = 8$, $PC = 24$.

   b $TB = 9$, $AB = 18$, $AP = 6$, $PC = 6$.

   c $AT = 4$, $AB = 12$, $AP = 6$, $AC = 15$.

   d $AT = 3$, $TB = 9$, $PC = 4$, $AC = 12$.

   e $AT = \frac{1}{3} \cdot AB$ and $PC = 2 \cdot AP$.

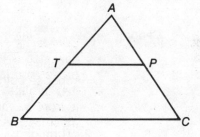

9. If $\overline{KW} \parallel \overline{EG}$, find the length of the indicated segments.

   a $HE = 20$, $KE = 12$, $WG = 9$, $HG = ?$

   b $KH = 7$, $KE = 14$, $HG = 12$, $HW = ?$

   c $HW = 4$, $WG = 12$, $HE = 28$, $KH = ?$

   d $KH = 9$, $KE = 12$, $HG = 42$, $WG = ?$

   e $KH = 2x - 15$, $KE = x$, $HW = 1$, $HG = 4$. Find KH and KE.

10. $\triangle GAL \sim \triangle SHE$. Name three pairs of congruent angles and three equal ratios.

11. The ratio of similitude of two similar polygons is $3:5$. If the length of the shortest side of the smaller polygon is 24, find the length of the shortest side of the larger polygon.

12. $\triangle ZAP \sim \triangle MYX$. If $ZA = 3$, $AP = 12$, $ZP = 21$, and $YX = 20$, find the lengths of the remaining sides of $\triangle MYX$.

13. Quadrilateral $ABCD \sim$ quadrilateral $RSTW$. The lengths of the sides of quadrilateral $ABCD$ are 3, 6, 9, and 15. If the length of the longest side of quadrilateral $RSTW$ is 20, find the perimeter of $RSTW$.

14. The longest side of a polygon exceeds twice the length of the longest side of a similar polygon by 3. If the ratio of similitude of the polygons is 4:9, find the lengths of the longest side of each polygon.

15. $\triangle RST \sim \triangle JKL$.

   a $\overline{RA}$ and $\overline{JB}$ are medians to sides $\overline{ST}$ and $\overline{KL}$, respectively. $RS = 10$ and $JK = 15$. If the length of JB exceeds the length of RA by 4, find the lengths of medians $\overline{JB}$ and $\overline{RA}$.

   b $\overline{SH}$ and $\overline{KO}$ are altitudes to sides $\overline{RT}$ and $\overline{JL}$, respectively. If $SH = 12$, $KO = 15$, $LK = 3x - 2$, and $TS = 2x + 1$, find the lengths of $\overline{LK}$ and $\overline{TS}$.

   c The perimeter of $\triangle RST$ is 25 and the perimeter of $\triangle JKL$ is 40. If $ST = 3x + 1$ and $KL = 4x + 4$, find the lengths of $\overline{ST}$ and $\overline{KL}$.

   d The ratio of similitude of $\triangle RST$ to $\triangle JKL$ is $3:x$. The length of altitude $\overline{SU}$ is $x - 4$ and the length of altitude $\overline{KV}$ is 15. Find the length of altitude $\overline{SU}$.

16. Supply the missing reasons to the following proof of the AA Theorem of Similarity (Theorem 10.3).

   GIVEN $\angle R \cong \angle L$, $\angle T \cong \angle P$.

   PROVE $\triangle RST \sim \triangle LMP$.

   PLAN $\angle S \cong \angle M$. Show sides are in proportion.

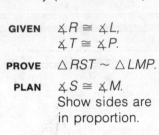
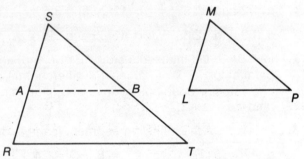

PROOF

Statements	Reasons
1. Assume $SR > ML$. On $\overline{SR}$ choose point A such that $SA = ML$.	1. At a given point on a line, a segment may be drawn equal in length to a given segment.
2. Through point A draw a line parallel to $\overline{RT}$, intersecting side $\overline{ST}$ at point B.	2. ?
3. $\dfrac{SR}{SA} = \dfrac{ST}{SB}$.	3. ?
4. $\dfrac{SR}{ML} = \dfrac{ST}{SB}$. Show $\triangle ASB \cong \triangle LMP$ in order to obtain $SB = MP$.	4. ?
5. $\angle SAB \cong \angle R$.	5. ?
6. $\angle R \cong \angle L$ and $\angle T \cong \angle P$.	6. ?
7. $\angle SAB \cong \angle L$.	7. ?

PROOF	Statements	Reasons
	8. $\angle S \cong \angle M$.	8. ?
	9. $\triangle ASB \cong \triangle LMP$.	9. ?
	10. $SB = MP$.	10. ?
	11. $\dfrac{SR}{ML} = \dfrac{ST}{MP}$.	11. ?

All that remains is to show that $\dfrac{ST}{MP} = \dfrac{RT}{LP}$. This can be accomplished by beginning the proof over again and this time locating a point C on side $\overline{RT}$ such that $TC = PL$, and so on. Since all pairs of corresponding angles are congruent and $\dfrac{SR}{ML} = \dfrac{ST}{MP} = \dfrac{RT}{LP}$, the reverse of the definition of similar polygons is satisfied. Thus, $\triangle RST \sim \triangle LMP$.

17. **GIVEN** $\overline{XW} \cong \overline{XY}$,
$\overline{HA} \perp \overline{WY}$, $\overline{KB} \perp \overline{WY}$.

PROVE $\triangle HWA \sim \triangle KYB$.

18. **GIVEN** $\triangle ABC \sim \triangle RST$,
$\overline{BX}$ bisects $\angle ABC$,
$\overline{SY}$ bisects $\angle RST$.

PROVE $\triangle BXC \sim \triangle SYT$.

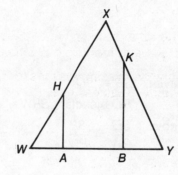

In Exercises 19 to 21, use the following diagram.

19. **GIVEN** $\triangle MCT \sim \triangle BAW$,
$\overline{SW}$ bisects $\angle AWM$.

PROVE $\triangle MCT \sim \triangle BCW$.

20. **GIVEN** $\overline{AW} \parallel \overline{ST}$,
$\overline{MS} \cong \overline{MW}$,
$\overline{WC} \cong \overline{WA}$.

PROVE $\triangle BCW \sim \triangle BTS$.

21. **GIVEN** $\overline{WB} \cong \overline{WC}$, $\overline{ST} \parallel \overline{AW}$,
$\overline{AT}$ bisects $\angle STW$.

PROVE $\triangle ABW \sim \triangle TCW$.

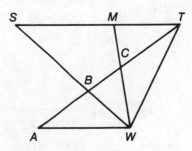

22. **GIVEN** $\overline{HW} \parallel \overline{TA}$,

$\overline{HY} \parallel \overline{AX}$.

PROVE $\dfrac{AX}{HY} = \dfrac{AT}{HW}$.

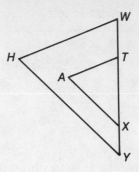

23. **GIVEN** $\overline{MN} \parallel \overline{AT}$,

$\angle 1 \cong \angle 2$.

PROVE $\dfrac{NT}{AT} = \dfrac{RN}{RT}$.

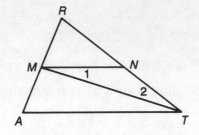

24. **GIVEN** $\overline{SR} \cong \overline{SQ}$,

$\overline{RQ}$ bisects $\angle SRW$.

PROVE $\dfrac{SQ}{RW} = \dfrac{SP}{PW}$.

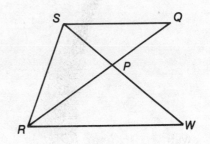

Use the following diagram for Exercises 25 and 26.

25. **GIVEN** $\overline{MC} \perp \overline{JK}$, $\overline{PM} \perp \overline{MQ}$,

$\overline{TP} \cong \overline{TM}$.

PROVE $\dfrac{PM}{MC} = \dfrac{PQ}{MK}$.

26. **GIVEN** T is the midpoint of $\overline{PQ}$,

$\overline{MP} \cong \overline{MQ}$, $\overline{JK} \parallel \overline{MQ}$.

PROVE $\dfrac{PM}{JK} = \dfrac{TQ}{JT}$.

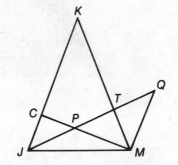

27. **GIVEN** $\overline{EF}$ is the median of
trapezoid $ABCD$.

PROVE $EI \times GH = IH \times EF$.

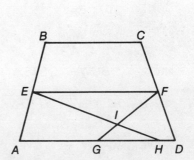

28. GIVEN $\overline{RS} \perp \overline{ST}$,
$\overline{SW} \perp \overline{RT}$.

PROVE $(ST)^2 = TW \times RT$.

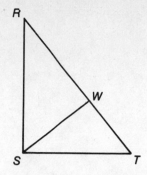

29. GIVEN $\overline{AF}$ bisects $\measuredangle BAC$,
$\overline{BH}$ bisects $\measuredangle ABC$,
$\overline{BC} \cong \overline{AC}$.

PROVE $AH \times EF = BF \times EH$.

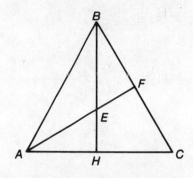

30. GIVEN $\overline{XY} \parallel \overline{LK}$,
$\overline{XZ} \parallel \overline{JK}$.

PROVE $JY \times ZL = XZ \times KZ$.

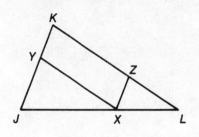

CHAPTER ELEVEN

THE RIGHT TRIANGLE

11.1 INTRODUCTION

Our main objective in this chapter is to see how the various parts of a right triangle are related. If we know the lengths of any two sides of a right triangle, can we determine the length of the remaining side? Or, can we use this information to determine the measure of an acute angle of the right triangle? If the length of a side of a right triangle and the measure of an acute angle are known, can the length of either one of the remaining sides of the triangle be determined? Are there any special relationships between the lengths of the sides in a right triangle whose acute angles measure 30 and 60, or which measure 45 and 45?

We may seek the answers to these questions by applying the properties of similar triangles. You may already be familiar with some of the ideas which result, such as the Pythagorean Theorem and trigonometric ratios. It will be interesting, however, to see how these relationships can be arrived at by using our knowledge of geometry.

Let's begin by reviewing some important terminology and introducing some new notation. In a right triangle, the side opposite the right angle is called the *hypotenuse*; each of the remaining sides is called a *leg*. It is sometimes convenient to refer to the length of a side of a triangle by using the lowercase letter of the vertex which lies opposite the side. See Figure 11.1.

11.2 PROPORTIONS IN A RIGHT TRIANGLE

The first set of properties of a right triangle that we will investigate involves the altitude drawn to the hypotenuse of a right triangle. See Figure 11.2. Notice that altitude $\overline{CD}$ divides hypotenuse $\overline{AB}$ into two segments, $\overline{AD}$ and $\overline{BD}$. The hypotenuse segment $\overline{AD}$ is adjacent to leg $\overline{AC}$ while hypotenuse segment $\overline{BD}$ is adjacent to leg $\overline{BC}$. How many triangles do you see in the

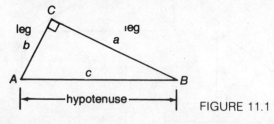

FIGURE 11.1

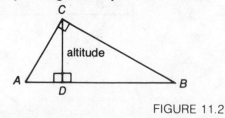

FIGURE 11.2

187

above diagram? There are *three* distinct triangles: the original right triangle ($\triangle ABC$) and the two smaller right triangles which are formed by the altitude on the hypotenuse ($\triangle ACD$ and $\triangle CBD$). As a result of the similarity relationships which exist between these triangles, a set of proportions can be written. Theorem 11.1 summarizes these relationships.

THEOREM 11.1 PROPORTIONS IN A RIGHT TRIANGLE

If in a right triangle the altitude to the hypotenuse is drawn, then:

1 The altitude separates the original triangle into two triangles which are similar to the original triangle and to each other (see Figure 11.2):

$$\triangle ACD \sim \triangle ABC$$

$$\triangle CBD \sim \triangle ABC$$

and $$\triangle ACD \sim \triangle CBD$$

2 The length of each leg is the mean proportional between the length of the hypotenuse segment adjacent to the leg and the length of the entire hypotenuse.
Since $\triangle ACD \sim \triangle ABC$,

$$\frac{AD}{AC} = \frac{AC}{AB}$$

Since $\triangle CBD \sim \triangle ABC$,

$$\frac{BD}{BC} = \frac{BC}{AB}$$

3 The length of the altitude is the mean proportional between the lengths of the segments it forms on the hypotenuse.
Since $\triangle ACD \sim \triangle CBD$,

$$\frac{AD}{CD} = \frac{CD}{DB}$$

OUTLINE OF PROOF $\triangle ACD \sim \triangle ABC$ since $\angle A \cong \angle A$ and $\angle ADC \cong \angle ACB$. $\triangle CBD \sim \triangle ABC$ since $\angle B \cong \angle B$ and $\angle CDB \cong \angle ACB$. $\triangle ACD \sim \triangle CBD$ since each triangle is similar to the same triangle they must be similar to each other. The indicated proportions follow from the principle that the lengths of corresponding sides of similar triangles are in proportion.

EXAMPLE 11.1 Find the value of x.

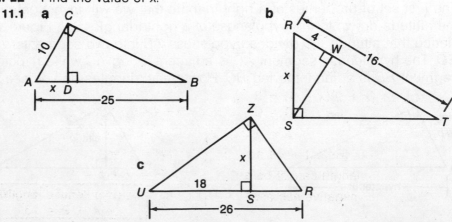

SOLUTION **a** $\dfrac{AD}{AC} = \dfrac{AC}{AB}$ $\left(\dfrac{\text{Hyp segment}}{\text{Leg}} = \dfrac{\text{leg}}{\text{hyp}}\right)$

$\dfrac{x}{10} = \dfrac{10}{25}$

$25x = 100$

$x = \dfrac{100}{25}$

$\boxed{x = 4}$

b $\dfrac{RW}{RS} = \dfrac{RS}{RT}$ $\left(\dfrac{\text{Hyp segment}}{\text{Leg}} = \dfrac{\text{leg}}{\text{hyp}}\right)$

$\dfrac{4}{x} = \dfrac{x}{16}$

$x^2 = 64$

$x = \sqrt{64}$

$\boxed{x = 8}$

c $\dfrac{US}{ZS} = \dfrac{ZS}{RS}$ $\left(\dfrac{\text{Hyp segment 1}}{\text{Altitude}} = \dfrac{\text{altitude}}{\text{hyp segment 2}}\right)$

$\dfrac{18}{x} = \dfrac{x}{8}$ NOTE: $RS = 26 - 18 = 8$

$x^2 = 144$

$x = \sqrt{144}$

$\boxed{x = 12}$

EXAMPLE 11.2 In right triangle JKL, angle K is a right angle. Altitude $\overline{KH}$ is drawn such that the length of $\overline{JH}$ exceeds the length of $\overline{HL}$ by 5. If $KH = 6$, find the length of the hypotenuse.

SOLUTION Let x = length of $\overline{LH}$.
Then $x + 5$ = length of $\overline{JH}$.

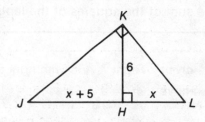

$\dfrac{x}{6} = \dfrac{6}{x + 5}$ $\left(\dfrac{\text{Hyp segment 1}}{\text{Altitude}} = \dfrac{\text{altitude}}{\text{hyp segment 2}}\right)$

$x(x + 5) = 36$

$x^2 + 5x = 36$

$x^2 + 5x - 36 = 36 - 36$

$x^2 + 5x - 36 = 0$

$(x + 9)(x - 4) = 0$

$x + 9 = 0$ or $x - 4 = 0$

$x = -9$ $x = 4$

(Reject since a $LH = x = 4$
length cannot be a $JH = x + 5 = 9$
negative number.)

$\boxed{\begin{array}{c} JL \text{ (hypotenuse length)} \\ = 4 + 9 = 13. \end{array}}$

┌─ SUMMARY ───
│ **PROPORTIONS IN A RIGHT TRIANGLE**
│
│ **1** $\dfrac{x}{b} = \dfrac{b}{c}$ and $\dfrac{y}{a} = \dfrac{a}{c}$
│
│ **2** $\dfrac{x}{h} = \dfrac{h}{y}$
│
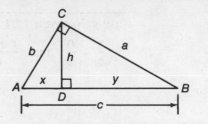
└──

11.3 THE PYTHAGOREAN THEOREM

One of the most famous and useful theorems in mathematics provides a means for finding the length of *any* side of a *right* triangle, given the lengths of the other two sides. The sides are related by the equation:

$$(\text{Hypotenuse})^2 = (\text{leg 1})^2 + (\text{leg 2})^2$$

This relationship is known as the *Pythagorean Theorem* and is named in honor of the Greek mathematician Pythagoras who is believed to have presented the first proof of this theorem in about 500 B.C. Since that time, many different proofs of this theorem have been offered. Here's the formal statement of the theorem and one such proof.

┌─ **THEOREM 11.2 THE PYTHAGOREAN THEOREM** ──────────
│ The square of the length of the hypotenuse of a right triangle is equal to
│ the sum of the squares of the lengths of the legs.
└──

GIVEN Right $\triangle ABC$ with right angle C.

PROVE $c^2 = a^2 + b^2$.

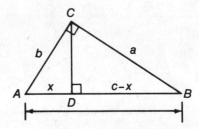

PLAN Draw the altitude from C to hypotenuse $\overline{AB}$, intersecting $\overline{AB}$ at point D. For convenience, refer to $\overline{AD}$ as x and $\overline{DB}$ as $c - x$. Apply Theorem 11.1.

PROOF

Statements	Reasons
1. Right $\triangle ABC$ with right angle C.	1. Given.
2. Draw the altitude from vertex C to hypotenuse $\overline{AB}$, intersecting $\overline{AB}$ at point D.	2. From a point not on a line, exactly one perpendicular may be drawn to the line.

PROOF	Statements	Reasons
	Apply Theorem 11.1:	
	3. $\dfrac{x}{b} = \dfrac{b}{c}$ and $\dfrac{c - x}{a} = \dfrac{a}{c}$.	3. If in a right triangle the altitude to the hypotenuse is drawn, then the length of each leg is the mean proportional between the length of the hypotenuse segment adjacent to the leg and the length of the entire hypotenuse.
	4. $cx = b^2$ and $c(c - x) = a^2$ which may be written as $c^2 - cx = a^2$.	4. In a proportion, the product of the means equals the product of the extremes.
	5. $\begin{aligned} c^2 - \cancel{cx} &= a^2 \\ + \qquad \cancel{cx} &= b^2 \\ \hline c^2 \qquad\quad &= a^2 + b^2 \end{aligned}$	5. Addition property.

EXAMPLE 11.3 Find the value of x.

a

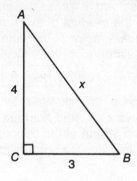

b

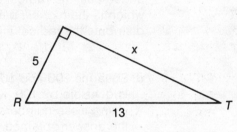

c

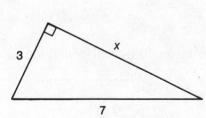

d

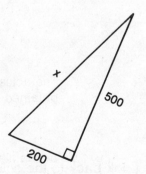

SOLUTION **a** Since x represents the length of the hypotenuse,

$$x^2 = 3^2 + 4^2$$
$$x^2 = 9 + 16$$
$$x^2 = 25$$
$$x = \sqrt{25}$$
$$\boxed{x = 5}$$

b Since x represents the length of a leg and the hypotenuse is 13,

$$13^2 = x^2 + 5^2$$
$$169 = x^2 + 25$$
$$169 - 25 = x^2 + 25 - 25$$
$$144 = x^2$$

or

$$x^2 = 144$$
$$x = \sqrt{144}$$
$$\boxed{x = 12}$$

c Since x represents the length of a leg and the hypotenuse is 7,

$$7^2 = x^2 + 3^2$$
$$49 = x^2 + 9$$
$$49 - 9 = x^2 + 9 - 9$$
$$40 = x^2$$

or

$$x^2 = 40$$
$$\boxed{x = \sqrt{40}}$$

REMARK To express a radical in simplest form, express the number underneath the radical sign as the product of two numbers, one of which is the *highest* perfect square factor of the number. Next, distribute the radical sign and simplify:

$$x = \sqrt{40} = \sqrt{4} \times \sqrt{10} = \boxed{2\sqrt{10}}$$

d Squaring 200 and 500 would be quite cumbersome. Since each is divisible by 100, we may work with the numbers 2 and 5. Using these numbers, find the length of the hypotenuse. *Multiply the answer obtained by 100 in order to compensate for dividing the original lengths by 100.*

$$c^2 = 2^2 + 5^2$$
$$c^2 = 4 + 25$$
$$c^2 = 29$$
$$c = \sqrt{29}$$

The actual value, x, may be found by multiplying the value obtained by 100:

$$x = 100c$$
$$\boxed{x = 100\sqrt{29}}$$

Look back to the solutions for Example 11.3. In parts (a) and (b) whole number values were obtained for the missing sides. Any set of three whole numbers x, y, and z is called a *Pythagorean triple* if the numbers satisfy the equation:

$$z^2 = x^2 + y^2$$

The set of numbers $\{3, 4, 5\}$ is an example of a Pythagorean triple. The sets $\{5, 12, 13\}$ and $\{8, 15, 17\}$ are also Pythagorean triples. There are many others.

If $\{x, y, z\}$ is a Pythagorean triple, then so is any whole number multiple of *each* member of this set. The set $\{6, 8, 10\}$ is a Pythagorean triple since each member was obtained by multiplying the corresponding member of the Pythagorean triple {3, 4, 5} by 2:

$$\{6, 8, 10\} = \{2 \cdot 3, 2 \cdot 4, 2 \cdot 5\}$$

To illustrate further, consider the following table:

PYTHAGOREAN TRIPLE	MULTIPLE OF A PYTHAGOREAN TRIPLE	MULTIPLYING FACTOR
{3, 4, 5}	{15, 20, 25}	5
{5, 12, 13}	{10, 24, 26}	2
{8, 15, 17}	{80, 150, 170}	10

You should *memorize* the sets {3, 4, 5}, {5, 12, 13}, and {8, 15, 17} as Pythagorean triples and be able to recognize their multiples. The need for this will become apparent as you work on the next set of examples.

EXAMPLE 11.4   The base (length) of a rectangle is 12 and its altitude (height) is 5. Find the length of a diagonal.

SOLUTION   Triangle *ABC* is a 5-12-13 right triangle. Diagonal *AC* = 13. If you didn't see this pattern, you could apply the Pythagorean Theorem:

$$x^2 = 5^2 + 12^2$$

$$x^2 = 25 + 144 = 169$$

$$x = \sqrt{169}$$

$$x = 13$$

EXAMPLE 11.5   The diagonals of a rhombus are 18 and 24. Find the length of a side of the rhombus.

SOLUTION   Recall that the diagonals of a rhombus bisect each other and intersect at right angles.

$\triangle AED$ is a multiple of a 3-4-5 right triangle. Each member of the triple is multiplied by 3. Hence, the length of a side is 3·5 or 15.

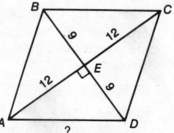

EXAMPLE 11.6   The lengths of the legs of an isosceles trapezoid are 17 each. The lengths of its bases are 9 and 39. Find the length of an altitude.

SOLUTION   Drop two altitudes, one from each of the upper vertices.

Quadrilateral *BEFC* is a rectangle (since $\overline{BE}$ and $\overline{CF}$ are congruent, parallel, and intersect $\overline{AD}$ at right angles). Hence *BC* = *EF* = 9. Since right $\triangle AEB \cong$ right $\triangle DFC$, *AE* = *DF* = 15. $\triangle AEB$ is an 8-15-17 right triangle. The length of an altitude is 8.

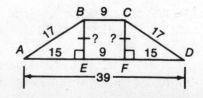

EXAMPLE 11.7 Find the length of a side of a square if a diagonal has a length of 8.

SOLUTION There are no Pythagorean triples here. Since a square is equilateral, we may represent the lengths of sides $\overline{AB}$ and $\overline{AD}$ using x and apply the Pythagorean Theorem in right $\triangle ABD$:

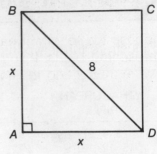

$$x^2 + x^2 = 8^2$$
$$2x^2 = 64$$
$$x^2 = 32$$
$$x = \sqrt{32} = \sqrt{16} \cdot \sqrt{2}$$
$$x = 4\sqrt{2}$$

The length of a side is $4\sqrt{2}$.

SUMMARY

PYTHAGOREAN TRIPLES

Some commonly encountered Pythagorean triples are:

$$\{3n, 4n, 5n\}$$
$$\{5n, 12n, 13n\}$$
$$\{8n, 15n, 17n\}$$

where n is any natural number ($n = 1, 2, 3, \ldots$). There are many other families of Pythagorean triples, including $\{7, 24, 25\}$ and $\{9, 40, 41\}$.

CONVERSE OF THE PYTHAGOREAN THEOREM

If the lengths of the sides of a triangle are known, then we can determine whether the triangle is a right triangle by applying the *converse* of the Pythagorean Theorem.

THEOREM 11.3 CONVERSE OF THE PYTHAGOREAN THEOREM

If in a triangle the square of the length of a side is equal to the sum of the squares of the lengths of the other two sides, then the triangle is a right triangle.

REMARK Suppose a, b, and c represent the lengths of the sides of a triangle and c is the largest of the three numbers. If $c^2 = a^2 + b^2$, then the triangle is a right triangle.

For example, to determine whether the triangle whose sides have lengths of 11, 60, and 61 is a right triangle, we test whether the square of the largest of the three numbers is equal to the sum of the squares of the other two numbers:

$$61^2 \stackrel{?}{=} 11^2 + 60^2$$
$$3721 \stackrel{?}{=} 121 + 3600$$
$$3721 \stackrel{\checkmark}{=} 3721 \quad \text{Hence, the triangle is a right triangle.}$$

11.4 SPECIAL RIGHT TRIANGLE RELATIONSHIPS

Suppose the length of each side of the equilateral triangle in Figure 11.3 is represented by 2*s*. If you drop an altitude from the vertex of this triangle, several interesting relationships materialize:

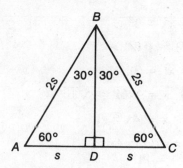

FIGURE 11.3

- Since $\triangle BDA \cong \triangle BDC$, $\overline{BD}$ bisects $\angle ABC$ and also bisects side $\overline{AC}$. Hence, $m\angle ABD = 30$. The acute angles of right triangle *ADB* measure 60 and 30. Also, $AD = \frac{1}{2}(AC) = \frac{1}{2}(2s) = s$.

- The length of side $\overline{BD}$ may be found using the Pythagorean Theorem:

$$(BD)^2 + (AD)^2 = (AB)^2$$
$$(BD)^2 + (s)^2 = (2s)^2$$
$$(BD)^2 + s^2 = 4s^2$$
$$(BD)^2 = 3s^2$$
$$BD = \sqrt{3s^2} = \sqrt{3}\cdot\sqrt{s^2} = \sqrt{3}\,s$$

Let's summarize the information we have gathered about $\triangle ADB$. First, $\triangle ADB$ is referred to as a *30-60 right triangle* since these numbers correspond to the measures of its acute angles. In a 30-60 right triangle the following relationships hold:

---

1 The length of the *shorter* leg (the side opposite the 30 degree angle) is one-half the length of the hypotenuse:

$$AD = \frac{1}{2}AB$$

2 The length of the *longer* leg (the side opposite the 60 degree angle) is one-half the length of the hypotenuse multiplied by $\sqrt{3}$:

$$BD = \frac{1}{2} \times AB \times \sqrt{3}$$

3 The length of the longer leg is equal to the length of the shorter leg multiplied by $\sqrt{3}$:

$$BD = AD \cdot \sqrt{3}$$

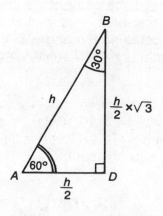

EXAMPLE 11.8   Fill in the following table:

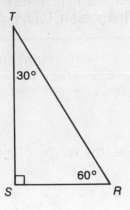

	RS	ST	RT
a	?	?	12
b	4	?	?
c	?	$7\sqrt{3}$	?

SOLUTION   **a** $RS = \frac{1}{2}(RT) = \frac{1}{2}(12) = 6$

$ST = RS \cdot \sqrt{3} = 6\sqrt{3}$

b $ST = RS \cdot \sqrt{3} = 4\sqrt{3}$

$RT = 2(RS) = 2(4) = 8$

c $RS = 7$
$RT = 2(RS) = 2(7) = 14$

EXAMPLE 11.9   In $\square RSTW$, $m\angle R = 30$ and $RS = 12$. Find the length of an altitude.

SOLUTION $SH = \frac{1}{2}(12) = 6$

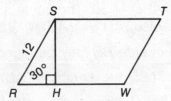

EXAMPLE 11.10   In $\triangle JKL$, $m\angle K = 120$ and $JK = 10$. Find the length of the altitude drawn from vertex J to side $\overline{LK}$ (extended if necessary).

SOLUTION   Since $\angle K$ is obtuse, the altitude falls in the exterior of the triangle, intersecting the extension of $\overline{KL}$, say at point H. $\triangle JHK$ is a 30-60 right triangle. Hence,

$$JH = \frac{1}{2} \cdot 10 \cdot \sqrt{3} = 5\sqrt{3}$$

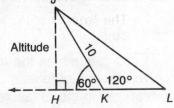

THE 45-45 RIGHT TRIANGLE

Another special right triangle is the isosceles right triangle. Since the legs of an isosceles right triangle are congruent, the angles opposite must also be congruent. This implies that the measure of each acute angle of an isosceles right triangle is 45. A *45-45 right triangle* is another name for an isosceles right triangle.

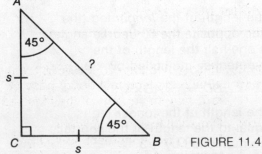

FIGURE 11.4

To determine the relationships between the lengths of the sides of a 45-45 right triangle, represent the length of each leg by the letter s (Figure 11.4).

Then the hypotenuse $\overline{AB}$ may be expressed in terms of s by applying the Pythagorean Theorem:

$$(AB)^2 = s^2 + s^2$$
$$(AB)^2 = 2s^2$$
$$AB = \sqrt{2s^2} = \sqrt{2} \times \sqrt{s^2}$$
$$AB = \sqrt{2} \times s$$

Thus, the length of the hypotenuse is $\sqrt{2}$ times the length of a leg.

$$\boxed{\text{Hypotenuse} = \sqrt{2} \times \text{ leg}}$$

By dividing each side of this equation by $\sqrt{2}$ we may find an expression for the length of a leg in terms of the length of the hypotenuse:

$$\text{Leg} = \frac{\text{hypotenuse}}{\sqrt{2}}$$

Rationalizing the denominator, we obtain:

$$\text{Leg} = \frac{\text{hypotenuse}}{\sqrt{2}} \times \frac{\sqrt{2}}{\sqrt{2}}$$

$$\boxed{\text{Leg} = \frac{\text{hypotenuse}}{2} \times \sqrt{2}}$$

In a 45-45 right triangle the following relationships hold:

---

1 The *lengths of the legs* are equal:

$$AC = BC$$

2 The *length of the hypotenuse* is equal to the length of either leg multiplied by $\sqrt{2}$:

$$AB = AC \text{ (or } BC\text{)} \cdot \sqrt{2}$$

3 The *length of either leg* is equal to one-half the length of the hypotenuse multiplied by $\sqrt{2}$:

$$AC \text{ (or } BC\text{)} = \tfrac{1}{2} AB \cdot \sqrt{2}$$

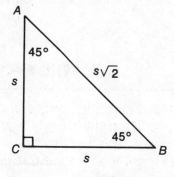

EXAMPLE 11.11 In isosceles trapezoid *ABCD*, the measure of a lower base angle is 45 and the length of upper base $\overline{BC}$ is 5. If the length of an altitude is 7, find the lengths of the legs, $\overline{AB}$ and $\overline{DC}$.

SOLUTION Drop altitudes from *B* and *C*, forming two congruent 45-45 right triangles. $AE = BE = 7$. Also, $FD = 7$. $AB = AE \times \sqrt{2} = 7\sqrt{2}$. The length of each leg is $7\sqrt{2}$.

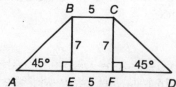

11.5 DEFINING TRIGONOMETRIC RATIOS

Each of the triangles pictured in Figure 11.5 is a 30-60 right triangle. The lengths of the sides in each of these right triangles must obey the relationships developed in the previous section.

In particular, we may write:

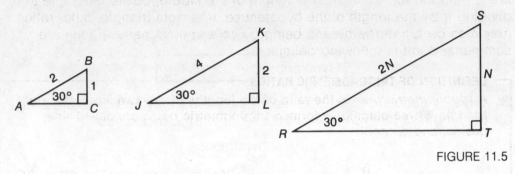

FIGURE 11.5

In △ABC	In △JKL	In △RST
$\dfrac{BC}{AB} = \dfrac{1}{2}$	$\dfrac{KL}{JK} = \dfrac{2}{4} = \dfrac{1}{2}$	$\dfrac{ST}{RS} = \dfrac{N}{2N} = \dfrac{1}{2}$

In each of these triangles the ratio of the length of the side opposite the 30 degree angle to the length of the hypotenuse is 1:2 or 0.5. This ratio will hold in *every* 30-60 right triangle no matter how large or small the triangle is. For other choices of congruent acute angles in right triangles, will this type of ratio also be constant? As the set of triangles in Figure 11.6 illustrates, the answer is yes! In each right triangle, the ratio of the length of the side opposite the angle whose measure is *x* degrees to the length of the hypotenuse must be the same.

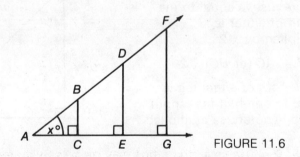

FIGURE 11.6

The ratios $\dfrac{BC}{AB}, \dfrac{DE}{AD}, \dfrac{FG}{AF}$ are equal since they represent the ratios of the lengths of corresponding sides of similar triangles. You should observe that in Figure 11.6 there are three right triangles each of which is similar to the other two. That is, $\triangle ABC \sim \triangle ADE \sim \triangle AFG$ by the AA Theorem of similarity since,

1 $\angle A \cong \angle A \cong \angle A$.

2 Each triangle includes a right angle and all right angles are congruent.

We are then entitled to write the following extended proportion:

$$\frac{BC}{AB} = \frac{DE}{AD} = \frac{FG}{AF} = \frac{\text{length of side opposite angle } A}{\text{length of the hypotenuse}}$$

For any given acute angle, this ratio is the same regardless of the size of the right triangle. It will be convenient to refer to this type of ratio by a special name—the *sine ratio*. The sine of an acute angle of a right triangle is the ratio formed by taking the length of the side opposite the angle and dividing it by the length of the hypotenuse. In a right triangle, other ratios may also be formed with each being given a special name. These are summarized in the following definition.

DEFINITION OF TRIGONOMETRIC RATIOS

A *trigonometric ratio* is the ratio of the lengths of *any* two sides of a *right* triangle. Three commonly formed trigonometric ratios are called sine, cosine, and tangent:

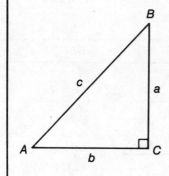

$$\text{Sine of } \angle A = \frac{\text{length of side opposite } \angle A}{\text{length of the hypotenuse}} = \frac{BC}{AB}$$

$$\text{Cosine of } \angle A = \frac{\text{length of side adjacent } \angle A}{\text{length of the hypotenuse}} = \frac{AC}{AB}$$

$$\text{Tangent of } \angle A = \frac{\text{length of side opposite } \angle A}{\text{length of side adjacent } \angle A} = \frac{BC}{AC}$$

REMARKS **1** In writing trigonometric ratios, the following abbreviations are used: sin A, cos A, and tan A. For example,

$$\sin A = \frac{a}{c}$$

$$\cos A = \frac{b}{c}$$

$$\tan A = \frac{a}{b}$$

2 Trigonometric ratios may be taken with respect to either of the acute angles of a right triangle:

$$\sin B = \frac{b}{c}$$

$$\cos B = \frac{a}{c}$$

$$\tan B = \frac{b}{a}$$

3 You should memorize these relationships.

EXAMPLE
11.12
In right $\triangle ABC$, angle C is a right angle. If $AC = 4$ and $BC = 3$, find the value of tan A, sin A, and cos A.

SOLUTION Since $\triangle ABC$ is a 3-4-5 right triangle, $AB = 5$.

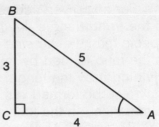

$$\tan A = \frac{\text{side opposite } \measuredangle A}{\text{side adjacent } \measuredangle A} = \frac{3}{4} = 0.75$$

$$\sin A = \frac{\text{side opposite } \measuredangle A}{\text{hypotenuse}} = \frac{3}{5} = 0.6$$

$$\cos A = \frac{\text{side adjacent } \measuredangle A}{\text{hypotenuse}} = \frac{4}{5} = 0.8$$

EXAMPLE
11.13
Express the values of sin 60°, cos 60°, and tan 60° in either radical or decimal form.

SOLUTION
$$\sin 60° = \frac{\sqrt{3} \times n}{2n} = \frac{\sqrt{3}}{2}$$

$$\cos 60° = \frac{n}{2n} = \frac{1}{2} = 0.5$$

$$\tan 60° = \frac{\sqrt{3} \times n}{n} = \sqrt{3}$$

TRIGONOMETRIC TABLES

In Example 11.13 we illustrated how to find the sine, cosine, and tangent of an angle having a measure of 60. Our approach was based on using the special relationships which exist between the lengths of the sides of a 30-60 right triangle. Suppose that an acute angle of a right triangle measures 38. How can we find the value of the sine, cosine, and tangent ratios of this angle? One approach would be to use a protractor to construct a right triangle having an acute angle of 38 degrees. A ruler would then be used to determine the lengths of each side of the right triangle. Calculating the ratios of appropriate lengths would give us the required trigonometric values.

This procedure is time consuming, not to mention inaccurate. Fortunately, the values of the trigonometric ratios for all angles between 0 and 90°, correct to the nearest ten-thousandth (fourth decimal place position), have been previously calculated and the results tabulated for us. Table 11.1 gives a partial listing of decimal approximations of trigonometric ratios for a group of angles. Let's concentrate on 38°. Notice that

$$\sin 38° = 0.6157$$

$$\cos 38° = 0.7880$$

$$\tan 38° = 0.7813$$

For what value of x, is cos x = 0.8387? Look in the cosine column, place your finger on 0.8387, and then move your finger to the left. You should now be pointing at 33 degrees. Let's do another example. For what value of x is sin x = 0.6751? Notice that the table does not include this number. It does, however, include numbers above and below this value:

$$\sin 42° = 0.6691$$
$$\text{difference} = 0.0060$$
$$\sin x \ = 0.6751$$
$$\text{difference} = 0.0069$$
$$\sin 43° = 0.6820$$

We must approximate the measure of the angle whose sine ratio is 0.6751. The value of x must be between 42 and 43 degrees. Which number is x closer to? To find out, take the absolute value of the difference between the values of the sines of 42 and x degrees, and then repeat the procedure for x degrees and 43 degrees. The measure of angle x will be closer to the angle which has the corresponding *smaller* calculated difference. In our example, the value of x, *approximated to the nearest degree*, is 42 degrees.

EXAMPLE Find the value of each of the following:

11.14   **a** tan 27°

      b sin 44°

      c cos 35°

SOLUTION   **a** tan 27° = 0.5095

      b sin 44° = 0.6947

      c cos 35° = 0.8192

EXAMPLE Find the measure of angle x.

11.15   **a** sin x = 0.7071

      b tan x = 0.6009

      c cos x = 0.7986

SOLUTION   **a** x = 45°

      b x = 31°

      c x = 37°

EXAMPLE Find the measure of angle A, correct to the nearest degree.

11.16   **a** tan A = 0.7413

      b cos A = 0.8854

      c sin A = 0.6599

SOLUTION   **a** A = 37°

      b A = 28°

      c A = 41°

See if you can observe any patterns in how the table values of the trigonometric ratios change as the angle increases. To get a more complete picture, turn to the trigonometric table which is located on page 202. You should observe that:

- As an angle increases in value, its sine and tangent ratio increases while the value of its cosine ratio decreases.

- The minimum value of sine, cosine, and tangent is 0. (*Note*: sin 0° = 0, cos 90° = 0, tan 0° = 0)

- The maximum value of cosine and sine is 1. (*Note*: cos 0° = 1 and sin 90° = 1) As the measure of an acute angle of a right triangle approaches 90 degrees, the value of its tangent ratio gets larger and larger, with *no* upper limit. We say that the tangent ratio is *unbounded* or undefined at 90 degrees. Consequently, there is no table entry for the value of tan 90°.

TABLE 11.1

ANGLE	SINE	COSINE	TANGENT
26°	.4384	.8988	.4877
27°	.4540	.8910	.5095
28°	.4695	.8829	.5317
29°	.4848	.8746	.5543
30°	.5000	.8660	.5774
31°	.5150	.8572	.6009
32°	.5299	.8480	.6249
33°	.5446	.8387	.6494
34°	.5592	.8290	.6745
35°	.5736	.8192	.7002
36°	.5878	.8090	.7265
37°	.6018	.7986	.7536
38°	**.6157**	**.7880**	**.7813**
39°	.6293	.7771	.8098
40°	.6428	.7660	.8391
41°	.6561	.7547	.8693
42°	.6691	.7431	.9004
43°	.6820	.7314	.9325
44°	.6947	.7193	.9657
45°	.7071	.7071	1.0000

11.6 INDIRECT MEASUREMENT IN THE RIGHT TRIANGLE

Trigonometric ratios may be used to arrive at the measure of a part of a right triangle which may be difficult, if not impossible, to calculate by direct measurement. For example, consider a plane which takes off from a runway, and climbs while maintaining a constant angle with the horizontal ground. Suppose that at the instant of time the plane has traveled 1000 meters, its altitude is 290 meters. Using our knowledge of trigonometry, it is possible to approximate the measure of the angle at which the plane has risen with respect to the horizontal ground.

A right triangle may be used to represent the situation we have just described. The hypotenuse corresponds to the path of the rising plane, the vertical leg of the triangle represents the plane's altitude, and the acute angle formed by the hypotenuse and the horizontal leg (the ground) is the desired angle whose measure we must determine.

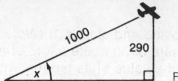

FIGURE 11.7

To find the value of x in Figure 11.7, we must first determine the appropriate trigonometric ratio. The sine ratio relates the three quantities under consideration:

$$\sin x = \frac{\text{side opposite} \sphericalangle}{\text{hypotenuse}}$$

$$\sin x = \frac{290}{1000} = 0.2900$$

To find the value of x to the nearest degree, the table of trigonometric values must be consulted. A search for the decimal number in the sine column that is closest to 0.2900 leads to the table value of 0.2924 which corresponds to 17 degrees. Hence, $x = 17°$, which is correct to the nearest degree.

EXAMPLE 11.17 Find $m\sphericalangle R$ to the nearest degree.

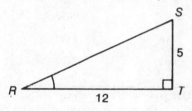

SOLUTION **1** Decide which trigonometric ratio to use and write the corresponding equation:

$$\tan R = \frac{\text{side opposite} \sphericalangle R}{\text{side adjacent} \sphericalangle R}$$

$$\tan R = \frac{5}{12}$$

2 Express the ratio in decimal form. If you have a calculator handy, press the following keys:

$$\boxed{5}\ \boxed{÷}\ \boxed{1}\ \boxed{2}\ \boxed{=}$$

The answer, rounded off to four decimal places, is 0.4167. Hence,

$$\tan R = 0.4167$$

3 Consult the appropriate table column (tangent in this case) in the trigonometric table. Find the decimal value that is closest to the calculated ratio that was determined in Step 2. Since 0.4245 (tan 23°) is closer to 0.4167 than 0.4040 (tan 22°), angle R equals 23°, correct to the nearest degree.

The previous example illustrates that given the lengths of any two sides of a right triangle, we may use an appropriate trigonometric ratio to find the measure of either acute angle:

GIVEN . . .	USE THE . . .
The lengths of the two legs	Tangent ratio
The length of the side opposite the desired angle and the length of the hypotenuse	Sine ratio
The length of the side adjacent to the desired angle and the length of the hypotenuse	Cosine ratio

The next two examples illustrate that given the length of a side of a right triangle and the measure of an acute angle, the length of either of the two remaining sides of the triangle may be found.

EXAMPLE   Find the value of x to the *nearest tenth*.
11.18

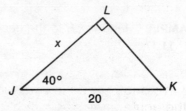

SOLUTION   1   Decide which trigonometric ratio to use and then write the corresponding equation:

$$\cos J = \frac{\text{side adjacent } \angle j}{\text{hypotenuse}}$$

$$\cos 40° = \frac{x}{20}$$

2   Consult the table of trigonometric values in order to evaluate the trigonometric ratio. Since cos 40° = 0.7660,

$$0.7660 = \frac{x}{20}$$

3   Solve the equation. It may help to think of the equation written in Step 2 as a proportion:

$$\frac{0.7660}{1} = \frac{x}{20}$$

Solve by cross-multiplying:

$$x = 20(0.7660) = 15.32$$

4   Round off the answer obtained in Step 3 to the desired accuracy. In this example we are asked to express the answer correct to the nearest tenth. Hence,

$$\boxed{x = 15.3}$$

EXAMPLE
11.19   Find the value of *x* to the *nearest tenth*.

SOLUTION   1

$$\tan \angle R = \frac{\text{side opposite } \angle R}{\text{side adjacent } \angle R}$$

$$\tan 75° = \frac{28}{x}$$

2

$$3.7321 = \frac{28}{x}$$

3

$$3.7321x = 28$$

$$\frac{3,7321x}{3.7321} = \frac{28}{3.7321}$$

$$x = 7.502$$

$$\boxed{x = 7.5}$$

4

In Step 3 we must divide a whole number by a decimal number. Without a calculator, this can be a messy business. This type of arithmetic operation can be avoided by working with the other acute angle of the right triangle. Since the acute angles of a right triangle are complementary, the measure of angle *S* is 15. We may therefore form the tangent ratio with respect to *angle S* by writing,

$$\tan 15° = \frac{x}{28}$$

$$0.2679 = \frac{x}{28}$$

$$x = 28(0.2679)$$

$$x = 7.5$$

This second approach has a slight computational advantage compared to the first method since most people find it easier to multiply a whole number by a decimal number than to divide a whole number by a decimal number.

EXAMPLE
11.20   The lengths of the diagonals of a rhombus are 12 and 16. Find to the nearest degree the measures of the angles of the rhombus.

SOLUTION   In △*AED*:

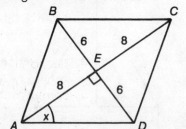

$$\tan x = \frac{\text{side opposite}}{\text{side adjacent}} = \frac{ED}{AE}$$

$$\tan x = \frac{6}{8} = 0.7500$$

$$x = 37° \text{ (to the nearest degree)}$$

Since the diagonals of a rhombus bisect its angles,

$$m\angle BAD = 2x = 2(37) = 74$$

Since consecutive angles of a rhombus are supplementary,

$$m\angle ABC = 180 - m\angle BAD$$
$$= 180 - 74$$
$$m\angle ABC = 106$$

Since opposite angles of a rhombus are equal in measure:

$$m\angle BCD = m\angle BAD = 74$$
$$m\angle ADC = m\angle ABC = 106$$

REVIEW EXERCISES FOR CHAPTER 11

In Exercises 1 to 3, find the values of r, s, and t.

1.

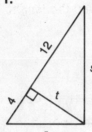

2.

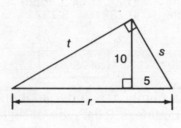

3.

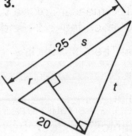

In Exercises 4 to 9, in right triangle JKL, angle JKL is a right angle and $\overline{KH} \perp \overline{JL}$.

4. If $JH = 4$ and $HL = 16$, find KH.

5. If $JH = 5$ and $HL = 4$, find KL.

6. If $JH = 8$, $JL = 20$, find KH.

7. If $KL = 18$, $JL = 27$, find JK.

8. If $JK = 14$, $HL = 21$, find JH.

9. If $KH = 12$, $JL = 40$, find JK (assume $\overline{JK}$ is the shorter leg of right $\triangle JKL$).

10. The altitude drawn to the hypotenuse of a right triangle divides the hypotenuse into segments such that their lengths are in the ratio of $1:4$. If the length of the altitude is 8, find the length of

a Each segment of the hypotenuse.　　**b** The longer leg of the triangle.

In Exercises 11 to 15, find the value of x.

11.

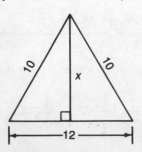

12.

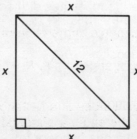

13.

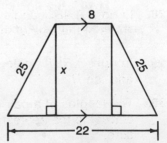

14.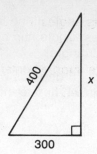

15.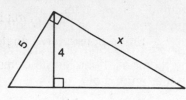

16. If the lengths of the diagonals of a rhombus are 32 and 24, find the perimeter of the rhombus.

17. If the perimeter of a rhombus is 164 and the length of the longer diagonal is 80, find the length of the shorter diagonal.

18. Find the length of the altitude drawn to a side of an equilateral triangle whose perimeter is 30.

19. The length of the base of an isosceles triangle is 14. If the length of the altitude drawn to the base is 5, find the length of each of the legs of the triangle.

20. The measure of the vertex angle of an isosceles triangle is 120 and the length of each leg is 8. Find the length of:

a The altitude drawn to the base.　　　**b** The base.

21. If the perimeter of a square is 24, find the length of a diagonal.

22. If the length of a diagonal of a square is 18, find the perimeter of the square.

23. Find the length of the altitude drawn to side $\overline{AC}$ of $\triangle ABC$ if $AB = 8$, $AC = 14$, and $m\angle A$ equals:

a 30　　　**b** 120　　　**c** 135

24. The lengths of the bases of an isosceles trapezoid are 9 and 25. Find the length of the altitude and each of the legs if the measure of each lower base angle is:

a 30　　　**b** 45　　　**c** 60

25. Find the value of x, y, and z.

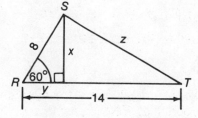

26. The lengths of two adjacent sides of a parallelogram are 6 and 14. If the measure of an included angle is 60, find the length of the shorter diagonal of the parallelogram.

27. The length of each side of a rhombus is 10 and the measure of an angle of the rhombus is 60. Find the length of the longer diagonal of the rhombus.

28. In right $\triangle ABC$, angle C is a right angle, $AC = 7$, and $BC = 24$. Find the value of the sine of the largest acute angle of the triangle.

29. In right $\triangle RST$, angle T is a right angle. If $\sin R = \frac{9}{41}$, find the value of $\cos R$ and $\tan R$.

30. The lengths of a pair of adjacent sides of a rectangle are 10 and 16. Find, correct to the nearest degree, the angle a diagonal makes with the longer side.

31. The measure of a vertex angle of an isosceles triangle is 72. If the length of the altitude drawn to the base is 10, find to the nearest whole number the length of the base and the length of each leg of the triangle.

32. The shorter diagonal of a rhombus makes an angle of 78 degrees with a side of the rhombus. If the length of the shorter diagonal is 24, find to the nearest tenth the lengths of:

a A side of the rhombus.　　**b** The longer diagonal.

33. Find the value of *x* and *y*, correct to the nearest tenth.

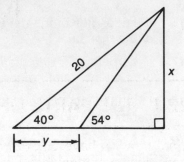

34. *ABCD* is a trapezoid. If *AB* = 14, *BC* = 10, and *m∡BCD* = 38, find the lengths of *AD* and *CD*, correct to the nearest tenth.

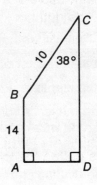

CHAPTER TWELVE
CIRCLES AND ANGLE MEASUREMENT

12.1 THE PARTS OF A CIRCLE

Until now we have considered only those geometric figures having line segments as sides (polygons). Curved-shaped figures represent an important class of figures in geometry as well as in everyday life. Imagine riding on a bicycle or in an automobile that didn't have circular-shaped wheels. It's difficult to think of a machine that doesn't include some circular-shaped parts.

Take a compass and using a fixed compass setting, draw a closed figure. (See Figure 12.1.) The resulting figure is a *circle*. The small puncture hole or impression that the metal compass point makes on the paper is called the *center* of the circle. The compass setting distance is called the *radius* of the circle. Since the compass setting (that is, radius) remained the same while drawing the circle with the compass, we may define a circle as follows:

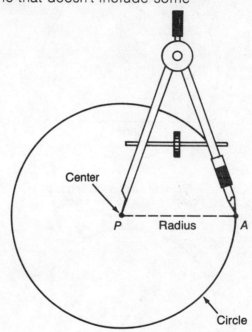

FIGURE 12.1

DEFINITION OF A CIRCLE

A *circle* is the set of all points in a plane having the same distance from a fixed point. The fixed point is called the *center* of the circle. The distance between the center of the circle and any point of the circle is called the *radius* (plural: radii) of the circle.

NOTATION: A capital letter is used to name the center of a circle. If the center of a circle is named by the letter *P*, then the circle is referred to as *circle P*. The shorthand notation ⊙*P* is commonly used, where the symbol that precedes the letter *P* is a miniature circle.

From the definition of a circle, it follows that *all radii of the same circle are congruent*. Two different circles are congruent if their radii are congruent. In circle *O* shown in Figure 12.2, several radii are drawn. Each of these radii must have the same length so that $\overline{OA} \cong \overline{OB} \cong \overline{OC} \cong \overline{OD} \cong$ ··· Circles *O* and *P* are congruent (written as $\odot O \cong \odot P$) if $\overline{OA} \cong \overline{PX}$.

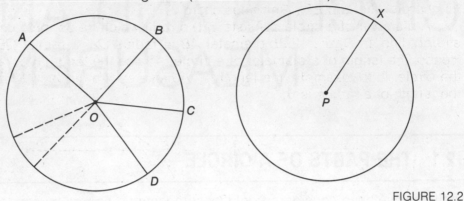

FIGURE 12.2

There are several fundamental terms associated with circles which you should be familiar with. These are illustrated in Figure 12.3.

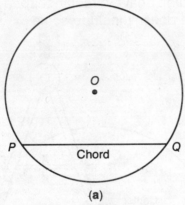

(a)

A *chord* is a segment whose end points lie on the circle.

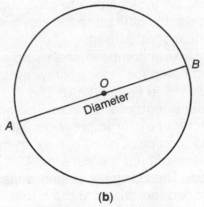

(b)

A *diameter* is a *chord* that passes through the center of the circle.

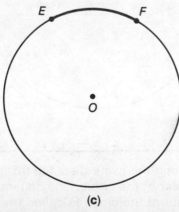

(c)

An *arc* is a portion of a circle that consists of two end points and the set of points on the circle that lie between these points. The arc illustrated is referred to as arc *EF*. Just as a line segment is named by placing a bar (–) over the end points of the segment, an arc is denoted by placing the symbol "⌢" over the end points of the arc. $\overset{\frown}{EF}$ is read as, "arc *EF*."

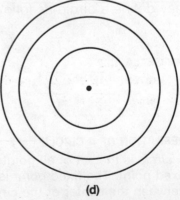

(d)

Concentric circles are circles which have the same center. Concentric circles are frequently encountered in the world around us. A metal doughnut-shaped washer illustrates concentric circles. When you toss a stone in a pond, a pattern of concentric wavelike ripples is generated.

FIGURE 12.3

Notice in Figure 12.3a chord $\overline{PQ}$ cuts off or *determines* an arc of the circle. See Figure 12.4 where we can say that chord $\overline{PQ}$ intercepts $\overset{\frown}{PQ}$ or has $\overset{\frown}{PQ}$ as its arc.

In Figure 12.3c one and only one chord can be drawn which joins the end points of $\overset{\frown}{EF}$. See Figure 12.5 where we can say that arc *EF has* $\overline{EF}$ as its chord or that arc *EF determines* chord $\overline{EF}$.

A diameter of a circle consists of two radii which lie on the same straight line. In Figure 12.3b diameter AB = radius OA + radius OB. Hence, the length of a diameter of a circle is twice the length of a radius of the circle. If, for example, the length of a diameter of a circle is 12, then the length of a radius is 6.

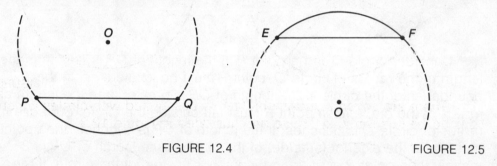

FIGURE 12.4   FIGURE 12.5

EXAMPLE 12.1   In circle O, radius $OA = 3n - 10$ and radius $OB = n + 2$. Find the length of a diameter of circle O.

SOLUTION   Since all radii of a circle are congruent,

$$OA = OB$$
$$3n - 10 = n + 2$$
$$3n = n + 12$$
$$2n = 12$$
$$n = 6$$
$$OB = OA = n + 2 = 6 + 2 = 8$$
$$\text{A diameter of } \odot O = 2 \times 8$$
$$= 16$$

EXAMPLE 12.2
  GIVEN $\overline{AB}$ and $\overline{CD}$ are diameters of $\odot O$.
  PROVE $\overline{AD} \cong \overline{BC}$.

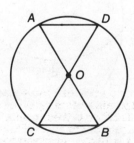

SOLUTION
  PLAN   Show $\triangle AOD \cong \triangle BOC$.

PROOF	Statements		Reasons
	1. $\overline{AB}$ and $\overline{CD}$ are diameters of $\odot O$.		1. Given.
	2. $\overline{OA} \cong \overline{OB}$.	(S)	2. All radii of a circle are congruent.
	3. $\angle AOD \cong \angle BOC$.	(A)	3. Vertical angles are congruent.
	4. $\overline{OD} \cong \overline{OC}$.	(S)	4. Same as reason 2.
	5. $\triangle AOD \cong \triangle BOC$.		5. SAS Postulate.
	6. $\overline{AD} \cong \overline{BC}$.		6. CPCTC.

Consider circle O and a point P. If the length of $\overline{OP}$ is less than the length of the radius of circle O, point P must be located within the boundaries of the circle. If the length of $\overline{OP}$ is greater than the length of the radius of the circle, then point P must fall outside the circle. Suppose the radius of circle O is 5 inches. If the length of $\overline{OP}$ is 6 inches, then point P must lie in the exterior (outside) of the circle. If the length of $\overline{OP}$ is 2 inches, then point P must lie in the interior (inside) of the circle. If the length of $\overline{OP}$ is 5 inches, then point P is a point on the circle.

DEFINITION OF INTERIOR AND EXTERIOR OF A CIRCLE

The *interior of a circle* is the set of all points whose distance from the center of the circle is less than the length of the radius of the circle.

The *exterior of a circle* is the set of all points whose distance from the center of the circle is greater than the length of the radius of the circle. See Figure 12.6.

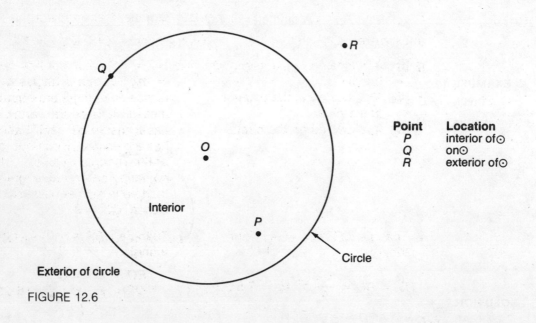

Point	Location
P	interior of $\odot$
Q	on $\odot$
R	exterior of $\odot$

FIGURE 12.6

EXAMPLE 12.3 $\overline{PQ}$ is a diameter.

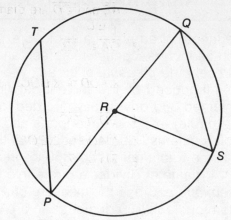

a What point is the center of the circle?

b Which segments are chords?

c Which segments are radii?

d Name all congruent segments.

e If $RS = 6$, find PQ.

f If $PQ = 15$, find RQ.

g For each of the following conditions, determine whether point X lies in the interior of the circle, the exterior of the circle, or on the circle:
(**i**) $RQ = 5$, and $RX = 4$.
(**ii**) $RP = 6$, and $RX = 8$.
(**iii**) $PQ = 14$, and $RX = 7$.

h If $m\measuredangle QRS = 60$ and $RQ = 9$, find QS.

i If $m\measuredangle QRS = 90$ and $QS = 15$, find RQ and RS.

j If $m\measuredangle QRS = 90$ and $RS = 6$, find RQ and QS.

SOLUTION

a Point R.

b $\overline{PT}$, $\overline{PQ}$, and $\overline{QS}$.

c $\overline{RP}$, $\overline{RQ}$, and $\overline{RS}$.

d $\overline{RP} \cong \overline{RQ} \cong \overline{RS}$ since all radii of a circle are congruent.

e Diameter $PQ = 2(6) = 12$.

f Radius $RQ = \frac{1}{2}(15) = 7.5$.

g (**i**) X is located in the interior of the circle.
(**ii**) X is located in the exterior of the circle.
(**iii**) X is located on the circle.

h $RQ = RS$ which implies $m\measuredangle RSQ$ $= m\measuredangle RQS$ (base angles of an isosceles triangle are equal in measure). Since the vertex angle has measure 60, each base angle has a measure of one-half of 120 or 60. Triangle RQS is equiangular. An equiangular triangle is also equilateral. Hence, $QS = 9$.

i Triangle QRS is a 45-45 right triangle.

$$RS = RQ = \frac{\sqrt{2}}{2} \times 15 = 7.5\sqrt{2}$$

j Triangle QRS is a 45-45 right triangle.

$$RQ = RS = 6$$
$$QS = RS \times \sqrt{2} = 6\sqrt{2}$$

12.2 ARCS AND CENTRAL ANGLES

The length of an arc of a circle is expressed in linear units of measurement such as inches, centimeters, feet, and so on. Degrees are used to represent the *measure* of an arc. A degree is a unit of measurement that was introduced by the ancient Babylonians who believed that a year contained 360 days. They decided to divide one complete revolution (that is, a circle) into 360 equal parts and to refer to each part as a degree. A circle contains 360 degrees so that the measure of an arc of a circle must be some fractional part of 360 degrees.

A diameter divides a circle into two equal arcs, each of which is called a *semicircle*. Since the measure of a circle is 360 degrees, the measure of a semicircle is 180 degrees. An arc whose measure is less than 180 is called a *minor arc*. A *major* arc has a measure greater than 180 but less than 360. See Figure 12.7.

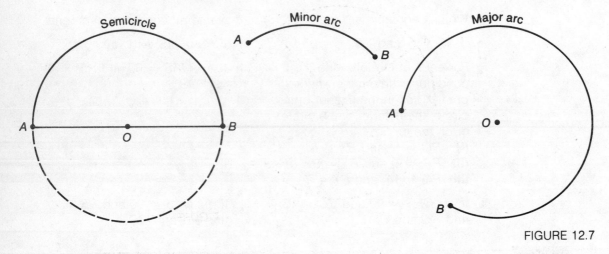

FIGURE 12.7

The measure of an arc of a circle may also be related to an angle whose vertex is the center of the circle and whose sides intercept acrs of the circle. See Figure 12.8 where angle *AOB* is called a *central angle*. The minor arc which lies in the interior of the angle is *defined* to have the same measure as its central angle, $\angle AOB$. For example, if $m\angle AOB = 83$, then $m\widehat{AB} = 83$.

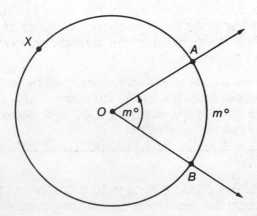

FIGURE 12.8

Notice that a central angle cuts off *two* arcs on a circle. The arc which lies in the exterior of the central angle is a major arc of the circle. Since its end points are the same as the end points of the corresponding minor arc, an additional point on the circle (the long way around from *A* to *B*) is needed in order to refer to the major arc. In Figure 12.8, $\widehat{AB}$ refers to *minor* arc *AB* while $\widehat{AXB}$ names *major* arc *AB*. Arc *AXB* includes end points *A* and *B* and the set of all points on the circle from *A* to *B*, taking a continuous path which includes point *X*. Three letters are generally used to name a major arc.

Naming arcs can be a bit tricky. Different letter combinations may name the same arc. For example, $\widehat{AXB}$ and $\widehat{BXA}$ name the same arc since all that was done was to interchange the positions of the end points of the arc. In Figure 12.9, $\widehat{KWR}$ and $\widehat{KAR}$ name the same major arc of circle *O* since points *W* and *A* determine the same path from end point *K* to end point *R*. On the other hand, $\widehat{AKR}$ and $\widehat{ANR}$ name *different* arcs. $\widehat{AKR}$ is a major arc, whereas $\widehat{ANR}$ is a minor arc.

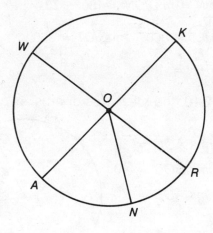

FIGURE 12.9

SUMMARY

- A *semicircle* is an arc of a circle whose end points are the end points of a diameter of the circle; its measure is 180 degrees.

- A *central angle* is an angle whose vertex is at the center of the circle.

- A *minor arc* of a circle is an arc which lies in the interior of the central angle that intercepts the arc. The measure of a minor arc is the same as the measure of its central angle.

- A *major arc* of a circle is an arc which lies in the exterior of the central angle which intercepts the arc. The degree measure of a major arc is equal to 360 minus the measure of the central angle which determines the end points of the arc.

- The *measure* of an arc, as with angle measurement, is denoted by preceding the name of the arc by the letter *m*:
  $m\widehat{AB} = 67$ is read as *the measure of arc AB is 67*. When this notation is used, the degree symbol is omitted.

At times we will have occasion to refer to a central angle and *its* *intercepted arc*. Although the sides of a central angle determine a minor *and* major arc, we will refer to the minor arc as *the* intercepted arc.

EXAMPLE 12.4

a Name four minor arcs of circle *O*.

b Name major $\overset{\frown}{TEJ}$ in three different ways. (Use three letters.)

c Name four different major arcs of circle *O*.

d If $m\angle JOT = 73$ and $\overline{NT}$ is a diameter of circle *O*, find:
  (i) $m\overset{\frown}{JT}$ **(ii)** $m\overset{\frown}{JN}$
  (iii) $m\overset{\frown}{JET}$ **(iv)** $m\overset{\frown}{TEN}$

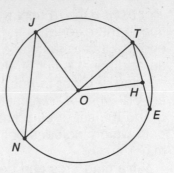

SOLUTION

a $\overset{\frown}{JT}$, $\overset{\frown}{JN}$, $\overset{\frown}{ET}$, and $\overset{\frown}{EN}$.

b $\overset{\frown}{JET}$, $\overset{\frown}{TNJ}$, and $\overset{\frown}{JNT}$.

c $\overset{\frown}{NTE}$, $\overset{\frown}{TNE}$, $\overset{\frown}{NTJ}$, and $\overset{\frown}{TNJ}$.

d (i) $m\overset{\frown}{JT} = m\angle JOT = 73$
  (ii) $m\overset{\frown}{JN} = m\angle JON = 180 - 73 = 107$
  (iii) $m\overset{\frown}{JET} = 360 - m\angle JOT = 360 - 73 = 287$
  (iv) $m\overset{\frown}{TEN} = 180$

In circle *O* of Figure 12.10, the measures of arcs *AC*, *CB*, and *AB* may be determined as follows:

$$m\overset{\frown}{AC} = m\angle AOC = 30$$
$$m\overset{\frown}{CB} = m\angle BOC = 40$$
$$m\overset{\frown}{AB} = m\angle AOB = 70$$

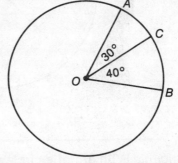

FIGURE 12.10

Notice that $m\overset{\frown}{AB} = m\overset{\frown}{AC} + m\overset{\frown}{CB}$. This result holds provided *AC* and *CB* are consecutive arcs, having exactly one point (point *C*) in common. This concept of arc addition is presented as Postulate 12.1 and is analogous to the concept of betweenness of points on line segments.

POSTULATE 12.1 ARC ADDITION POSTULATE

If point *C* is on $\overset{\frown}{AB}$, then $m\overset{\frown}{AC} + m\overset{\frown}{CB} = m\overset{\frown}{AB}$.

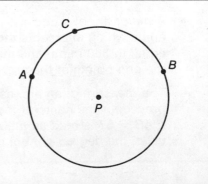

For convenience, we postulate that the same arc measure may be added or subtracted from a pair of equal arcs.

┌─ **POSTULATE 12.2 ARC SUM AND ARC DIFFERENCE POSTULATE** ─┐

- If points C and D are on $\overset{\frown}{AB}$ and $m\overset{\frown}{AC} = m\overset{\frown}{BD}$, then $m\overset{\frown}{AD} = m\overset{\frown}{BC}$. (Arc Sum Postulate)

- If points C and D are on $\overset{\frown}{AB}$ and $m\overset{\frown}{AD} = m\overset{\frown}{BC}$, then $m\overset{\frown}{AC} = m\overset{\frown}{BD}$. (Arc Difference Postulate)

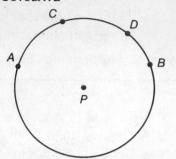

Are arcs which have the same degree measure congruent (that is, equal in length)? If we are working in the same circle the answer is yes. But what if we are comparing two arcs having the same measure in two different circles? Consider two circles which have the same center (called *concentric* circles).

In Figure 12.11, although $m\overset{\frown}{AB} = 40$ and $m\overset{\frown}{XY} = 40$, arcs AB and XY are *not* congruent. This observation will help us to frame the definition of congruent arcs.

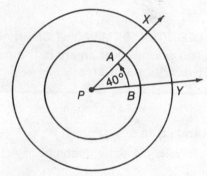

FIGURE 12.11

┌─ **DEFINITION OF CONGRUENT ARCS** ─┐

Congruent arcs are arcs in the *same or in congruent circles* which have the same degree measure.

It follows from the definition of congruent arcs that in the same or in congruent circles, *congruent central angles intercept congruent arcs* and *congruent arcs have congruent central angles*. See Figure 12.12.

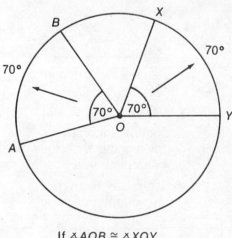

If $\angle AOB \cong \angle XOY$, then $\overset{\frown}{AB} \cong \overset{\frown}{XY}$.

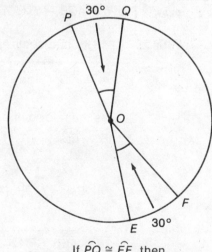

If $\overset{\frown}{PQ} \cong \overset{\frown}{EF}$, then $\angle POQ \cong \angle EOF$.

FIGURE 12.12

┌─ THEOREM 12.1 ≅ CENTRAL ANGLES IMPLIES ≅ ARCS AND VICE VERSA ─┐
In the same or congruent circles,

 ■ Congruent central angles intercept congruent arcs.

 ■ Congruent arcs have congruent central angles.

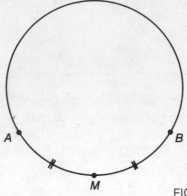

If point M is located on $\overarc{AB}$ such that $\overarc{AM} \cong \overarc{MB}$, then point M is the midpoint of $\overarc{AB}$, as shown in Figure 12.13.

FIGURE 12.13

DEFINITION OF MIDPOINT OF AN ARC
The *midpoint of an arc* is the point of the arc which divides the arc into two congruent arcs.

EXAMPLE 12.5

  GIVEN B is the midpoint of $\overarc{AC}$.

  PROVE $\triangle AOB \cong \triangle COB$.

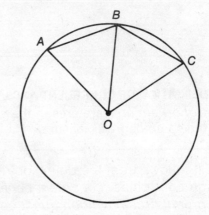

SOLUTION

  PLAN   Prove $\triangle AOB \cong \triangle COB$ by SAS.

  PROOF

Statements	Reasons
1. $\overline{OA} \cong \overline{OC}$. (Side)	1. Radii of the same circle are congruent.
2. B is the midpoint of $\overarc{AC}$.	2. Given.
3. $\overarc{AB} \cong \overarc{BC}$.	3. A midpoint of an arc divides the arc into two congruent arcs.
4. $\angle AOB \cong \angle COB$. (Angle)	4. In the same circle, congruent arcs have congruent central angles.
5. $\overline{OB} \cong \overline{OB}$. (Side)	5. Reflexive property of congruence.
6. $\triangle AOB \cong \triangle COB$.	6. SAS Postulate.

EXAMPLE 12.6

GIVEN $\overset{\frown}{AB} \cong \overset{\frown}{CD}$.

PROVE $\overline{AB} \cong \overline{CD}$.

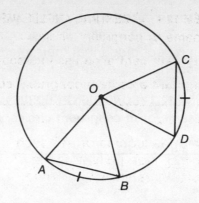

SOLUTION

PLAN Show $\triangle AOB \cong \triangle DOC$.

PROOF Statements	Reasons
1. $\overline{AO} \cong \overline{DO}$. (Side)	1. Radii of the same circle are congruent.
2. $\overset{\frown}{AB} \cong \overset{\frown}{CD}$.	2. Given.
3. $\angle AOB \cong \angle DOC$. (Angle)	3. In the same circle, congruent arcs have congruent central angles.
4. $\overline{OB} \cong \overline{OC}$. (Side)	4. Same as reason 1.
5. $\triangle AOB \cong \triangle DOC$.	5. SAS Postulate.
6. $\overline{AB} \cong \overline{CD}$.	6. CPCTC.

Example 12.6 establishes the following theorem.

THEOREM 12.2 CONGRUENT ARCS IMPLY CONGRUENT CHORDS

In the same or in congruent circles, congruent arcs have congruent chords.

In Figure 12.14, if circle O is congruent to circle P and $\overset{\frown}{AB} \cong \overset{\frown}{CD}$, then Theorem 12.2 entitles us to conclude that $\overline{AB} \cong \overline{CD}$.

Theorem 12.2 provides a convenient method for proving a pair of chords are congruent; show that the chords to be proven congruent cut off

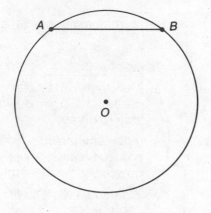

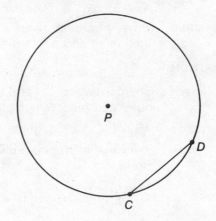

FIGURE 12.14

congruent arcs in the same or in congruent circles. The converse of Theorem 12.2 states that if you know that a pair of chords are congruent, then you may conclude that their intercepted arcs are congruent, provided you are working in the same or in congruent circles.

THEOREM 12.3 CONGRUENT CHORDS IMPLY CONGRUENT ARCS

In the same or in congruent circles, congruent chords intercept congruent arcs.

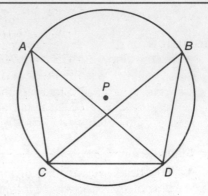

EXAMPLE 12.7

GIVEN $\overline{CA} \cong \overline{DB}$.

PROVE $\angle A \cong \angle B$.

SOLUTION

PLAN   Show $\triangle ACD \cong \triangle BDC$ by SSS Postulate.

PROOF

Statements	Reasons
1. $\overline{CD} \cong \overline{CD}$. (Side)	1. Reflexive property of $\cong$.
2. $\overline{CA} \cong \overline{DB}$. (Side)	2. Given.
3. $m\overset{\frown}{CA} = m\overset{\frown}{DB}$.	3. In the same circle, congruent chords intercept equal arcs.
4. $m\overset{\frown}{ACD} = m\overset{\frown}{BDC}$.	4. Arc Sum Postulate (Postulate 12.2).
5. $\overline{AD} \cong \overline{BC}$. (Side)	5. In the same circle, equal arcs have congruent chords.
6. $\triangle ACD \cong \triangle BDC$.	6. SSS Postulate.
7. $\angle A \cong \angle B$.	7. CPCTC.

EXAMPLE 12.8   Prove if two chords are the perpendicular bisectors of each other, then each chord is a diameter.

SOLUTION

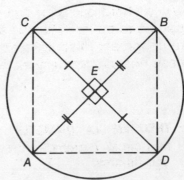

GIVEN   Chords $\overline{AB}$ and $\overline{CD}$ are $\perp$ to each other and bisect each other.

PROVE $\overline{AB}$ and $\overline{CD}$ are diameters.

PLAN   Show that each chord divides the circle into two congruent arcs.

To prove that $\overline{AB}$ is a diameter show that $m\overset{\frown}{ACB} = m\overset{\frown}{ADB}$.

- Draw $\overline{AC}$, $\overline{AD}$, $\overline{BC}$, and $\overline{BD}$.

- *Prove* $\triangle AEC \cong \triangle AED$ by SAS. $\overline{AC} \cong \overline{AD}$ which implies that $m\overset{\frown}{AC} = m\overset{\frown}{AD}$.

- Prove $\triangle BEC \cong \triangle BED$ by SAS. $\overline{BC} \cong \overline{BD}$ which implies that $m\overset{\frown}{BC} = m\overset{\frown}{BD}$.

- $\begin{aligned} m\overset{\frown}{AC} &= m\overset{\frown}{AD} \\ + \quad m\overset{\frown}{BC} &= m\overset{\frown}{BD} \end{aligned}$

  $m\overset{\frown}{ACB} = m\overset{\frown}{ADB}$ which implies $\overline{AB}$ is a diameter since it divides the circle into two equal arcs. A similar approach can be used to show that $m\overset{\frown}{CAD} = m\overset{\frown}{CBD}$ which establishes that $\overline{CD}$ is a diameter

---

SUMMARY

- To prove *arcs* are congruent in the same or congruent circles, show one of the following statements is true:

    1 The central angles which intercept the arcs are congruent.

    2 The chords which cut off the arcs are congruent.

- To prove *central angles* are congruent in the same or in congruent circles, show that their intercepted arcs are congruent.

- To prove *chords* are congruent in the same or in congruent circles, show that the chords have congruent arcs.

- To prove a *chord is a diameter* show that the chord divides the circle into two congruent arcs.

12.3 DIAMETERS AND CHORDS

Draw a chord and then draw a diameter perpendicular to the chord. What appears to be true? Notice that the resulting chord segments look like they are congruent. Also, corresponding pairs of arcs seem to be congruent. These observations are stated in the next theorem.

THEOREM 12.4   DIAMETER ⊥ CHORD THEOREM

In a circle, a diameter drawn perpendicular to a chord bisects the chord and its arcs.

GIVEN ⊙O with diameter

$\overline{AB} \perp \overline{CD}$.

PROVE **a** $\overline{CX} \cong \overline{DX}$.

b $\overparen{CB} \cong \overparen{DB}$.

$\overparen{CA} = \overparen{DA}$.

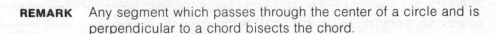

- *Draw radii* $\overline{OC}$ *and* $\overline{OD}$.

- *Prove* △$OXC \cong$ △OXD *by Hy-Leg*.

- $\overline{CX} \cong \overline{DX}$ *and* ∢1 $\cong$ ∢2 *by CPCTC*.

- $\overparen{CB} \cong \overparen{DB}$ *since* $\cong$ *central angles intercept* $\cong$ *arcs*.

- ∢3 $\cong$ ∢4 *since supplements of congruent angles are congruent*.

- $\overparen{CA} \cong \overparen{DA}$.

REMARK Any segment which passes through the center of a circle and is perpendicular to a chord bisects the chord.

EXAMPLE 12.9 The length of a diameter of circle O is 20 and the length of chord $\overline{AB}$ is 16. What is the distance between the chord and the center of the circle?

SOLUTION Recall that the distance from a point (that is, the center of the circle) to a line segment (that is, the chord) is the length of the perpendicular segment from the point to the line segment. We therefore choose to draw a diameter which is perpendicular to chord $\overline{AB}$.

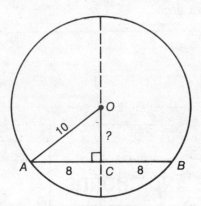

The distance between the chord and the center of the circle is represented by the length of $\overline{OC}$. By drawing radius $\overline{OA}$ a right triangle is formed. Since the diameter is 20, the radius is 10. Applying Theorem 12.4, we deduce that the length of $\overline{AC}$ is 8. Hence, △OCA is a right triangle, the lengths of whose sides form a Pythagorean triple (3-4-5). The length of $\overline{OC}$ is 6.

After drawing several pairs of parallel chords in a circle, a pattern emerges, as shown in Figure 12.15.

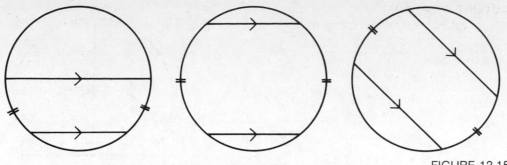

FIGURE 12.15

Notice that the arcs that the parallel chords cut off (and which lie between the chords) appear to be congruent. Here's the formal statement of the corresponding theorem and its proof.

THEOREM 12.5 PARALLEL CHORDS AND CONGRUENT ARCS

In a circle, parallel chords cut off equal arcs.

GIVEN ☉O with $\overline{AB} \parallel \overline{CD}$.

PROVE $m\overparen{AC} = m\overparen{BD}$.

PLAN Draw diameter $\overline{XY} \perp \overline{AB}$. $\overline{XY}$ will also be $\perp \overline{CD}$. Apply Theorem 12.4.

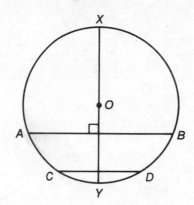

PROOF

Statements	Reasons
1. ☉O with $\overline{AB} \parallel \overline{CD}$.	1. Given.
2. Draw $\overline{XY}$ through point O and perpendicular to $\overline{AB}$.	2. Through a point not on a line, exactly one perpendicular may be drawn from the point to the line.
3. $\overline{XY} \perp \overline{CD}$.	3. A line perpendicular to one of two parallel lines is perpendicular to the other line.
4. $m\overparen{AY} = m\overparen{BY}$. $m\overparen{CY} = m\overparen{DY}$.	4. A diameter perpendicular to a chord bisects its arcs.
5. $m\overparen{AY} - m\overparen{CY} = m\overparen{BY} - m\overparen{DY}$.	5. Subtraction property of equality.
6. $m\overparen{AC} = m\overparen{BD}$.	6. Substitution.

EXAMPLE 12.10 If chord $\overline{CD}$ is parallel to diameter $\overline{AB}$, and $m\overset{\frown}{CD} = 40$, find the measures of arcs AC and BD.

SOLUTION Since $\overline{CD} \parallel \overline{AB}$, $m\overset{\frown}{AC} = m\overset{\frown}{BD} = x$.

$$x + 40 + x = 180$$
$$2x + 40 = 180$$
$$2x = 140$$
$$x = 70$$
$$m\overset{\frown}{AC} = m\overset{\frown}{BD} = 70$$

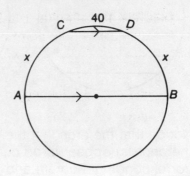

12.4 TANGENTS AND SECANTS

A line may intersect a circle at no points, one point, or two points. See Figure 12.16.

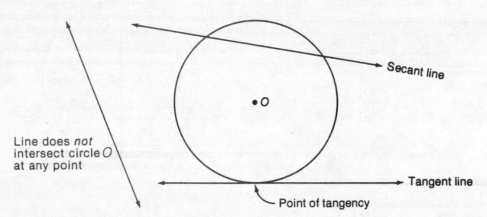

Line does *not* intersect circle *O* at any point

Secant line

•*O*

Tangent line

Point of tangency

FIGURE 12.16

DEFINITION OF TANGENT AND SECANT

- A *tangent* line is a line which intersects a circle in exactly one point. The point of contact is called the *point of tangency*.

- A *secant* line is a line which intersects a circle in two different points. (Every secant line includes a chord of the circle.)

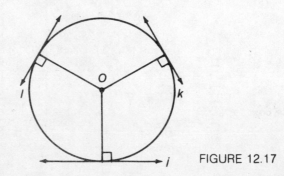

FIGURE 12.17

In Figure 12.17, lines *j*, *k*, and *l* are tangent to circle *O*. Radii have been drawn to their points of tangency. In each case the radius appears to be perpendicular to the tangent.

THEOREM 12.6 RADIUS ⊥ TANGENT

A radius drawn to a point of tangency is perpendicular to the tangent.

GIVEN $\overleftrightarrow{AB}$ is tangent to ⊙*P* at point *A*.

PROVE $\overline{PA} \perp \overleftrightarrow{AB}$.

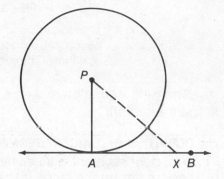

OUTLINE OF PROOF When congruent or similar triangles do not seem appropriate, look to apply an indirect proof.

■ Assume $\overline{PA}$ is *not* perpendicular to $\overleftrightarrow{AB}$. If this is the case, then there must be another segment, say $\overline{PX}$, which can be drawn from *P* perpendicular to $\overleftrightarrow{AB}$.

■ Since $\overline{PX} \perp \overleftrightarrow{AB}$, *PX* represents the *shortest* distance from *P* to $\overleftrightarrow{AB}$. Thus, *PX* is *less than PA*.

■ But point *X* lies in the exterior of circle *P*. Since $\overline{PA}$ is a radius, *PX* must be *greater than PA*. This contradicts our previous assertion that *PX* is less than *PA*. Our assumption that $\overline{PA}$ is not perpendicular to $\overleftrightarrow{AB}$ must be false. $\overline{PA}$ is therefore perpendicular to $\overleftrightarrow{AB}$.

The converse of this theorem is given in Theorem 12.7 and may also be proven indirectly.

THEOREM 12.7 RADIUS ⊥ LINE IMPLIES A TANGENT

If a radius is perpendicular to á line at the point at which the line intersects a circle, then the line is a tangent.

GIVEN ⊙*P* with $\overline{PA} \perp l$ at point *A*.

PROVE *l* is tangent to ⊙*P*.

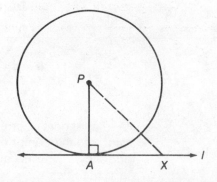

Use an indirect proof.

- Assume *l* is not a tangent line. It must therefore intersect ⊙*P* at another point, say point *X*.

- Since $\overline{PX}$ is the hypotenuse of right triangle *PAX*, *PX* is greater than *PA*.

- But $\overline{PA}$ is a radius which implies that point *X* must lie in the *exterior* of circle *P*.

- The assumption that there is a second point *on the circle* at which line *l* intersects is false. Since there is exactly one point at which *l* intersects the circle, *l* is a tangent line.

The next definition makes a distinction between a tangent *line* and a tangent *segment*.

DEFINITION OF TANGENT SEGMENT

A *tangent segment* is a line segment that has a point on the tangent line and the point of tangency as end points.

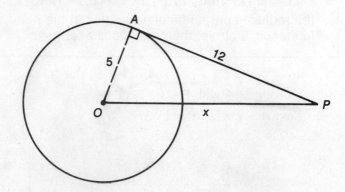

EXAMPLE
12.11
The length of a tangent segment drawn from point *P* to circle *O* is 12. If the radius of circle *O* is 5, find the distance from point *P* to the center of the circle.

$\overline{PA}$ is a tangent segment and *PO* represents the distance from point *P* to the center of the circle. Draw radius to the point of tangency. $\overline{OA} \perp \overline{PA}$ and *x* = 13 since triangle *OAP* is a 5-12-*13* right triangle.

INSCRIBED AND CIRCUMSCRIBED POLYGONS

Quadrilateral *ABCD* in Figure 12.18 is *circumscribed about circle O*.
Pentagon *ABCDE* in Figure 12.19 is *inscribed in* circle *O*.

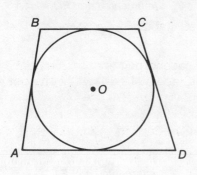

FIGURE 12.18

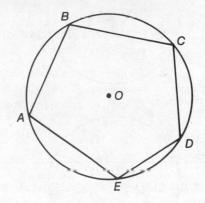

FIGURE 12.19

DEFINITIONS OF CIRCUMSCRIBED AND INSCRIBED POLYGONS

A polygon is *circumscribed about* a circle if each of its sides is tangent to the circle.

A polygon is *inscribed in* a circle if each of its vertices lie on the circle.

EXAMPLE 12.12   A quadrilateral is inscribed in a circle such that the sides of the quadrilateral divide the circle into arcs whose measures have the ratio 1:2:3:4. How many degrees are there in each arc?

SOLUTION   Since a circle contains 360°, the sum of the measures of the arcs must equal 360:

$$1x + 2x + 3x + 4x = 360°$$

$$10x = 360°$$

$x = 36°$
$2x = 72°$
$3x = 108°$
$4x = 144°$

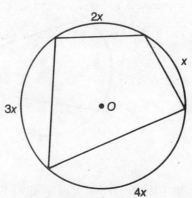

12.5 ANGLE MEASUREMENT: VERTEX ON THE CIRCLE

The vertex of an angle whose sides intercept arcs on a circle may be located in any one of the positions shown in Figure 12.20.

In each of these cases a relationship exists between the measures of the intercepted arcs and the measure of the angle. We have already seen, for example, that the measure of any angle whose vertex is at the center of

At the center of the circle:

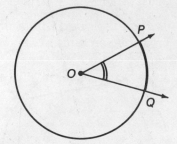

Intercepted arc: $\overset{\frown}{PQ}$

On the circle:

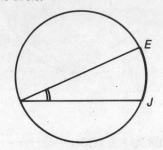

Intercepted arc: $\overset{\frown}{EJ}$

In the interior of the circle but not at the center:

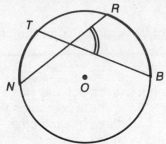

Intercepted arcs: $\overset{\frown}{NT}$ and $\overset{\frown}{RB}$

In the exterior of the circle:

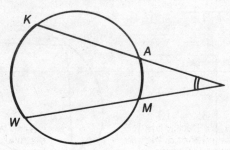

Intercepted arcs: $\overset{\frown}{KW}$ and $\overset{\frown}{AM}$

FIGURE 12.20

the circle (a central angle) is equal to the measure of its intercepted arc. Let's now consider the situation in which the vertex of the angle is a point on the circle and the sides of the angle are chords (or secants) of the circle. In Figure 12.21, angle ABC is called an *inscribed angle* and $\overset{\frown}{AC}$ is its intercepted arc.

DEFINITION OF INSCRIBED ANGLE

An *inscribed angle* is an angle whose vertex lies on a circle and whose sides are chords (or secants) of the circle.

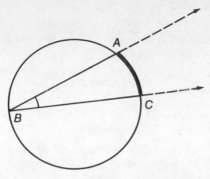

FIGURE 12.21

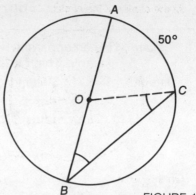

FIGURE 12.22

In Figure 12.22, suppose $m\widehat{AC} = 50$. Then $m\angle AOC = 50$. Since $\angle AOC$ is an exterior angle of $\triangle BOC$, $m\angle AOC = m\angle B + m\angle C$. Since $m\angle B = m\angle C$ (base angles of an isosceles triangle), $m\angle B = 25$. How do the measures of inscribed angle ABC and its intercepted arc ($\widehat{AC}$) compare? Observe that $m\angle ABC = \frac{1}{2}m\widehat{AC} = \frac{1}{2}(50) = 25$.

THEOREM 12.8   THE INSCRIBED ANGLE THEOREM

The measure of an inscribed angle is equal to one-half the measure of its intercepted arc.

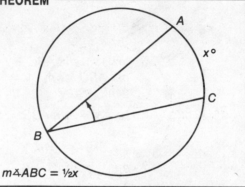

$m\angle ABC = \frac{1}{2}x$

In our example, we assumed that one of the sides of the inscribed angle passed through the center of the circle. In general, this need not be true. The center of the circle may lie on a side of the inscribed angle, in the interior of the angle, or in the exterior of the angle. The proof of the inscribed angle theorem must treat each of these three cases separately.

The three cases of Theorem 12.8 to be proved are shown in Figure 12.23.

CASE 1: **Center lies on a side of the inscribed angle.**

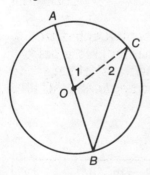

CASE 2: **Center lies in the interior of the inscribed angle.**

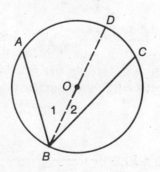

CASE 3: **Center lies in the exterior of the inscribed angle.**

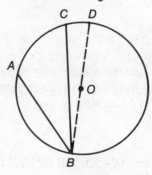

FIGURE 12.23

GIVEN $\odot O$ with inscribed angle ABC.

PROVE $m\angle ABC = \frac{1}{2}m\widehat{AC}$.

PROOF OF CASE 1

PLAN   Use an approach which parallels the solution to the numerical example which was used to motivate this theorem.

PROOF   Statements   Reasons

Statements	Reasons
1. Draw radius $\overline{OC}$.	1. Two points determine a line.
2. $m\angle 1 = m\angle ABC + m\angle 2$.	2. The measure of an exterior angle of a triangle is equal to the sum of the measures of the two nonadjacent interior angles of the triangle.
3. $\overline{OB} \cong \overline{OC}$.	3. All radii of the same circle are congruent.
4. $m\angle 2 = m\angle ABC$.	4. If two sides of a triangle are congruent, then the angles opposite these sides are equal in measure.
5. $m\angle 1 = m\angle ABC + m\angle ABC$, or $m\angle 1 = 2\,m\angle ABC$.	5. Substitution.
6. $\frac{1}{2}m\angle 1 = m\angle ABC$ or $m\angle ABC = \frac{1}{2}m\angle 1$.	6. Division property.
7. $m\angle 1 = m\overparen{AC}$.	7. The measure of a central angle is equal to the measure of its intercepted arc.
8. $m\angle ABC = \frac{1}{2}mAC$.	8. Substitution.

OUTLINE OF PROOF FOR CASE 2

Draw diameter $\overline{BD}$.
Apply the result established in Case 1:

$$m\angle 1 = \tfrac{1}{2}m\overparen{AD}$$

$$+\ \underline{m\angle 2 = \tfrac{1}{2}m\overparen{DC}}$$

$$m\angle 1 + m\angle 2 = \tfrac{1}{2}(m\overparen{AD} + m\overparen{DC})$$

Using substitution,

$$m\angle ABC = \tfrac{1}{2}m\overparen{AC}$$

OUTLINE OF PROOF FOR CASE 3

Draw diameter $\overline{BD}$.
Apply the result established in Case 1:

$$m\angle ABD = \tfrac{1}{2}m\overparen{ACD}$$

$$-\ \underline{m\angle CBD = \tfrac{1}{2}m\overparen{CD}}$$

$$m\angle ABD - m\angle CBD = \tfrac{1}{2}(m\overparen{ACD} - m\overparen{CD})$$

Using substitution,

$$m\angle ABC = \tfrac{1}{2}m\overparen{AC}$$

EXAMPLE
12.13
In each of the following, find the value of *x*.

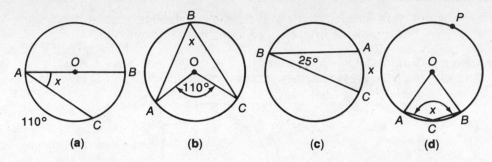

(a) (b) (c) (d)

SOLUTION **a** $\overset{\frown}{ACB}$ is a semicircle so that $m\overset{\frown}{BC} = 180 - 110 = 70$.

$$x = \tfrac{1}{2}m\overset{\frown}{BC} = \tfrac{1}{2}(70)$$

$$x = 35$$

b $\angle AOC$ is a central angle so that $m\overset{\frown}{AC} = 110$.

$$m\angle ABC = \tfrac{1}{2}m\overset{\frown}{AC} = \tfrac{1}{2}(110)$$

$$m\angle ABC = 55$$

c The measure of the arc intercepted by an inscribed angle must be twice the measure of the inscribed angle. Hence,

$$x = 2(m\angle ABC) = 2(25)$$

$$x = 50$$

d $\angle AOB$ is a central angle so that $m\overset{\frown}{ACB} = 70$.

$$m\overset{\frown}{APB} = 360 - 70 = 290$$

$$x = \tfrac{1}{2}m\overset{\frown}{APB} = \tfrac{1}{2}(290)$$

$$x = 145$$

EXAMPLE
12.14
A triangle is inscribed in a circle so that its sides divide the circle into arcs whose measure have the ratio $2:3:7$. Find the measure of the largest angle of the triangle.

SOLUTION First, determine the measures of the arcs of the circle.

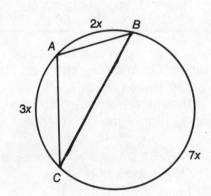

$$m\overset{\frown}{AB} + m\overset{\frown}{AC} + m\overset{\frown}{BC} = 360$$

$$2x + 3x + 7x = 360$$

$$12x = 360$$

$$x = 30$$

The measures of the arcs of the circle are

$$m\overset{\frown}{AB} = 2x = 60$$

$$m\overset{\frown}{AC} = 3x = 90$$

$$m\overset{\frown}{BC} = 7x = 210$$

ANGLE MEASUREMENT: VERTEX ON THE CIRCLE **231**

Each angle of the triangle is an inscribed angle. The largest angle lies opposite the arc having the greatest measure. Hence,

$$m \angle A = \tfrac{1}{2} m \overset{\frown}{BC} = \tfrac{1}{2}(210)$$

$$m \angle A = 105$$

EXAMPLE 12.15 Quadrilateral *ABCD* is inscribed in circle *O*. If $m \angle A = x$, and $m \angle B = y$, express the measures of angles of *C* and *D* in terms of *x* and *y*.

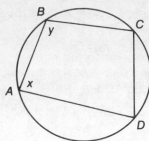

SOLUTION

To find $m \angle C$:	To find $m \angle D$:
$m \overset{\frown}{BCD} = 2(x) = 2x$	$m \overset{\frown}{ADC} = 2(y) = 2y$
$m \overset{\frown}{BAD} = 360 - 2x$	$m \overset{\frown}{ABC} = 360 - 2y$
$m \angle C = \tfrac{1}{2}(360 - 2x)$	$m \angle D = \tfrac{1}{2}(360 - 2y)$
$m \angle C = 180 - x$	$m \angle D = 180 - y$

Example 12.15 establishes the following theorem.

THEOREM 12.9 ANGLES OF AN INSCRIBED QUADRILATERAL

Opposite angles of an inscribed quadrilateral are supplementary.

$$a° + c° = 180°$$
$$b° + d° = 180°$$

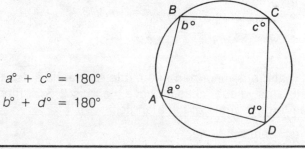

A chord which has as one of its end points the point of tangency of a tangent to a circle forms an angle whose vertex is on the circle. In Figure 12.24, the sides of $\angle ABC$ are chord $\overline{AB}$ and tangent ray $\overrightarrow{BC}$. The end points of the intercepted arc are the vertex of the angle (point *B*) and the other end point of the chord (point *A*).

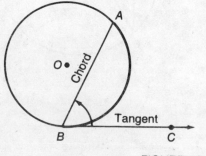

FIGURE 12.24

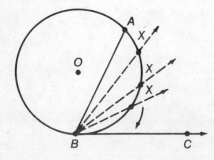

FIGURE 12.25

In Figure 12.25, consider inscribed angle *ABX* and suppose point *X* slides down the circle so that $\overrightarrow{BX}$ gets closer and closer to tangent $\overrightarrow{BC}$. At each position, inscribed angle *ABX* is measured by one-half of its intercepted arc, $\overset{\frown}{AX}$. As point *X* approaches point *B*, the measure of $\overset{\frown}{AX}$ gets closer and closer to the measure of $\overset{\frown}{AB}$. Hence, when $\overrightarrow{BX}$ finally coincides with tangent $\overrightarrow{BC}$, the angle formed is measured by one-half the measure of $\overset{\frown}{AB}$.

THEOREM 12.10 CHORD-TANGENT ANGLE THEOREM

The measure of an angle formed by a tangent and a chord drawn to the point of tangency is equal to one-half the measure of the intercepted arc.

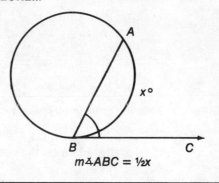

$m\angle ABC = \tfrac{1}{2}x$

OUTLINE OF PROOF

GIVEN $\overrightarrow{BC}$ is tangent to $\odot O$ at point *B*. $\overline{AB}$ is drawn.

PROVE $m\angle 1 = \tfrac{1}{2}m\overset{\frown}{AB}$.

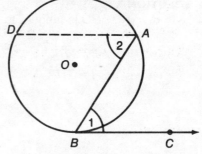

Through point *A* draw a chord parallel to $\overline{BC}$, intersecting the circle at point *D*. Then,

- $m\angle 1 = m\angle 2$ (Alternate interior angles)

- $m\angle 2 = \tfrac{1}{2}m\overset{\frown}{BD}$ (Inscribed Angle Theorem)

- $m\angle 1 = \tfrac{1}{2}m\overset{\frown}{BD}$ (Substitution)

- $m\overset{\frown}{BD} = m\overset{\frown}{AB}$ (Since $\overline{AD} \parallel \overrightarrow{BC}$)

- $m\angle 1 = \tfrac{1}{2}m\overset{\frown}{AB}$ (Substitution)

EXAMPLE 12.16 Find the value of *x* if $m\overset{\frown}{ACB} = 250$.

SOLUTION $m\overset{\frown}{AB} = 360 - 250 = 110$

$x = \tfrac{1}{2}m\overset{\frown}{AB} = \tfrac{1}{2}(110)$

$x = 55$

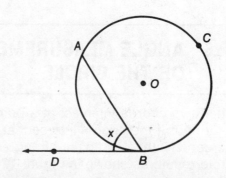

EXAMPLE **12.17**	In circle O, $m\overset{\frown}{AC} = 110$. Find the measures of each of the numbered angles.
SOLUTION	$m\angle 1 = \frac{1}{2}m\overset{\frown}{AC} = \frac{1}{2}(110)$ $= 55$

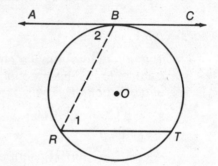

Since $m\overset{\frown}{BC} = 180 - 110$
    $= 70$,

$m\angle 2 = \frac{1}{2}m\overset{\frown}{BC} = \frac{1}{2}(70) = 35$

$m\angle 3 = \frac{1}{2}m\overset{\frown}{BC} = \frac{1}{2}(70) = 35$

$m\angle 4 = m\overset{\frown}{AC} = 110$

$m\angle 5 = \frac{1}{2}m\overset{\frown}{AC} = \frac{1}{2}(110) = 55$

$m\angle 6 = \frac{1}{2}m\overset{\frown}{AB} = \frac{1}{2}(180) = 90$

NOTE: Angles 1 and 2 are formed by a chord and a tangent; angle 4 is a central angle; and angles 3, 5, and 6 are inscribed angles.

EXAMPLE **12.18**	Prove that a tangent drawn parallel to a chord of a circle cuts off congruent arcs between the tangent and the chord.
SOLUTION **GIVEN**	$\overleftrightarrow{ABC}$ is tangent to $\odot O$ at point B, $\overleftrightarrow{ABC} \parallel \overline{RT}$.
PROVE	$\overset{\frown}{BR} \cong \overset{\frown}{BT}$.

OUTLINE OF **PROOF**	■ $m\angle 1 = \frac{1}{2}m\overset{\frown}{BT}$ and $m\angle 2 = \frac{1}{2}m\overset{\frown}{BR}$. ■ Since $\overleftrightarrow{ABC} \parallel \overline{RT}$, $m\angle 1 = m\angle 2$ so that $\frac{1}{2}m\overset{\frown}{BT} = \frac{1}{2}m\overset{\frown}{BR}$. ■ Multiplying each side of this equation by 2 yields $m\overset{\frown}{BT} = m\overset{\frown}{BR}$, or $\overset{\frown}{BT} \cong \overset{\frown}{BR}$.
REMARK	Parallel chords, parallel tangents, and a tangent parallel to a chord cut off congruent arcs on a circle.

12.6 ANGLE MEASUREMENT: VERTEX IN THE INTERIOR OF THE CIRCLE

When two chords intersect at a point on the circle, then an inscribed angle is formed. If the chords intersect at a point in the interior of a circle (not at the center), then two pairs of angles whose sides intercept arcs on the circle result, as shown in Figure 12.26. With respect to angle 1 (and its

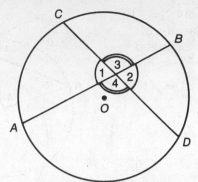

FIGURE 12.26

vertical angle 2), the intercepted arcs are the arcs which lie opposite the vertical angle pair: $\overset{\frown}{AC}$ and $\overset{\frown}{BD}$. With respect to the vertical angle pair formed by angles 3 and 4, the intercepted arcs are $\overset{\frown}{AD}$ and $\overset{\frown}{BC}$. For convenience we will refer to angles 1, 2, 3, and 4 as *chord-chord* angles since they are formed by intersecting chords.

A chord-chord angle is measured by the *average* of its two intercepted arcs:

$$m \angle 1 = m \angle 2 = \tfrac{1}{2}(m\overset{\frown}{AC} + m\overset{\frown}{BD})$$

and

$$m \angle 3 = m \angle 4 = \tfrac{1}{2}(m\overset{\frown}{AD} + m\overset{\frown}{BC})$$

THEOREM 12.11 CHORD-CHORD ANGLE THEOREM

The measure of an angle formed by two chords intersecting in the interior of a circle is equal to one-half the sum of the measures of the two intercepted arcs.

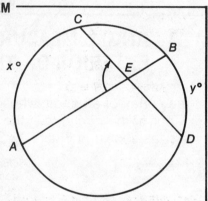

$$m \angle AEC = \frac{1}{2}(x° + y°)$$

REMARK The sides of the angle may be secants which intersect in the interior of a circle since secants contain chords of the circle.

OUTLINE OF PROOF

GIVEN In $\odot O$ chords $\overline{AB}$ and $\overline{CD}$ intersect at point E.

PROVE $m \angle 1 = \tfrac{1}{2}(m\overset{\frown}{AC} + m\overset{\frown}{BD})$.

Draw $\overline{AD}$. Angle 1 is an exterior angle of $\triangle AED$. Hence,

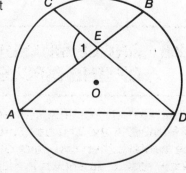

$$m \angle 1 = m \angle D + m \angle A$$
$$m \angle 1 = \tfrac{1}{2}m\overset{\frown}{AC} + \tfrac{1}{2}m\overset{\frown}{BD}$$
$$m \angle 1 = \tfrac{1}{2}(m\overset{\frown}{AC} + m\overset{\frown}{BD})$$

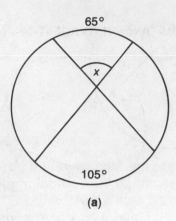

(a)

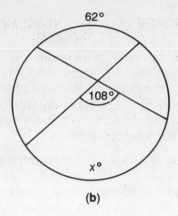

(b)

SOLUTION **a** $x = \frac{1}{2}(65 + 105)$

$x = \frac{1}{2}(170)$

$x = 85$

b $108 = \frac{1}{2}(x + 62)$

or

$x + 62 = 2(108)$

$x + 62 = 216$

$x = 216 - 62$

$x = 154$

12.7 ANGLE MEASUREMENT: VERTEX IN THE EXTERIOR OF THE CIRCLE

If the vertex of an angle whose sides intercept arcs on the circle lies in the exterior of the circle, then the angle may be formed in one of three possible ways, as shown in Figure 12.27.

1. Each side is a secant.

2. One side is a secant and the other side is a tangent.

3. Each side is a tangent.

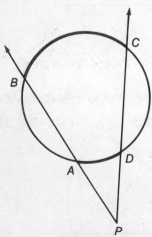

Intercepted arcs:
$\overset{\frown}{BC}$ and $\overset{\frown}{AD}$

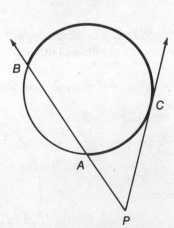

Intercepted arcs:
$\overset{\frown}{BC}$ and $\overset{\frown}{AC}$

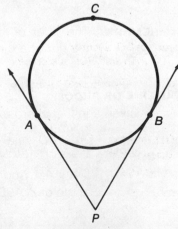

Intercepted arcs:
$\overset{\frown}{ACB}$ and $\overset{\frown}{AB}$

FIGURE 12.27

In each of these three cases, the angle at vertex P is measured by one-half the *difference* of the intercepted arcs.

THEOREM 12.12 SECANT-SECANT, SECANT-TANGENT, TANGENT-TANGENT THEOREM

The measure of an angle formed by two secants, a secant and a tangent, or two tangents intersecting in the exterior of a circle is equal to one-half the difference of the measures of the intercepted arcs.

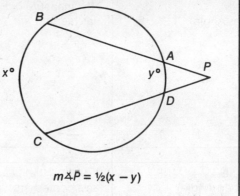

$$m\angle P = \tfrac{1}{2}(x - y)$$

The three cases of Theorem 12.12 to be proved are shown in Figure 12.28.

CASE 1 **Secant-Secant**

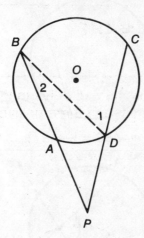

GIVEN In ⊙O, secants $\overline{PAB}$ and $\overline{PDC}$ are drawn

PROVE $m\angle P = \tfrac{1}{2}(m\widehat{BC} - m\widehat{AD})$

CASE 2 **Secant-Tangent**

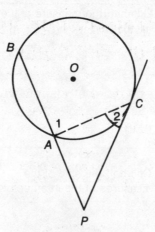

GIVEN In ⊙O, secant $\overline{PAB}$ and tangent $\overline{PC}$ are drawn.

PROVE $m\angle P = \tfrac{1}{2}(m\widehat{BC} - m\widehat{AC})$

CASE 3 **Tangent-Tangent**

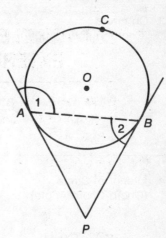

GIVEN In ⊙O, tangents $\overline{PA}$ and $\overline{PB}$ are drawn

PROVE $m\angle P = \tfrac{1}{2}(m\widehat{ACB} - m\widehat{AB})$

FIGURE 12.28

OUTLINE OF PROOF OF CASE 1 Draw $\overline{BD}$. Angle 1 is an exterior angle of $\triangle DBP$. Hence,

$$m\angle 1 = m\angle 2 + m\angle P$$

or

$$m\angle P = m\angle 1 - m\angle 2$$
$$m\angle P = \tfrac{1}{2}m\widehat{BC} - \tfrac{1}{2}m\widehat{AD}$$
$$m\angle P = \tfrac{1}{2}(m\widehat{BC} - m\widehat{AD})$$

The proofs of Cases 2 and 3 follow a similar pattern.

VERTEX OF ANGLE	MEASURE OF ANGLE EQUALS	MODEL
1. At center (Central Angle)	The measure of the intercepted arc.	$m \angle AOB = m \widehat{AB}$
2. On circle (Inscribed Angle)	One-half the measure of the intercepted arc.	$m \angle ABC = \frac{1}{2} m \widehat{AC}$
3. On circle (Chord-Tangent Angle)	One-half the measure of the intercepted arc.	$m \angle ABC = \frac{1}{2} m \widehat{AB}$
4. Interior of Circle (Chord-Chord Angle)	One-half the sum of the measures of the intercepted arcs.	$m \angle AED = \frac{1}{2}(m \widehat{AD} + m \widehat{BC})$
5. Exterior of Circle (Secant-Secant, Secant-Tangent, and Tangent-Tangent Angles)	One-half the difference of the measures of the intercepted arcs.	$m \angle APC = \frac{1}{2}(m \widehat{AC} - m \widehat{AB})$

EXAMPLE
12.20
Find the value of x

(a)
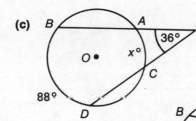

(b)
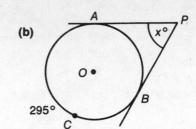

(c)

(d)
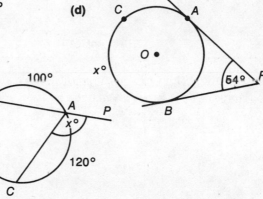

(e)

SOLUTION

a $m\widehat{AC} = 360 - (76 + 84)$

$\qquad = 360 - 160$

$m\widehat{AC} = 200$

$\qquad x = \frac{1}{2}(m\widehat{AC} - m\widehat{AB})$

$\qquad\quad = \frac{1}{2}(200 - 76)$

$\qquad\quad = \frac{1}{2}(124)$

$\boxed{x = 62°}$

c $m\measuredangle P = \frac{1}{2}(m\widehat{BD} = m\widehat{AC})$

$\quad 36 = \frac{1}{2}(88 - x)$

Multiplying both sides by 2,

$\quad 72 = 88 - x$

$x + 72 = 88$

$\quad x = 88 - 72$

$\boxed{x = 16°}$

b $m\widehat{AB} = 360 - 295 = 65$

$\qquad x = \frac{1}{2}(m\widehat{ACB} - m\widehat{AB})$

$\qquad\quad = \frac{1}{2}(295 - 65)$

$\qquad\quad = \frac{1}{2}(230)$

$\boxed{x = 115°}$

d $m\widehat{AB} = 360 - m\widehat{ACB}$

$\quad m\widehat{AB} = 360 - x$

$\quad m\measuredangle P = \frac{1}{2}(m\widehat{ACB} - m\widehat{AB})$

$\qquad 54 = \frac{1}{2}[x - (360 - x)]$

Multiplying both sides by 2,

$\quad 108 = x - 360 + x$

$\quad 108 = 2x - 360$

$\quad 468 = 2x$

$\boxed{x = 234°}$

e A common *error* is to find the $m\measuredangle CAP$ by taking $\frac{1}{2}m\widehat{AC}$. This is incorrect since the sides of the angle are a chord and *part of a secant*. Since angles *CAP* and *CAB* are supplementary, we first determine the measure of inscribed angle *CAB*:

$$m\measuredangle CAB = \frac{1}{2}m\widehat{BC} = \frac{1}{2}(140) = 70$$

$$m\measuredangle CAP = 180 - 70$$

$$\boxed{x = 110°}$$

EXAMPLE
12.21
$\overrightarrow{PA}$ is tangent to circle O at point A. Secant $\overline{PBC}$ is drawn. Chords $\overline{CA}$ and $\overline{BD}$ intersect at point E. If $m\widehat{AD}:m\widehat{AB}:m\widehat{DC}:m\widehat{BC} = 2:3:4:6$, find the measures of each of the numbered angles.

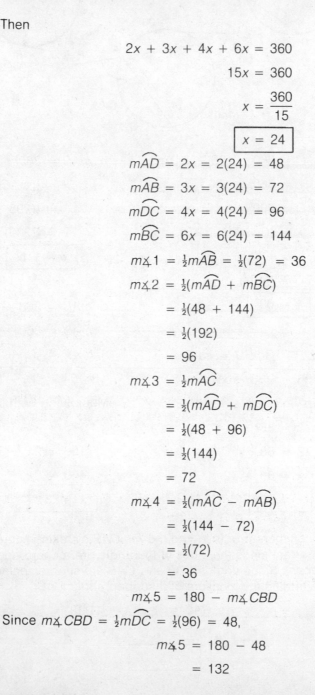

SOLUTION Let

$$m\widehat{AD} = 2x$$
$$m\widehat{AB} = 3x$$
$$m\widehat{DC} = 4x$$
$$m\widehat{BC} = 6x$$

Then

$$2x + 3x + 4x + 6x = 360$$
$$15x = 360$$
$$x = \frac{360}{15}$$
$$\boxed{x = 24}$$

$$m\widehat{AD} = 2x = 2(24) = 48$$
$$m\widehat{AB} = 3x = 3(24) = 72$$
$$m\widehat{DC} = 4x = 4(24) = 96$$
$$m\widehat{BC} = 6x = 6(24) = 144$$
$$m\angle 1 = \tfrac{1}{2}m\widehat{AB} = \tfrac{1}{2}(72) = 36$$
$$m\angle 2 = \tfrac{1}{2}(m\widehat{AD} + m\widehat{BC})$$
$$= \tfrac{1}{2}(48 + 144)$$
$$= \tfrac{1}{2}(192)$$
$$= 96$$
$$m\angle 3 = \tfrac{1}{2}m\widehat{AC}$$
$$= \tfrac{1}{2}(m\widehat{AD} + m\widehat{DC})$$
$$= \tfrac{1}{2}(48 + 96)$$
$$= \tfrac{1}{2}(144)$$
$$= 72$$
$$m\angle 4 = \tfrac{1}{2}(m\widehat{AC} - m\widehat{AB})$$
$$= \tfrac{1}{2}(144 - 72)$$
$$= \tfrac{1}{2}(72)$$
$$= 36$$
$$m\angle 5 = 180 - m\angle CBD$$

Since $m\angle CBD = \tfrac{1}{2}m\widehat{DC} = \tfrac{1}{2}(96) = 48$,

$$m\angle 5 = 180 - 48$$
$$= 132$$

12.8 USING ANGLE MEASUREMENT THEOREMS

Angle measurement theorems can be used to help establish that angles, arcs, and chords of a circle are congruent. We first pause to formally state some observations which you have probably made while working on numerical applications of these theorems.

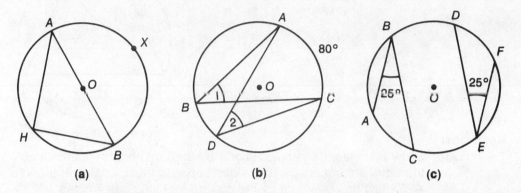

(a)   (b)   (c)

FIGURE 12.29

In Figure 12.29a $\overline{AB}$ is a diameter of circle O. Angle AHB is said to be *inscribed in* semicircle $\widehat{AHB}$ since the end points of the chords which form the angle coincide with the end points of a diameter. What is the measure of $\angle AHB$? Angle AHB is an inscribed angle having semicircle $\widehat{AXB}$ as its intercepted arc. Hence, $m\angle AHB = \frac{1}{2}(180) = 90$. *An angle inscribed in a semicircle is a right angle.*

In Figure 12.29b the measures of inscribed angles 1 and 2 are each equal to one-half of 80° (or 40°). *Inscribed angles which intercept the same arc (or congruent arcs) are congruent.* The inscribed angles which are shown in Figure 12.29c are given as congruent (each has a measure of 25). It follows that $m\widehat{AC} = 50$ and $m\widehat{DF} = 50$. *If a pair of inscribed angles are congruent, then their intercepted arcs are congruent.*

Each of the preceding observations were based on a direct application of the Inscribed Angle Theorem (Theorem 12.8) and are therefore referred to as *corollaries.*

COROLLARIES TO THE INSCRIBED ANGLE THEOREM

■ An angle inscribed in a semicircle is a right angle.

■ If inscribed angles (or angles formed by a tangent and a chord) intercept the same or congruent arcs, then they are congruent.

■ If inscribed angles are congruent, then their intercepted arcs are congruent.

EXAMPLE 12.22

GIVEN   In ⊙O, $\overline{CD} \parallel \overline{AB}$. Chords $\overline{AC}$ and $\overline{BD}$ are extended to meet at point P.

PROVE $\overline{AP} \cong \overline{BP}$.

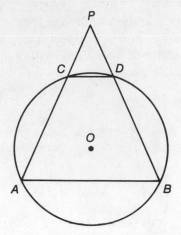

SOLUTION

PLAN   Show that inscribed angles A and B intercept congruent arcs and therefore are congruent. The sides opposite these angles in triangle APB must be congruent by the converse of the Base Angles Theorem.

PROOF	Statements	Reasons
	1. $\overline{CD} \parallel \overline{AB}$.	1. Given.
	2. $m\overset{\frown}{AC} = m\overset{\frown}{BD}$.	2. Parallel chords intercept arcs equal in measure.
	3. $m\overset{\frown}{ACD} = m\overset{\frown}{BDC}$.	3. Arc Sum Postulate.
	4. $m\angle B = \frac{1}{2}m\overset{\frown}{ACD}$. $m\angle A = \frac{1}{2}m\overset{\frown}{BDC}$.	4. The measure of an inscribed angle is one-half the measure of its intercepted arc.
	5. $\frac{1}{2}m\overset{\frown}{ACD} = \frac{1}{2}m\overset{\frown}{BDC}$.	5. Halves of equals are equal.
	6. $m\angle B = m\angle A$.	6. Substitution.
	7. $\overline{AP} \cong \overline{BP}$.	7. If two angles of a triangle are equal in measure, then the sides opposite are congruent.

EXAMPLE 12.23

GIVEN   In ⊙O, $\overline{AB}$ is a diameter. $\overset{\frown}{AC} \cong \overset{\frown}{BD}$.

PROVE $\triangle ABC \cong \triangle BAD$.

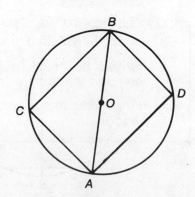

SOLUTION

PLAN   Prove triangles are congruent by Hy-Leg.

PROOF	Statements	Reasons
	1. In $\odot O$, $\overline{AB}$ is a diameter.	1. Given.
	2. Angles C and D are right angles.	2. An angle inscribed in a semicircle is a right angle.
	3. Triangles ABC and BAD are right triangles.	3. A triangle which contains a right angle is a right triangle.
	4. $\overline{AB} \cong \overline{AB}$. (Hy)	4. Reflexive property of congruence.
	5. $\overarc{AC} \cong \overarc{BD}$.	5. Given.
	6. $\overline{AC} \cong \overline{BD}$. (Leg)	6. In a circle, congruent arcs have congruent chords.
	7. $\triangle ABC \cong \triangle BAD$.	7. Hy-Leg.

REMARK   The solution presented for this example is not unique. For example, the AAS theorem could have been used:

  (A) $\angle C \cong \angle D$ (Right $\angle$s are $\cong$.)

  (A) $\angle ABC \cong \angle BAD$ (Inscribed angles which intercept $\cong$ arcs are $\cong$.)

  (S) $\overline{AB} \cong \overline{AB}$

REVIEW EXERCISES FOR CHAPTER 12

In Exercises 1 to 13, find the value of x.

1.

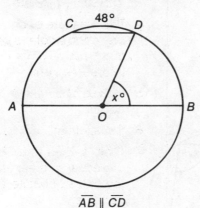

$\overline{AB} \parallel \overline{CD}$

2.

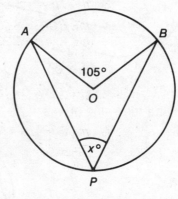

3.

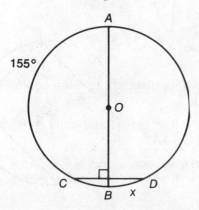

4.

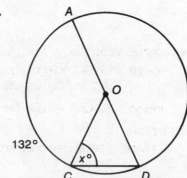

5.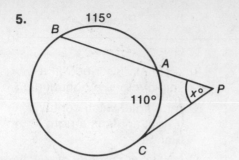

6.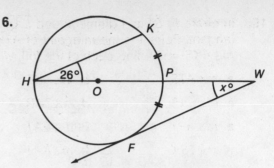

P is the midpoint of $\overset{\frown}{KPF}$.

7.

8.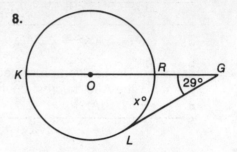

9.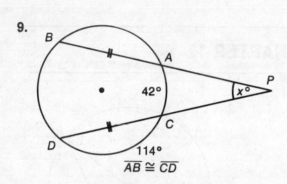

$\overline{AB} \cong \overline{CD}$

10.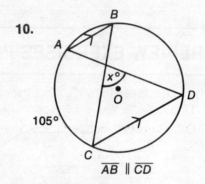

$\overline{AB} \parallel \overline{CD}$

11. Find the value of x.

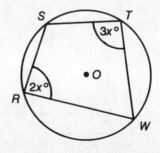

12. The length of a chord of a circle is 24 and its distance from the center is 5. Find the length of a diameter of the circle.

13. Tangents $\overline{AX}$ and $\overline{AY}$ are drawn to circle P from an exterior point A. Radii $\overline{PX}$ and $\overline{PY}$ are drawn. If $m\angle XPY = 74$, find $m\angle XAY$.

14. The length of tangent segment $\overline{PA}$ drawn from exterior point P to circle O is 24. If the radius of the circle is 7, find the distance from point P to the center of the circle.

15. In circle P, $\overline{ST}$ is a diameter and $\overleftrightarrow{LAB}$ is a tangent. Point T is the midpoint of $\overparen{ATK}$. If $m\angle AKS = 74$, find each of the following:

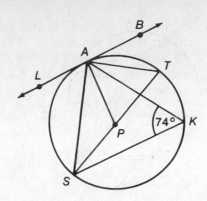

a $m\angle ATS$ b $m\angle AST$

c $m\angle LAS$ d $m\angle TAB$

e $m\angle APT$ f $m\angle SAT$

g $m\angle TSK$ h $m\angle KAT$

i $m\angle TAP$

16. Fill in the following table:

	$m\overparen{BD}$	$m\overparen{AC}$	$m\angle DEB$	$m\angle ABC$	$m\angle AED$
a	118°	62°	?	?	?
b	102°	?	?	28°	?
c	115°	?	83°	?	?
d	?	?	?	41°	64°

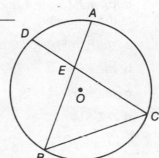

17. Find the values of x, y, and z.

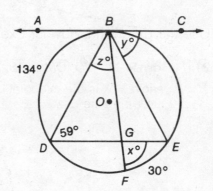

18. A is the midpoint of $\overparen{DAB}$. Fill in the following table.

	$m\overparen{ADC}$	$m\overparen{AB}$	$m\angle CAQ$	$m\angle DBC$	$m\angle AEB$	$m\angle CPQ$
a	120	40	?	?	?	?
b	132	?	?	?	?	37
c	?	64	?	?	82	?
d	?	?	71	?	?	49

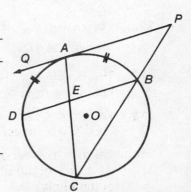

19. In circle O, tangent $\overrightarrow{PW}$ and secant $\overline{PST}$ are drawn. Chord $\overline{WA}$ is parallel to chord $\overline{ST}$. Chords $\overline{AS}$ and $\overline{WT}$ intersect at point B. If $m\widehat{WA}:m\widehat{AT}:m\widehat{ST} = 1:3:5$, find each of the following:

a $m\widehat{WA}$, $m\widehat{AT}$, $m\widehat{ST}$, and $m\widehat{SW}$

b $m\angle WTS$

c $m\angle TBS$

d $m\angle TWP$

e $m\angle WPT$

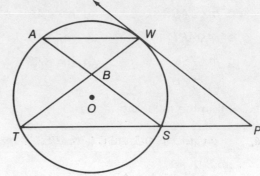

20. In circle P, $\overline{KM}$ is a diameter and $\overline{LFK}$ and $\overline{LHJ}$ are secants. Point F is the midpoint of $\widehat{KFH}$. If $m\widehat{KJ}:m\widehat{JM} = 5:4$ and $m\angle HEM = 64$, find each of the following:

a $m\widehat{KF}$, $m\widehat{FM}$, $m\widehat{JK}$, $m\widehat{JM}$, and $m\widehat{HM}$

b $m\angle KPF$

c $m\angle KJH$

d $m\angle KLJ$

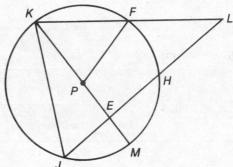

Use the diagram below for Exercises 21 and 22.

21. **GIVEN** In $\odot O$, $\overline{OM} \perp \overline{AB}$.

PROVE X is the midpoint of $\widehat{AB}$.

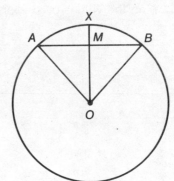

22. **GIVEN** In $\odot O$, $\widehat{RST} \cong \widehat{WTS}$.

PROVE $\angle RTS \cong \angle WST$.

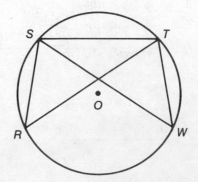

23. **GIVEN** In ⊙O, OA > AC.

   PROVE $m\overset{\frown}{BC} > m\overset{\frown}{AC}$.

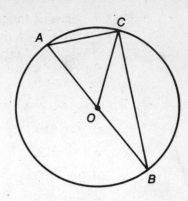

24. **GIVEN** In ⊙O, $m\overset{\frown}{BC} > m\overset{\frown}{AC}$.

   PROVE OA > AD.

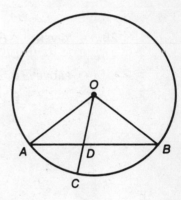

25. **GIVEN** ⊙X ≅ ⊙Y.
   $\overleftrightarrow{PQ}$ is tangent to ⊙X at P
   and tangent to ⊙Y at Q.

   PROVE Point M is the midpoint of $\overline{XY}$.

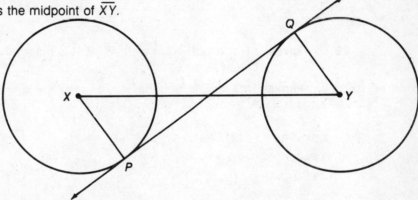

26. **GIVEN** △EFG is inscribed in ⊙P,
   $\overleftrightarrow{AB}$ is tangent to F
   $\overline{FE} \cong \overline{FG}$.

   PROVE $\overline{AB} \parallel \overline{EG}$.

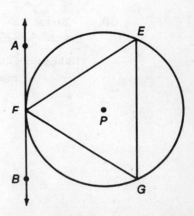

27. **GIVEN** In ⊙O, $\overline{AB}$ is a diameter, point M is the midpoint of $\overline{BD}$, chords $\overline{BM}$ and $\overline{AC}$ are extended to meet at point D.

PROVE M is the midpoint of $\overset{\frown}{BMC}$.

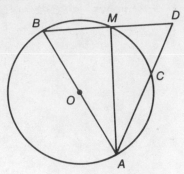

28. **GIVEN** △RST is inscribed in ⊙O and chords $\overline{SW}$ and $\overline{WK}$ are drawn, $\overline{SW}$ bisects ∡RST, $\overset{\frown}{RW} \cong \overset{\frown}{SK}$.

PROVE **a** △NWS is isosceles.

b △NTK is isosceles.

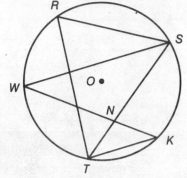

29. **GIVEN** Trapezoid JKLM with $\overline{JK} \parallel \overline{LM}$ is inscribed in ⊙O, $m\overset{\frown}{LM} > m\overset{\frown}{KL}$.

PROVE $m∡MKL > m∡JKM$.

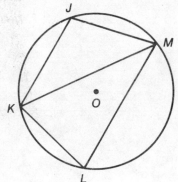

30. **GIVEN** $\overline{SR} \cong \overline{TW}$, $\overline{MA} \cong \overline{MT}$.

PROVE Quadrilateral RSTW is a parallelogram.

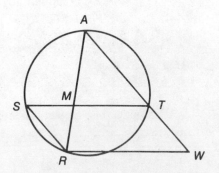

CHAPTER THIRTEEN

CHORD, TANGENT, AND SECANT SEGMENTS

13.1 EQUIDISTANT CHORDS

In Figure 13.1, $\overline{OC}$ is drawn perpendicular to chord $\overline{AB}$. The length of $\overline{OC}$ represents the *distance* of chord $\overline{AB}$ from the center (point O) of the circle. If the radius of circle O is 5 and the length of $\overline{AB}$ is 8, what is the length of $\overline{OC}$? If a line passes through the center of a circle and is perpendicular to a chord, then it bisects the chord (see Theorem 12.4). Hence, $AC = 4$. If we now draw radius $\overline{OA}$, a 3-4-5 right triangle is formed, where $OC = 3$. (See Figure 13.2.)

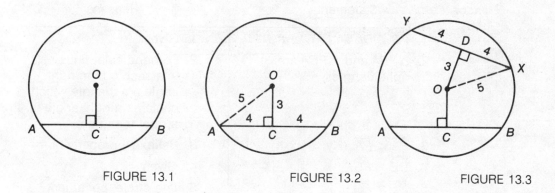

FIGURE 13.1   FIGURE 13.2   FIGURE 13.3

Suppose we now draw another chord in circle O, say $\overline{XY}$, which also has a length of 8. How far is it from the center of the circle? (See Figure 13.3.)

Using a similar analysis, we find that chord $\overline{XY}$ is also 3 units from the center of the circle. What do chords $\overline{AB}$ and $\overline{XY}$ have in common? What conclusion follows? The answers are stated in Theorem 13.1.

> **THEOREM 13.1 CONGRUENT CHORDS AND DISTANCE FROM THE CENTER**
>
> In the same or in congruent circles, congruent chords are equidistant (the same distance) from the center(s) of the circle(s).

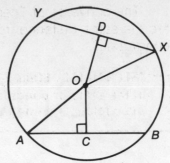

GIVEN In $\odot O$, $\overline{AB} \cong \overline{XY}$, $\overline{OC} \perp \overline{AB}$, $\overline{OD} \perp \overline{XY}$.

PROVE $OC = OD$.

Draw $\overline{OA}$ and $\overline{OX}$. Prove $\triangle OCA \cong \triangle ODX$ by Hy-Leg:

$$\overline{OA} \cong \overline{OX} \text{ (Hy)}$$

$$\overline{AC} \cong \overline{XD} \text{ (Leg)}$$

$AC = XD$ since $AC = \frac{1}{2}AB$ and $XD = \frac{1}{2}XY$, and halves of equals (AB and XY) are equal (AC and XD). $\overline{OC} \cong \overline{OD}$ by CPCTC from which it follows that $OC = OD$.

EXAMPLE 13.1

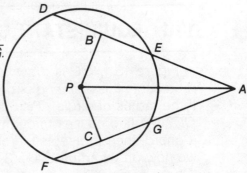

GIVEN In $\odot P$, $\overline{PB} \perp \overline{DE}$, $\overline{PC} \perp \overline{FG}$, $\overline{DE} \cong \overline{FG}$.

PROVE $\overline{PA}$ bisects $\measuredangle FAD$.

SOLUTION

PLAN Show right $\triangle PBA \cong$ right $\triangle PCA$.

PROOF

Statements	Reasons
1. $\overline{PB} \perp \overline{DE}$ and $\overline{PC} \perp \overline{FG}$.	1. Given.
2. $\triangle PBA$ and $\triangle PCA$ are right triangles.	2. Perpendicular lines intersect to form right angles; a triangle which contains a right angle is a right triangle.
3. $\overline{PA} \cong \overline{PA}$. (Hy)	3. Reflexive property of $\cong$.
4. $\overparen{DE} \cong \overparen{FG}$.	4. Given.
5. $\overline{DE} \cong \overline{FG}$.	5. In a circle, congruent arcs have congruent chords.
6. $PB = PC$.	6. In a circle, congruent chords are equidistant from the center of the circle.
7. $\overline{PB} \cong \overline{PC}$. (Leg)	7. Segments equal in length are congruent.
8. $\triangle PBA \cong \triangle PCA$.	8. Hy-Leg.
9. $\measuredangle PAB \cong \measuredangle PAC$.	9. CPCTC.
10. $\overline{PA}$ bisects $\measuredangle FAD$.	10. If a line divides an angle into two congruent angles, it bisects the angle.

The converse of Theorem 13.1 states that if we know that two chords are the same distance from the center of a circle, then they must be congruent.

THEOREM 13.2 EQUIDISTANT CHORDS

In the same or congruent circles, chords equidistant from the center(s) of the circle(s) are congruent.

EXAMPLE 13.2

GIVEN $\odot O$ with $\overline{OX} \perp \overline{AB}$, $\overline{OY} \perp \overline{CB}$, $\angle OXY \cong \angle OYX$.

PROVE $\overparen{AB} \cong \overparen{CB}$.

SOLUTION

PLAN Prove that $\overline{AB}$ and $\overline{CB}$ are the same distance from the center of the circle by showing that $OX = OY$. By Theorem 13.2, $\overline{AB} \cong \overline{CB}$ which implies $\overparen{AB} \cong \overparen{CB}$.

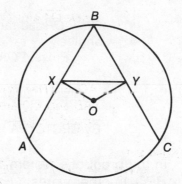

PROOF

Statements	Reasons
1. $\overline{OX} \perp \overline{AB}$ and $\overline{OY} \perp \overline{CB}$.	1. Given.
2. $\angle OXY \cong \angle OYX$.	2. Given.
3. $OX = OY$.	3. If two angles of a triangle are congruent, then the sides opposite are equal in length.
4. $\overline{AB} \cong \overline{CB}$.	4. In the same circle, chords equidistant from the center of the circle are congruent.
5. $\overparen{AB} \cong \overparen{CB}$.	5. In the same circle, congruent chords intercept congruent arcs.

13.2 TANGENTS AND CIRCLES

In the same circle, chords which are the same distance from the center of the circle are congruent. If two tangent segments are drawn from the same exterior point to a circle, then they are congruent.

THEOREM 13.3 CONGRUENT TANGENT SEGMENTS

If two tangent segments are drawn to a circle from the same exterior point, then they are congruent.

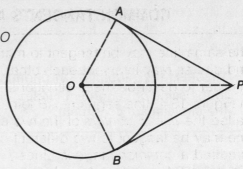

GIVEN $\overline{PA}$ and $\overline{PB}$ are tangent to $\odot O$ at points A and B, respectively.

PROVE $\overline{PA} \cong \overline{PB}$.

Draw $\overline{OP}$ and radii $\overline{OA}$ and $\overline{OB}$. Angles OAP and OBP are right angles.

Prove $\triangle OAP \cong \triangle OBP$ by Hy-Leg:

$$\overline{OP} \cong \overline{OP} \quad \text{(Hy)}$$

$$\overline{OA} \cong \overline{OB} \quad \text{(Leg)}$$

By CPCTC, $\overline{PA} \cong \overline{PB}$.

In the proof of Theorem 13.3, notice that since $\triangle OAP \cong \triangle OBP$, $\angle APO \cong \angle BPO$. In other words, $\overline{OP}$ bisects the angle formed by the two tangent segments.

EXAMPLE 13.3 Find the value of x.

a

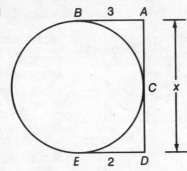

b

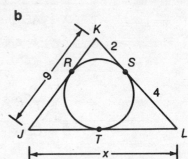

SOLUTION

a $AC = AB = 3$

$DC = DE = 2$

$x = AC + DC = 5$

b $KR = KS = 2$

$JR = 9 - KR = 9 - 2 = 7$

$JT = JR = 7$

$LT = LS = 4$

$x = JT + LT = 7 + 4 = 11$

EXAMPLE 13.4 Find the values of x and y.

SOLUTION $PK = PJ$ and $PK = PL$.

Hence, $PJ = PL$.

$$2x - 7 = x + 3$$

$$2x = x + 10$$

$$x = 10$$

$$PL = x + 3 = 13$$

$$y = PK = PL = 13$$

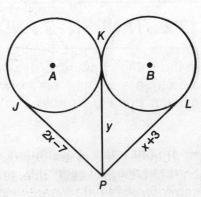

COMMON TANGENTS AND TANGENT CIRCLES

The same line may be tangent to more than one circle (a common tangent), and circles may intersect each other in exactly one point (tangent circles). In order to describe these situations, some terminology must be introduced. In Figure 13.4, line segment $\overline{AB}$ joins the centers of circles A and B and is called the *line of centers* of the two circles. Figure 13.5 illustrates that a line may be tangent to two different circles. A line which has this property is called a *common tangent*. Lines j, k, l, and m are common tangents. In Figure 13.6 two different circles are shown to be tangent to the same line at the same point. These circles are known as *tangent circles*. Circles A and B are tangent to line l at point P. Circles A and C are tangent to line m at point Q.

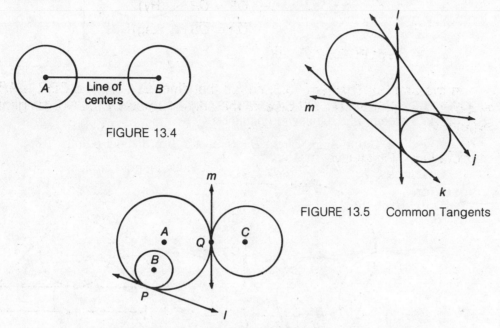

FIGURE 13.4

FIGURE 13.5   Common Tangents

FIGURE 13.6   Tangent Circles

We may further distinguish between types of common tangents and types of tangent circles. A common tangent may be either a common *internal* or a common *external* tangent, as shown in Figure 13.7.

Common internal tangents

Common external tangents

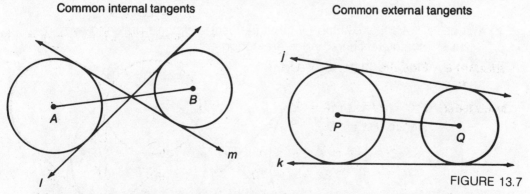

FIGURE 13.7

Lines l and m are common *internal* tangents since each is tangent to both circles and each intersects their line of centers. Lines j and k are common *external* tangents since each is tangent to both circles and each does *not* intersect their line of centers.

Tangent circles may be tangent either *internally* or *externally* to each other, as shown in Figure 13.8. Circles *A* and *B* are tangent *internally* since they lie on the *same* side of their common tangent. Circles *P* and *Q* are tangent *externally* since they lie on *opposite* sides of their common tangent.

Internally Tangent Circles

Externally Tangent Circles

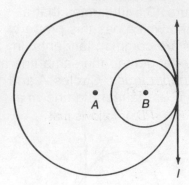
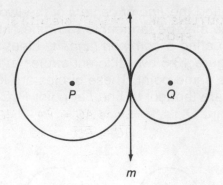

FIGURE 13.8

EXAMPLE 13.5 Determine the number of common tangents which can be drawn for each of the following situations:

a Circle *A* and circle *B* intersect in two distinct points.

b Circle *A* and circle *B* are externally tangent circles.

SOLUTION

(a)

(b)

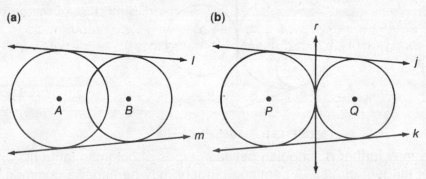

Two common external tangents

Two common external tangents and one common internal tangent

EXAMPLE 13.6 Prove that common internal tangent segments drawn to two nonintersecting circles are congruent.

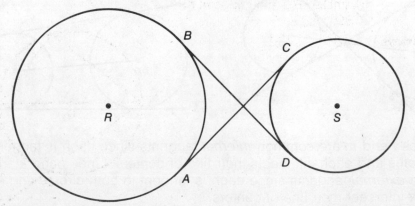

SOLUTION Draw two nonintersecting circles with their common internal tangent segments drawn. Let *P* represent the point at which the tangent segments intersect.

GIVEN Nonintersecting circles *R* and *S*, common internal tangent segments $\overline{AC}$ and $\overline{BD}$, intersecting at point *P*.

PROVE $\overline{AC} \cong \overline{BD}$.

OUTLINE OF PROOF Apply Theorem 13.3:

$$PA = PB$$
$$\underline{+\, PC = PD}$$
$$\overline{PA + PC = PB + PD}$$

Since $AC = PA + PC$ and $BD = PB + PD$, it follows that $\overline{AC} \cong \overline{BD}$.

13.3 SIMILAR TRIANGLES AND CIRCLES

The sides of a pair of triangles may intercept arcs of a circle so that the measures of some of the angles of the triangles may be determined by the measures of the intercepted arcs. The following facts about angle measurement will be found useful in proving that these triangles are *similar*.

- Inscribed angles (or angles formed by a tangent and a chord) which intercept the same or congruent arcs are congruent.

- An angle inscribed in a semicircle is a right angle.

- An angle formed by a radius drawn to the point of tangency is a right angle.

Let's look at an actual problem that illustrates these concepts.

GIVEN $\overline{EB}$ is tangent to $\odot O$ at *B*, $\overline{AB}$ is a diameter, *B* is the midpoint of $\overarc{CBD}$.

PROVE $\triangle ABC \sim \triangle AEB$.

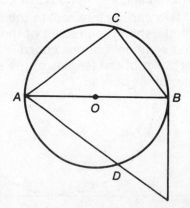

In planning the proof, note that (Figure 13.9):

- Angles *ACB* and *ABE* are congruent since they are both right angles.

- Angles *CAB* and *EAB* are congruent since they are inscribed angles which intercept congruent arcs ($\overset{\frown}{BC} \cong \overset{\frown}{BD}$).

- $\triangle ABC \sim \triangle AEB$ by the AA Theorem of Similarity.

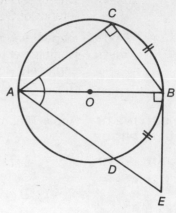

FIGURE 13.9

PROOF	Statements	Reasons
	1. *B* is the midpoint $\overset{\frown}{CBD}$.	1. Given.
	2. $\overset{\frown}{BC} \cong \overset{\frown}{BD}$.	2. A midpoint of an arc divides the arc into two congruent arcs.
	3. $\angle CAB \cong \angle EAB$. (Angle)	3. Inscribed angles of a circle which intercept congruent arcs are congruent.
	4. $\angle ACB$ is a right angle.	4. An angle inscribed in a semicircle is a right angle.
	5. $\angle ABE$ is a right angle.	5. An angle formed by a radius drawn to the point of tangency is a right angle.
	6. $\angle ACB \cong \angle ABE$. (Angle)	6. All right angles are congruent.
	7. $\triangle ABC \sim \triangle AEB$.	7. AA Theorem of Similarity.

The properties of similar triangles may be used to establish a special relationship between the segments formed by two chords which intersect in the interior of a circle.

THEOREM 13.4 PRODUCTS OF LENGTHS OF SEGMENTS OF INTERSECTING CHORDS

If two chords intersect in the interior of a circle, then the product of the lengths of the segments of one chord is equal to the product of the lengths of the segments of the other chord.

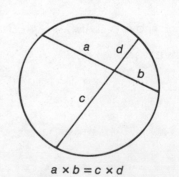

$a \times b = c \times d$

GIVEN Chords $\overline{AB}$ and $\overline{CD}$ intersect in the interior of $\odot O$ at point E.

PROVE $AE \cdot EB = CE \cdot ED$.

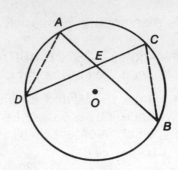

OUTLINE OF PROOF

■ Draw $\overline{AD}$ and $\overline{CB}$.

■ $AE \cdot EB = CE \cdot ED \Longrightarrow \dfrac{AE}{CE} = \dfrac{ED}{EB} \Longrightarrow \triangle AED \sim \triangle CEB$

■ $\triangle AED$ is similar to $\triangle CEB$ since:

$\measuredangle AED \cong \measuredangle BEC$ (Vertical angles)

$\measuredangle ADC \cong \measuredangle CBA$ (Inscribed angles which intercept the same arc)

EXAMPLE 13.7   Find the value of x.

(a)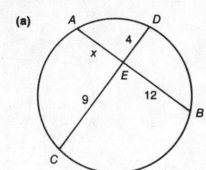

(b)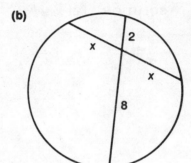

(c)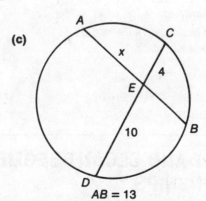

$AB = 13$

SOLUTION   **a** $AE \cdot EB = CE \cdot ED$

$(x)(12) = (4)(9)$

$12x = 36$

$x = \dfrac{36}{12}$

$\boxed{x = 3}$

b $(x)(x) = (8)(2)$

$\qquad x^2 = 16$

$\qquad x = \sqrt{16}$

$\qquad \boxed{x = 4}$

c If $x = AE$, then $13 - x = EB$.

$$(x)(13 - x) = (10)(4)$$

$$13x - x^2 = 40$$

Writing the quadratic equation in standard form,

$$x^2 - 13x + 40 = 0$$

$$(x - 8)(x - 5) = 0$$

$$x - 8 = 0 \quad or \quad x - 5 = 0$$

$$\boxed{x = 8} \quad or \quad \boxed{x = 5}$$

If $AE = 8$ then $EB = 5$. Alternatively, AE may equal 5, in which case $EB = 8$.

EXAMPLE 13.8 A diameter divides a chord of a circle into two segments whose lengths are 7 and 9. If the length of the shorter segment of the diameter is 3, find the length of a radius of the circle.

SOLUTION Let $x = PB$.

$$(3)(x) = (7)(9)$$

$$3x = 63$$

$$x = \frac{63}{3}$$

$$\boxed{x = 21}$$

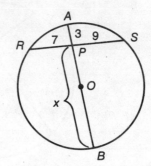

Diameter $AB = 3 + 21 = 24$

The length of a radius of the circle is $\frac{1}{2}(24)$ or 12.

13.4 TANGENT AND SECANT SEGMENT RELATIONSHIPS

In Figure 13.10, $\overline{PAB}$ is called a secant *segment*. Its end points are a point in the exterior of the circle (point P) and the point on the circle furthest from point P at which the secant intersects the circle (point B). The circle divides the secant segment into two segments: an *internal* secant segment (chord $\overline{AB}$) and an external secant segment ($\overline{AP}$).

When two secant segments are drawn to a circle from the same exterior point, then a special relationship exists between the lengths of the secant segments and the lengths of their external segments.

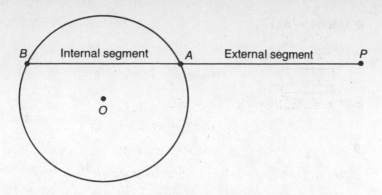

Internal segment   External segment

FIGURE 13.10

THEOREM 13.5   SECANT-SECANT SEGMENT PRODUCTS

If two secant segments are drawn to a circle from the same exterior point, then the product of the lengths of one secant segment and its *external* segment is equal to the product of the lengths of the other secant segment and its *external* segment.

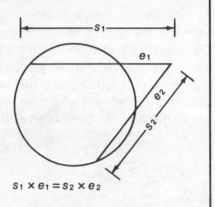

$$s_1 \times e_1 = s_2 \times e_2$$

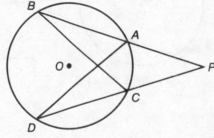

GIVEN $\overline{PB}$ and $\overline{PD}$ are secant segments drawn to $\odot O$.

PROVE $PB \cdot PA = PD \cdot PC$.

OUTLINE OF PROOF

- Draw $\overline{AD}$ and $\overline{CB}$.

- $PB \cdot PA = PD \cdot PC \;\Rightarrow\; \dfrac{PB}{PD} = \dfrac{PC}{PA} \;\Rightarrow\;$ Show $\triangle PBC \sim \triangle PDA$.

- $\triangle PBC \sim \triangle PDA$ since $\begin{array}{l} \measuredangle P \cong \measuredangle P \\ \measuredangle PBC \cong \measuredangle PDA \end{array}$

EXAMPLE 13.9   Find the value of x.

(a)

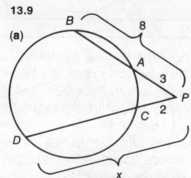

(b)

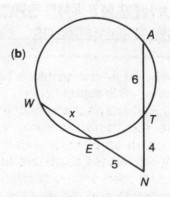

TANGENT AND SECANT SEGMENT RELATIONSHIPS   **259**

SOLUTION **a** $PA \cdot PB = PC \cdot PD$

$3 \cdot 8 = 2 \cdot x$

$24 = 2x$

$\boxed{12 = x}$

b $NE \cdot NW = NT \cdot NA$, where

$NE = 5$

$NW = x + 5$

$NT = 4$

$NA = 6 + 4 = 10$

$5 \cdot (x + 5) = 4 \cdot 10$

$5x + 25 = 40$

$5x = 15$

$\boxed{x = 3}$

When a *tangent* segment and a secant segment are drawn to a circle from the same exterior point, a relationship analogous to the one stated in Theorem 13.5 results. Figure 13.11 illustrates that when secant $\overline{PAB}$ is rotated clockwise, it will eventually be tangent to the circle. This analysis suggests that we replace the secant segment length with the tangent segment length $\overline{PA}$ and also replace the external portion of the secant segment with tangent segment length $\overline{PA}$.

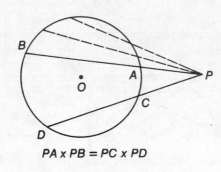

$PA \times PB = PC \times PD$

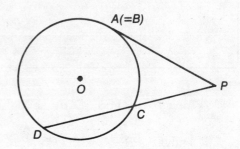

$PA \times PB = PC \times PD$

$PA \times PA = PC \times PD$ or $(PA)^2 = PC \times PD$

FIGURE 13.11

THEOREM 13.6 TANGENT-SECANT SEGMENT PRODUCTS

If a tangent segment and a secant segment are drawn to a circle from the same exterior point, then the square of the length of the tangent segment is equal to the product of the lengths of the secant segment and its external segment.

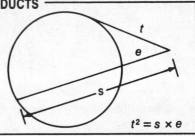

$t^2 = s \times e$

GIVEN $\overline{PA}$ is tangent to $\odot O$ at A and $\overline{PC}$ is a secant.

PROVE $(PA)^2 = PC \cdot PB$.

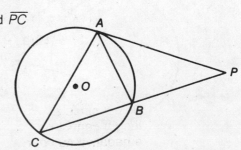

OUTLINE OF PROOF

- Draw $\overline{CA}$ and $\overline{BA}$.

- $(PA)^2 = PC \cdot PB \Longrightarrow \dfrac{PA}{PC} = \dfrac{PB}{PA} \Longrightarrow$ Show $\triangle PAB \sim \triangle PCA$.

- $\triangle PAB \sim \triangle PCA$ since:

$$\angle P \cong \angle P$$

$$\angle PCA \cong \angle PAB$$

(NOTE: Both angles are measured by $\frac{1}{2}\overset{\frown}{AB}$.)

EXAMPLE 13.10

Find the value of x.

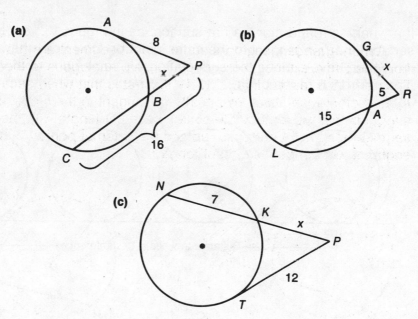

(a)

(b)

(c)

SOLUTION

a $(PA)^2 = PB \cdot PC$

$(8)^2 = x \cdot 16$

$64 = 16x$

$\boxed{4 = x}$

b $(RG)^2 = RA \cdot RL$

$x^2 = 5(15 + 5)$

$x^2 = 5 \cdot 20$

$x^2 = 100$

$x = \sqrt{100}$

$\boxed{x = 10}$

c $(PT)^2 = PK \cdot PN$

$(12)^2 = x(x + 7)$

$144 = x^2 + 7x$

Writing the quadratic equation in standard form,

$$x^2 + 7x - 144 = 0 \quad or \quad\quad\quad x - 9 = 0$$

$$(x + 16)(x - 9) = 0$$

$$x + 16 = 0$$

$$x = -16 \quad\quad\quad \boxed{x = 9}$$

Reject since a
length cannot be
a negative number

13.5 CIRCUMFERENCE AND ARC LENGTH

The distance around a polygon is referred to as the *perimeter* of the polygon. The perimeter of a circle is given a special name, *circumference*.

> **DEFINITION OF CIRCUMFERENCE**
>
> The *circumference* of a circle is the distance around the circle, expressed in linear units of measurement (e.g., inches, centimeters, feet).

If a wheel is rolled along a flat surface, as in Figure 13.12, so that the surface is again tangent to the same point on the circle, then the distance the wheel travels along the surface has the same numerical value as the circumference of the circle.

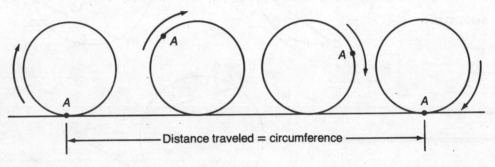

Distance traveled = circumference

FIGURE 13.12

The longer the diameter of a circle, the larger the circle's circumference. Interestingly, however, if the circumference of a circle is divided by the length of its diameter, then the value obtained will be the same regardless of the size of the circle:

$$\frac{\text{Circumference}}{\text{Diameter}} = 3.1415926\ldots$$

The three dots which follow the decimal number indicate that this value is a nonterminating (never-ending) decimal number. This constant value is referred to as *pi* and is denoted by the Greek letter π. We may therefore write

$$\frac{\text{Circumference}}{\text{Diameter}} = \pi \qquad \text{or} \qquad \text{Circumference} = \pi \cdot \text{Diameter}$$

where π is *approximately equal to* 3.14 or *approximately equal to* the improper fraction $\frac{22}{7}$.

> **THEOREM 13.7 CIRCUMFERENCE OF A CIRCLE**
>
> The circumference of a circle is equal to the product of π and the length of its diameter: $C = \pi D$.

REMARKS   **1.** $\pi \approx 3.14$ and $\pi \approx \frac{22}{7}$, where the symbol $\approx$ is read as *is approximately equal to.*

2. Since the length of a diameter is numerically equal to twice the length of the radius, we may write Circumference = $\pi \cdot 2 \cdot$Radius. In writing a formula it is a common practice to write a digit (2 in this case) before any symbols. We therefore write Circumference = $2 \cdot \pi \cdot$Radius, or $C = 2\pi R$.

3. The choice of which approximation is to be used for π (3.14 or $\frac{22}{7}$) will usually be specified in the statement of the problem. Frequently, you will be asked *not* to substitute an approximation for π, and instead, to express the answer in terms of π. This is illustrated in Example 13.11.

EXAMPLE   Find the circumference of a circle if:
13.11   **a**   Diameter = 10 (Use π = 3.14.)

b   Radius = 14 (Use $\pi = \frac{22}{7}$.)

c   Diameter = 29 (Express answer in terms of π.)

SOLUTION   **a** $C = \pi D$

$$= 3.14 \times 10$$

$$\boxed{C = 31.4}$$

b $C = 2\pi R$

$$= 2 \times \frac{22}{\cancel{7}_1} \times \cancel{14}^2$$

$$\boxed{C = 88}$$

c $C = \pi D$

$$= \pi 29 \quad \text{or}$$

$$\boxed{C = 29\pi}$$

REMARK   When an answer is expressed in terms of π, the symbol for π is usually written last.

EXAMPLE   Rectangle *ABCD* has a width of 5 and a length of 12 and is
13.12   inscribed in circle *O*. Find the circumference of the circle.

SOLUTION   Since $\angle BAD$ is a right angle, diagonal $\overline{BD}$ must coincide with a diameter of the circle. Right $\triangle BAD$ is a 5-12-*13* right triangle where diagonal BD = 13. Hence,

$$C = \pi D = 13\pi$$

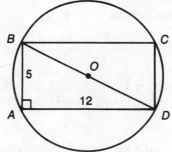

Circumference represents the distance around the *entire* circle. Our next concern is how to determine the length (in linear units) of an *arc* of the circle. Since a circle contains 360°, the circumference of a circle represents the length of a 360° arc of the circle (Figure 13.13). The ratio of the length of an arc to the circumference must be equal to the ratio of the degree measure of the arc to 360°:

$$\frac{\text{Arc measurement}}{\text{Circle measurement}} = \frac{\text{arc length}}{2\pi R} = \frac{n°}{360°}$$

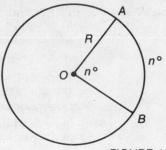

FIGURE 13.13

If you consider the circle to be made of a string, then the length of arc *AB* corresponds to the number arrived at by taking the section of the string from *A* to *B*, stretching it out, and then using a ruler to measure its length.

THEOREM 13.8 ARC LENGTH PROPORTION

$$\frac{\text{Length of arc}}{\text{Circumference}} = \frac{\text{degree measure of arc}}{360°}$$

EXAMPLE 13.13 In a circle having a radius of 10, find the length of an arc whose degree measure is 72°. (Leave answer in terms of π.)

SOLUTION

$$\frac{\text{Length of arc}}{\text{Circumference}} = \frac{\text{degree measure of arc}}{360°}$$

$$\frac{L}{2\pi \cdot 10} = \frac{72°}{360°}$$

Simplify each ratio before cross-multiplying:

$$\frac{L}{20\pi} = \frac{1}{5}$$

Cross-multiply:

$$5L = 20\pi$$

$$L = 4\pi$$

EXAMPLE 13.14 In a circle a 40° arc has a length of 8π. Determine the radius of the circle.

SOLUTION

$$\frac{\text{Length of arc}}{\text{Circumference}} = \frac{\text{degree measure of arc}}{360°}$$

$$\frac{8\pi}{2\pi R} = \frac{40°}{360°}$$

Simplifying each ratio *before* cross-multiplying:

$$\frac{4}{R} = \frac{1}{9}$$

Cross-multiply:

$$\boxed{R = 36}$$

EXAMPLE 13.15 Right triangle *ABC* is inscribed in circle *O* so that $\overline{AB}$ is a diameter and has a length of 27. If $m\angle CAB = 50$, find the length of $\overset{\frown}{AC}$, expressed in terms of π.

SOLUTION In order to find the degree measure of $\overset{\frown}{AC}$, draw $\overline{OC}$ and determine the measure of central angle *AOC*. Since $\overline{OA} \cong \overline{OC}$, $m\angle OCA = m\angle CAB = 50$. Using the fact that the sum of the angles of a triangle is 180, $m\angle AOC = 80$. Since a central angle and its intercepted arc have the same measure, $m\overset{\frown}{AC} = 80$.

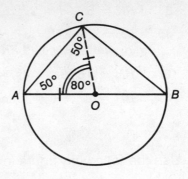

$$\frac{\text{Length of arc}}{\text{Circumference}} = \frac{\text{degree measure of arc}}{360°}$$

$$\frac{L}{27\pi} = \frac{80°}{360°}$$

$$\frac{L}{27\pi} = \frac{2}{9}$$

$$9L = 54\pi$$

$$\boxed{L = 6\pi}$$

REVIEW EXERCISES FOR CHAPTER 13

1. For each of the given situations, determine the number of common internal tangents and the number of common external tangents which can be drawn.

   a Circle *P* lies in the exterior of circle *Q* and has no points in common with circle *Q*.

   b Circle *P* lies in the interior of circle *Q* and has no points in common with circle *Q*.

   c Circle *P* and circle *Q* are tangent internally.

2. Find the value of *x*.

a

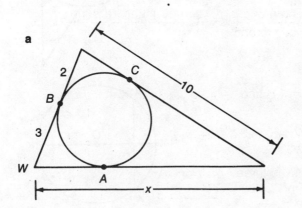

b

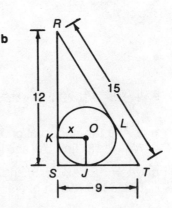

Use the following figure for
Exercises 3 to 6.

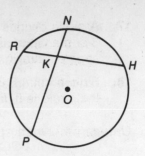

3. If $RK = 5$, $KH = 9$, $PK = 15$, find KN.

4. If $PK = 27$, $KN = 3$, and K is the
midpoint of RH, find RK.

5. If $RH = 16$, $RK = 4$, $PK = 8$, find PN.

6. If $RH = 22$, $PK = 7$, and $KN = 3$, find RK.

Use the following figure for Exercises 7
to 11.

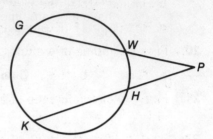

7. If $PW = 5$, $PG = 8$, $PH = 2$, find KH.

8. If $GW = 7$, $PW = 3$, $PK = 15$, find PH.

9. If W is the midpoint of $\overline{GP}$ and $PH =$
5, $KH = 35$, find PG.

10. If $PW = 6$, $WG = 9$, $PH = 9$, find KH.

11. If $GW = 11$, $PH = 8$, $KH = 2$, find PW.

Use the following figure for Exercises
12 to 14.

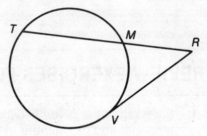

12. If $RV = 9$, $RM = 3$, find RT.

13. If $MT = 24$, $RM = 1$, find RV.

14. If $RV = 8$, $RM = 4$, find MT.

15. $RE = 2$
$RA = 14.5$
$ZG = 6$
$ZF = 8$
$SF = x$
E is the midpoint of $\overline{BG}$. Find SZ.

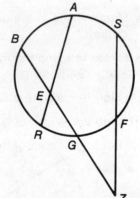

16. a If $NA = 4$, $JA = 8$, $WA = 11$, find
OK.

b If A is the midpoint of $\overline{JW}$, $AK = 32$,
$NA = 18$, find JW and OA.

c $JA = 12$, $AW = 9$, and AK is three
times the length of $\overline{NA}$. Find OK.

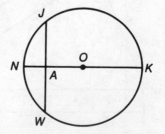

17. A circle divides a secant segment into an internal segment having a length of 8 and an external segment having a length of 2. Find the length of a tangent segment drawn to the circle from the same exterior point.

18. From a point 2 units from a circle, a tangent segment is drawn. If the radius of the circle is 8, find the length of the tangent segment.

Unless otherwise specified, answers may be left in terms of π.

19. Find the circumference of a circle if:

a Diameter = 21 (Use $\pi = \frac{22}{7}$.)

b Radius = 50 (Use $\pi = 3.14$.)

c Radius = 17 (Express answer in terms of π.)

20. Find the radius of a circle if its circumference is:

a 32π **b** 15π **c** 23

21. Find the circumference of a circle that is inscribed in a square whose side is 9.

22. Find the circumference of a circle which is circumscribed about a square whose perimeter is 36.

23. Fill in the following table:

	RADIUS OF CIRCLE	DEGREE MEASURE OF ARC	LENGTH OF ARC
a	9	120°	?
b	12	?	3π
c	?	72°	4π

24. A square having a side of 12 inches is inscribed in a circle. Find the length of an arc of the circle intercepted by one of the square's sides.

25. In a pulley system of a certain machine a belt of negligible thickness is wrapped around two identical wheels having a radius of 9. See the figure provided for this example. The belt criss-crosses at a point P between the two circles such that the measure of angle NPJ is 60. $\overline{NPS}$ and $\overline{JPK}$ may be considered to be tangent segments.

a Find the degree measure of *minor* arc NJ.

b Find the length of *major* arc NJ.

c Find the length of the belt.

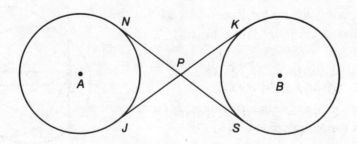

Use the following figure for Exercises 26 and 27.

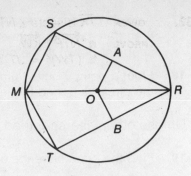

26. **GIVEN** In $\odot O$, $\overline{MR}$ is a diameter, $\overline{OA} \parallel \overline{MS}$,
     $\overline{OB} \parallel \overline{MT}$, $\overline{AR} \cong \overline{BR}$.
     PROVE $\overline{SR} \cong \overline{TR}$.

27. **GIVEN** In $\odot O$, $\overline{SR} \cong \overline{TR}$, $\overline{OA} \parallel \overline{MS}$, $\overline{OB} \parallel \overline{MT}$.
     PROVE $\angle AOM \cong \angle BOM$.

28. **GIVEN** In $\odot O$, quadrilateral $OXEY$ is a square.
     PROVE $\overparen{QP} \cong \overparen{JT}$.

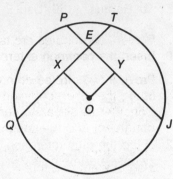

29. **GIVEN** $\triangle HBW$ is inscribed in $\odot O$,
     tangent segment $\overline{AB}$ is tangent at point
     B, $ABLM$ is a parallelogram.
     PROVE $\dfrac{BL}{BW} = \dfrac{BM}{BH}$.

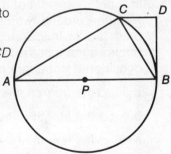

30. **GIVEN** In $\odot P$, $\overline{AB}$ is a diameter, $\overline{DB}$ is tangent to
     $\odot P$ at B, $\overline{CD} \perp \overline{DB}$.
     PROVE BC is the mean proportional between CD
     and AB.

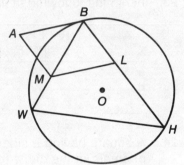

Use the following figure for Exercises 31 and 32.

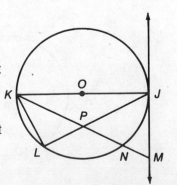

31. **GIVEN** In $\odot O$, $\overline{KJ}$ is a diameter, $\overline{MJ}$ is tangent at
     point J, N is the midpoint of $\overparen{LNJ}$.
     PROVE $KL : KJ = KP : KM$.

32. **GIVEN** In $\odot O$, $\overline{KJ}$ is a diameter, $\overline{MJ}$ is tangent at
     point J, $\overline{JP} \cong \overline{JM}$.
     PROVE $KL \cdot KM = JK \cdot LP$.

33. **GIVEN** $\overleftrightarrow{TK}$ bisects $\angle NTW$, $\overline{WK} \cong \overline{WT}$.

PROVE **a** $\overleftrightarrow{NTP} \parallel \overleftrightarrow{KW}$.

b $(TW)^2 = JT \cdot TK$.

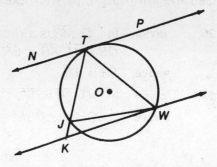

34. Prove if two circles are tangent externally, then the common internal tangent bisects a common external tangent.

35. Prove if two circles do not intersect, with one circle lying in the exterior of the other, then their common external tangent segments are congruent. (HINT: Consider as separate cases the situations in which the two circles are congruent and not congruent.)

36. Lines *l* and *m* are parallel and are each tangent to circle *P*. Prove if line *k* is also tangent to circle *P*, then the segments determined by the points at which line *k* intersects lines *l* and *m* and point *P*, intersect at right angles at *P*.

CHAPTER FOURTEEN
AREA OF POLYGONS AND CIRCLES

14.1 AREA OF A RECTANGLE, SQUARE, AND PARALLELOGRAM

When we speak of the *area of a figure* we simply mean the number of square boxes which the figure can enclose. If a figure can enclose a total of 30 square boxes, then its area is said to be 30 square *units*. If the length of a side of each square box is 1 centimeter (cm), then the area of the figure is 30 cm² (cm² is read as *square centimeters*). If the length of a side of each square box is 1 in., then the area of the figure under consideration would be expressed as 30 square in.

Figure 14.1 shows a rectangle that encloses a total of 36 square boxes. The area of this rectangle is 36 square units.

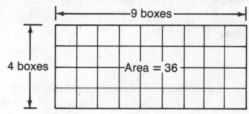

FIGURE 14.1

The value of 36 can be arrived at by multiplying the number of boxes along the length by the number of boxes along the width: Area = 9 × 4 = 36. Our goal in this chapter is to develop formulas which will enable us to calculate the areas of figures which we have previously investigated: rectangle, square, parallelogram, triangle, rhombus, trapezoid, regular polygon, and circle. We begin by summarizing some fundamental area postulates.

AREA POSTULATES

POSTULATE 14.1   For any given closed region and unit of measurement, there is a positive number which represents the area of the region.

POSTULATE 14.2   If two figures are congruent, then they have equal areas.

POSTULATE 14.3   The area A of a rectangle is equal to the product of its length (l) and width (w).

$$A = lw$$

Sometimes one side of a rectangle is referred to as the *base* and an adjacent side (which is perpendicular to the base) is referred to as the *altitude*. The formula for the area of a rectangle may be expressed as $A = bh$, where b represents the length of the base and h represents the length of the altitude. The terms *base* and *altitude* will recur in our investigations of the areas of other figures. In each instance, the altitude will always be a segment that is perpendicular to a side which is specified to be the base. Also, when the terms base and altitude are used in connection with area, we will always understand these terms to mean the *length* of the base and the *length* of the altitude.

EXAMPLE 14.1 Find the area of a rectangle if its base is 12 cm and its diagonal has a length of 13 cm.

SOLUTION Triangle *ABD* is a 5-12-13 right triangle where 12 is the base and 5 is the altitude. Hence,

$$A = bh$$
$$= 12 \times 5$$
$$A = 60 \text{ cm}^2$$

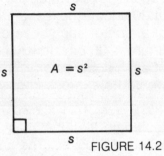

A *square* is a rectangle so that the area formula for a rectangle also applies to a square. See Figure 14.2. If the length of a side of a square is represented by s, then the area of a square is given by the relationship $A = s \times s$ or $A = s^2$.

FIGURE 14.2

EXAMPLE 14.2 A certain rectangle and square are *equivalent* (have the same area). The base of the rectangle exceeds three times its altitude by 4. If the length of a side of the square is 8, find the dimensions of the rectangle.

SOLUTION Let x = altitude of rectangle.
Then $3x + 4$ = base of rectangle.

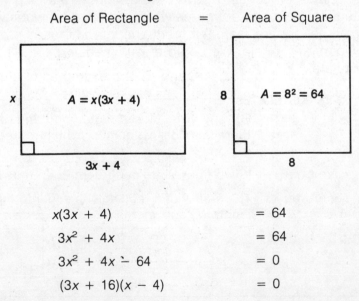

$$
\begin{aligned}
x(3x + 4) &= 64 \\
3x^2 + 4x &= 64 \\
3x^2 + 4x - 64 &= 0 \\
(3x + 16)(x - 4) &= 0
\end{aligned}
$$

$$3x + 16 = 0 \quad \textit{or} \quad x - 4 = 0$$

$$3x = -16$$

$$\frac{3x}{3} = -\frac{16}{3}$$

x = altitude = 4
$3x + 4$ = base = $3(4) + 4$
$\quad\quad\quad\quad = 16$

Reject this solution since x
(the length of a side) cannot
be negative.

It will sometimes be convenient to subdivide a region into component regions. For example, in Figure 14.3, diagonals $\overline{AC}$ and $\overline{AD}$ separate pentagon $ABCDE$ into three triangular regions such that:
Area pentagon $ABCDE$ = Area $\triangle ABC$
+ Area $\triangle ACD$ + Area $\triangle ADE$. The
generalization of this notion is presented as
Postulate 14.4.

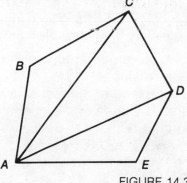

FIGURE 14.3

AREA ADDITION POSTULATE (14.4)

The area of a closed region is equal to the sum of the areas of any nonoverlapping division of that region.

A rectangle is a special type of parallelogram. It therefore seems reasonable that there may be some relationship between the area formulas for these figures. We will look to establish that a parallelogram has the same area as a rectangle having the same base and altitude as the parallelogram. We begin our analysis by relating the concept of base and altitude to a parallelogram.

Any side of a parallelogram may be identified as the *base* of the parallelogram. An *altitude* of a parallelogram is a segment drawn perpendicular to the base from any point on the side opposite the base.

In Figure 14.4, $\overline{SJ}$, $\overline{AK}$, and $\overline{BL}$ are examples of altitudes, each drawn to base $\overline{RW}$. Since parallel lines are everywhere equidistant, altitudes drawn to a given base of a parallelogram are equal. That is, $SJ = AK = BL = \cdots$

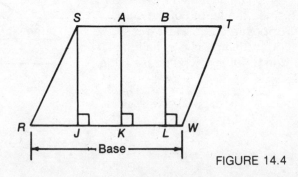

FIGURE 14.4

Next consider □ABCD with altitude $\overline{BH}$ drawn to base $\overline{AD}$ (Figure 14.5). If right triangle AHB were cut off the figure and then slid over to the right so that $\overline{AB}$ and $\overline{CD}$ were made to coincide, then the resulting figure would be a rectangle. This is illustrated in Figure 14.6, where right angle DKC corresponds to right angle AHB.

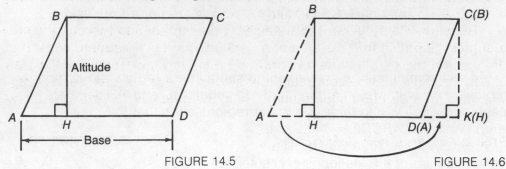

FIGURE 14.5   FIGURE 14.6

The area of rectangle BHKC (see Figure 14.6) is given by the relationship $A = HK \times BH$, where $\overline{BH}$ is the altitude and $\overline{HK}$ is the base of the rectangle. Since we have not thrown away or created any additional area, the area of the orignal parallelogram equals the area of the newly formed rectangle. Hence, the area of □ABCD also equals $HK \times BH$. Since $\triangle AHB \cong \triangle DKC$, $\overline{AH} \cong \overline{DK}$. This implies that $AD = HK$. Substituting AD for HK in the area relationship we obtain the following:

$$\text{Area of } □ABCD = HK \cdot BH$$

$$= AD \cdot BH$$

Note that $\overline{AD}$ is the base of the parallelogram and $\overline{BH}$ is an altitude drawn to that base. This relationship is stated formally in Theorem 14.1

THEOREM 14.1   AREA OF A PARALLELOGRAM

The area of a parallelogram is equal to the product of the base and the altitude drawn to that base.

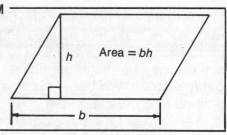

EXAMPLE 14.3   A pair of adjacent sides of a parallelogram are 6 and 10 centimeters. If the measure of their included angle is 30, find the area of the parallelogram.

SOLUTION   In □ABCD, altitude BH = 3 since the length of the side opposite a 30° angle in a 30-60 right triangle is one-half the length of the hypotenuse (side $\overline{AB}$).

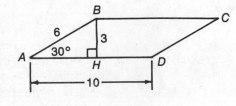

$$\text{Area } □ABCD = bh$$

$$= AD \cdot BH$$

$$= 10 \cdot 3$$

$$= 30 \text{ cm}^2$$

14.2 AREA OF A TRIANGLE AND TRAPEZOID

The formula to find the area of a triangle can be deduced easily from the area of a parallelogram relationship. Once we know how to find the area of a triangle, we may derive a formula to find the area of a trapezoid.

Recall that a diagonal separates a parallelogram into two congruent triangles. Suppose that the area of a parallelogram is 60 square units. If a diagonal of the parallelogram is drawn (Figure 14.7), what will be the area of each triangle? Since congruent triangles have equal areas, each triangle will have an area of 30 square units. Thus, the area of each triangle is one-half the area of the parallelogram.

Area $\triangle ABD = \frac{1}{2}$ area $\square ABCD$

$\qquad = \frac{1}{2}$ base $\times$ height

$\qquad = \frac{1}{2} AD \times BH$

This analysis suggests Theorem 14.2

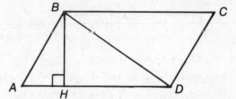

FIGURE 14.7

THEOREM 14.2 AREA OF A TRIANGLE

The area of a triangle is equal to one-half the product of the base and the altitude drawn to that base.

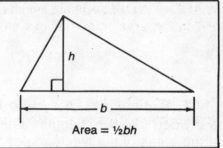

Area = ½bh

EXAMPLE 14.4 Find the area of each of the following triangles.

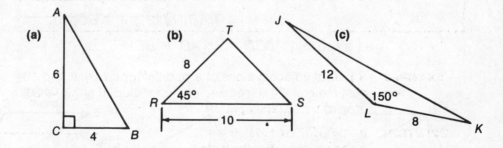

SOLUTION **a** In a right triangle either leg may be considered the base while the other leg then becomes the altitude. The area of a *right* triangle is therefore equal to one-half the product of the lengths of the legs:

$$A = \tfrac{1}{2}(\text{leg}_1)(\text{leg}_2)$$

$$= \tfrac{1}{2}(6)(4)$$

$$= 12$$

b From vertex T, drop an altitude to base $\overline{RS}$. Since $\overline{TH}$ is the side opposite a 45° angle in a 45-45 right triangle, the length of $\overline{TH}$ is equal to one-half the length of the hypotenuse ($\overline{RT}$) multiplied by $\sqrt{2}$: $TH = \frac{1}{2}(8)\sqrt{2} = 4\sqrt{2}$.

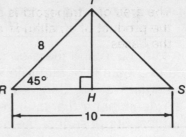

$$A = \frac{1}{2}(RS)(TH)$$
$$= \frac{1}{2}(10)(4\sqrt{2})$$
$$= 5(4\sqrt{2})$$
$$= 20\sqrt{2}$$

c From vertex J, drop an altitude to base $\overline{LK}$, extended.

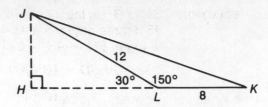

Angle JLH has measure 30 so that the length of altitude $\overline{JH}$ (the side opposite the 30° angle in a 30-60 right triangle) is equal to one-half the length of the hypotenuse ($\overline{JL}$). Hence, $JH = \frac{1}{2}(12) = 6$.

$$A = \frac{1}{2}(LK)(JH)$$
$$= \frac{1}{2}(8)(6)$$
$$= 24 \text{ square units}$$

To find the area of a *trapezoid*, draw a diagonal so that the trapezoid is divided into two triangles. The area of the trapezoid is given by the sum of the areas of the two triangles. See Figure 14.8. Notice that the lengths of perpendicular segments $\overline{BX}$ and $\overline{DY}$ both represent the distance between bases $\overline{AD}$ and $\overline{BC}$. Since parallel lines are everywhere equidistant, $DY = BX$. For convenience, let us refer to the length of the altitude by the letter h so that $h = DY = BX$. We may then write

$$\text{Area trapezoid } ABCD = \text{area } \triangle ABD \quad + \quad \text{area } \triangle BCD$$
$$= \frac{1}{2}(AD)(h) \qquad + \frac{1}{2}(BC)(h)$$

$$\text{Area trapezoid } ABCD = \frac{1}{2}h(AD + BC)$$

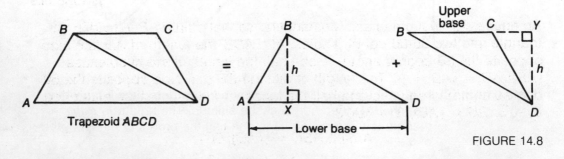

FIGURE 14.8

In this relationship, the terms inside the parentheses represent the lengths of the bases of the trapezoid and the letter h represents the length of an altitude of the trapezoid. This result is stated formally in Theorem 14.3.

THEOREM 14.3   AREA OF A TRAPEZOID

The area of a trapezoid is equal to one-half the product of an altitude and the sum of the bases.

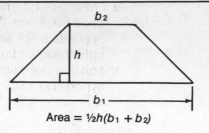

Area $= \frac{1}{2}h(b_1 + b_2)$

EXAMPLE 14.5   Find the area of an isosceles trapezoid whose bases are 8 and 20 and whose lower base angle has a measure of 45:

SOLUTION   Since $\overline{BH}$ is the side opposite the 45 degree angle in a 45-45 right triangle, $BH = AH = 6$.

$$\text{Area } ABCD = \frac{1}{2}BH(AD + BC)$$
$$= \frac{1}{2}(6)(20 + 8)$$
$$= 3(28)$$
$$= 84 \text{ square units}$$

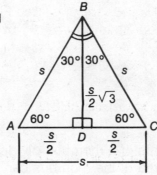

The area of a triangle formula may also be used to derive convenient formulas for finding the areas of equilateral triangles and rhombuses. The formula for the area of an equilateral triangle may be expressed exclusively in terms of the length of a side s of the triangle by representing the length of the altitude in terms of s and then applying the area of a triangle relationship.

Let's illustrate by considering equilateral triangle ABC and drawing altitude $\overline{BD}$ to side $\overline{AC}$ (Figure 14.9).

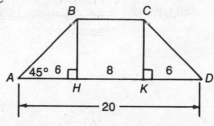

FIGURE 14.9

An equilateral triangle is also equiangular so that altitude $\overline{BD}$ divides the triangle into two 30-60 right triangles. In $\triangle ADB$, the length of $\overline{AD}$ (the side opposite the 30 degree angle) is one-half the length of the hypotenuse ($\overline{AB}$): $AD = \frac{1}{2}AB = \frac{1}{2}s$. The length of altitude $\overline{BD}$ (the side opposite the 60 degree angle) is equal to one-half the length of the hypotenuse multiplied by $\sqrt{3}$: $BD = \frac{1}{2}AB\sqrt{3} = \frac{1}{2}s\sqrt{3}$.

$$\text{Area } \triangle ABC = \frac{1}{2}(AC)(BD)$$
$$= \frac{1}{2}(s)(\frac{1}{2}s\sqrt{3})$$
$$= \frac{1}{4}s^2\sqrt{3}$$

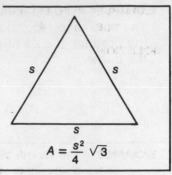

THEOREM 14.4 AREA OF AN EQUILATERAL TRIANGLE

The area of an equilateral triangle is equal to one-fourth of the square of the length of a side multiplied by $\sqrt{3}$.

$$A = \frac{s^2}{4}\sqrt{3}$$

EXAMPLE 14.6 Find the area of an equilateral triangle whose perimeter is 24.

SOLUTION If the perimeter of the triangle is 24, then each side is 8.

$$A = \frac{s^2}{4}\sqrt{3}$$

$$= \frac{8^2}{4}\sqrt{3}$$

$$= \frac{64}{4}\sqrt{3}$$

$$= 16\sqrt{3} \text{ square units}$$

EXAMPLE 14.7 Find the length of a side of an equilateral triangle that has an area of $25\sqrt{3}$ square centimeters.

SOLUTION

$$A = \frac{s^2}{4}\sqrt{3}$$

$$25\sqrt{3} = \frac{s^2}{4}\sqrt{3}$$

$$25 = \frac{s^2}{4}$$

$$s^2 = 100$$

$$s = \sqrt{100} = 10 \text{ cm}$$

The diagonals of a rhombus divide the rhombus into four congruent triangles. The area of the rhombus can be obtained by determining the area of one of these triangles and then multiplying it by four. As a matter of convenience, we will refer to the lengths of the two diagonals of a rhombus as d_1 and d_2. It can be shown that the area of any one of the four triangles formed by the diagonals is $\frac{1}{8}(d_1)(d_2)$. To find the area of the rhombus, multiply the expression $\frac{1}{8}d_1 \cdot d_2$ by 4, which leads to the expression $\frac{1}{2}d_1 \cdot d_2$.

THEOREM 14.5 AREA OF A RHOMBUS

The area of a rhombus is equal to one-half the product of the lengths of the diagonals.

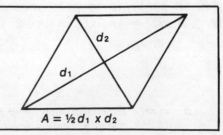

$$A = \tfrac{1}{2}d_1 \times d_2$$

EXAMPLE 14.8 Find the area of a rhombus if the lengths of its diagonals are 10 and 14.

SOLUTION
$$A = \tfrac{1}{2}d_1 \cdot d_2$$
$$= \tfrac{1}{2}(10)(14)$$
$$= \tfrac{1}{2}(140)$$
$$A = 70 \text{ square units}$$

EXAMPLE 14.9 The length of one diagonal of a rhombus is three times the length of the other diagonal. If the area of the rhombus is 54 square units, find the length of each diagonal.

SOLUTION Let x = length of the shorter diagonal.
Then $3x$ = length of the longer diagonal.

$$A - \tfrac{1}{2}d_1 \cdot d_2$$
$$54 = \tfrac{1}{2}(3x)(x)$$
$$108 = 3x^2$$
$$x^2 = \tfrac{108}{3} = 36$$
$$x = \sqrt{36}$$
$$x = \text{length of shorter diagonal} = 6$$
$$3x = \text{length of longer diagonal} = 18$$

14.3 COMPARING AREAS

The following theorem is easy to prove and will prove helpful in establishing that triangles have equal areas.

> **THEOREM 14.6 EQUIVALENT TRIANGLES**
> If two triangles have equal bases and equal altitudes, then their areas are equal.

The following demonstration problem will illustrate the usefulness of this theorem.

GIVEN $\overline{AD}$ and $\overline{BC}$ are bases of trapezoid $ABCD$.

PROVE **a** Area $\triangle AED$ = area $\triangle AFD$.
 b Area $\triangle AXE$ = area $\triangle DXF$.

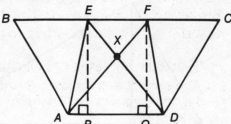

a PLAN Both triangles have side $\overline{AD}$ as a base and equal altitudes since parallel lines are everywhere equidistant.

PROOF	Statements	Reasons
	1. $\overline{AD}$ and $\overline{BC}$ are bases of trapezoid $ABCD$.	1. Given.
	2. $AD = AD$. (Base)	2. Reflexive property of equality.
	3. $\overline{AD} \parallel \overline{BC}$.	3. The bases of a trapezoid are parallel.
	4. From points E and F draw segments perpendicular to $\overline{AD}$, intersecting $\overline{AD}$ at points P and Q, respectively.	4. From a given point not on a line, exactly one segment may be drawn perpendicular to the line.
	5. $EP = FQ$. (Altitude)	5. Parallel lines are everywhere equidistant.
	6. Area $\triangle AED =$ area $\triangle AFD$.	6. If two triangles have equal bases and equal altitudes, then their areas are equal.

b **PLAN** Subtract the area of $\triangle AXD$ from the area of triangle AED and from the area of triangle AFD. This leads to the desired result.

	7. Area $\triangle AXD =$ area $\triangle AXD$.	7. Reflexive property of equality.
	8. Area $\triangle AED -$ area $\triangle AXD$ $=$ area $\triangle AFD -$ area $\triangle AXD$.	8. Subtraction property.
	9. Area $\triangle AXE =$ area $\triangle DXF$.	9. Substitution.

COMPARING AREAS OF SIMILAR POLYGONS

Recall that if two triangles are similar, then the ratio of the lengths of any pair of corresponding sides is the same as the ratio of the lengths of any pair of corresponding altitudes (or corresponding medians and angle bisectors). Suppose $\triangle ABC \sim \triangle RST$ and that their ratio of similitude is $3:1$. For example, an interesting relationship arises when we take the ratio of the areas of the two similar triangles in Figure 14.10:

$$\frac{\text{Area } \triangle ABC}{\text{Area } \triangle RST} = \frac{\frac{1}{2}(AC)(BX)}{\frac{1}{2}(RT)(SY)} = \frac{\frac{1}{2}(15)(3)}{\frac{1}{2}(5)(1)} = \frac{9}{1}$$

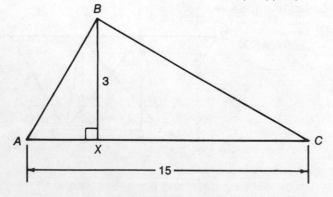
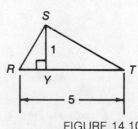

FIGURE 14.10

How does the ratio of the areas of the triangles compare with the ratio of the lengths of a pair of corresponding sides? Since the ratio of similitude is 3:1,

$$\frac{\text{Area } \triangle ABC}{\text{Area } \triangle RST} = \frac{9}{1} = \left(\frac{3}{1}\right)^2$$

$$= \left(\frac{\text{length of a side in } \triangle ABC}{\text{length of the corresponding side in } \triangle RST}\right)^2$$

THEOREM 14.7 COMPARING AREAS OF SIMILAR TRIANGLES

If two triangles are similar, then the ratio of their areas is equal to the square of the ratio of the lengths of any pair of corresponding sides:

$$\frac{\text{Area } \triangle I}{\text{Area } \triangle II} = \left(\frac{\text{side}_I}{\text{side}_{II}}\right)^2$$

where $\triangle I$ and $\triangle II$ refer to a pair of similar triangles and side_I and side_{II} represent the lengths of a pair of corresponding sides in triangles I and II, respectively.

REMARKS

1. Theorem 14.7 may also be generalized so that it applies to any pair of similar *polygons*.

2. Since the lengths of corresponding sides, altitudes, medians, and angle bisectors are in proportion, the ratio of the areas of a pair of similar triangles is equal to the square of the ratio of any pair of these corresponding segments.

3. The perimeters of similar polygons have the same ratio as the lengths of any pair of corresponding sides. The ratio of the areas of similar polygons is therefore equal to the square of the ratio of their perimeters.

EXAMPLE 14.10   The ratio of similitude of two similar triangles is 2:3. If the area of the smaller triangle is 12, find the area of the larger triangle.

SOLUTION

$$\frac{\text{Area of smaller } \triangle}{\text{Area of larger } \triangle} = \left(\frac{2}{3}\right)^2$$

Let x = area of the larger triangle.

$$\frac{12}{x} = \frac{4}{9}$$

$$4x = 108$$

$$x = \frac{108}{4}$$

$$\boxed{x = 27}$$

EXAMPLE 14.11   The areas of two similar polygons are 25 and 81. If the length of a side of the larger polygon is 72, find the length of the corresponding side of the smaller polygon.

SOLUTION Let $x =$ the length of the corresponding side of the smaller polygon.

$$\frac{\text{Area of smaller polygon}}{\text{Area of larger polygon}} = \left(\frac{x}{72}\right)^2$$

$$\frac{25}{81} = \left(\frac{x}{72}\right)^2$$

Solve for x by first taking the square root of both sides of the proportion:

$$\sqrt{\frac{25}{81}} = \frac{x}{72}$$

$$\frac{5}{9} = \frac{x}{72}$$

$$9x = 360$$

$$x = \frac{360}{9}$$

$$\boxed{x = 40}$$

14.4 AREA OF A REGULAR POLYGON

We have intentionally restricted our attention to finding the areas of familiar three-sided and four-sided polygons. As a general rule, it is difficult to develop convenient formulas for other types of polygons. There is, however, one notable exception. A formula for the area of a *regular* polygon can be derived by subdividing the regular polygon into a set of congruent triangles and then summing the areas of these triangles. Throughout our analysis we shall assume that *circles having the same center can be inscribed and circumscribed about a regular polygon.* Figures 14.11 and 14.12 identify some important terms relating to regular polygons and their circumscribed and inscribed circles. This information will be needed when we eventually turn our attention to finding the area of a regular polygon.

In Figure 14.11:

- Point O is the center of the circle inscribed and the circle circumscribed about regular pentagon $ABCDE$. Point O is referred to as the *center* of a regular polygon.

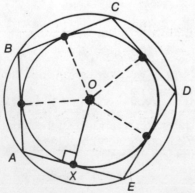

FIGURE 14.11

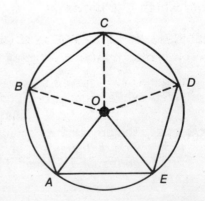

FIGURE 14.12

- $\overline{OX}$ is a segment drawn from the center of the regular pentagon to the point at which the inscribed circle is tangent to side $\overline{AE}$. $\overline{OX}$ is called an *apothem*. Apothems may be drawn to each of the other sides.

In Figure 14.12:

- $\overline{OA}$ and $\overline{OE}$ are examples of *radii* of regular pentagon *ABCDE*. A *radius* of a regular polygon is a segment drawn from the center to any vertex of the polygon.

- Angle *AOE* is an example of a *central angle* of regular pentagon *ABCDE*. A *central angle* of a regular polygon is formed by drawing two radii to consecutive vertices of the polygon.

DEFINITIONS RELATING TO REGULAR POLYGONS

- The *center* of a regular polygon is the common center of its inscribed and circumscribed circles.

- An *apothem* of a regular polygon is a segment whose end points are the center of the polygon and a point at which the inscribed circle is tangent to a side. An apothem of a regular polygon is also a radius of the *inscribed* circle.

- A *radius* of a regular polygon is a segment whose end points are the center of the polygon and a vertex of the polygon. A radius of a regular polygon is also a radius of the *circumscribed* circle.

- A *central angle* of a regular polygon is an angle whose vertex is the center of the polygon and whose sides are radii drawn to the end points of the same side of the polygon. An *n*-sided regular polygon will have *n* central angles.

In Figure 14.12, the sides of the regular pentagon divide the circle into five congruent arcs. It follows that each central angle must have the same measure. Since there are five central angles, the measure of each central angle may be found by dividing 360 by 5, obtaining a measure of 72 for each of the central angles. In general, the radii of a regular polygon divide the polygon into triangles which may be proven congruent to one another by SAS, where the congruent sides are the radii and the congruent included angles are the central angles. Using the CPCTC principle, we may further conclude that each radius bisects the angle located at the vertex to which it is drawn.

THEOREM 14.8   ANGLES OF A REGULAR POLYGON

- The central angles of a regular polygon are congruent.

- The measure of a central angle of a regular polygon is equal to 360 divided by the number of sides of the polygon.

- The radii of a regular polygon bisect the interior angles of the regular polygon.

EXAMPLE
14.12
The length of a side of a regular hexagon is 14. Find each of the following:

a The measure of a central angle.

b The length of a radius of the hexagon.

c The length of an apothem.

SOLUTION

a $m \angle AOB = \dfrac{360}{n}$

$= \dfrac{360}{6}$

$m \angle AOB = 60$

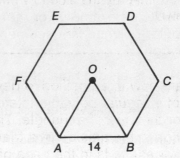

b Since the central angle has measure 60, and $\overline{OA} \cong \overline{OB}$, the measure of angles OAB and OBA must be congruent and therefore have measure 60. $\triangle AOB$ is equiangular which means it is also equilateral. Hence, the radius of the hexagon must have the same length as $\overline{AB}$ and is therefore *14*.

c Let us focus on triangle AOB and draw the apothem to side $\overline{AB}$.

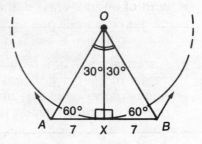

Since an apothem is drawn to a point at which the inscribed circle is tangent to a side of the polygon, the apothem must be perpendicular to the side to which it is drawn. Furthermore, it bisects the side to which it is drawn. As an illustration observe that apothem $\overline{OX}$ divides triangle AOB into two congruent triangles. $\triangle AOX \cong \triangle BOX$ by Hy-leg since

$$\overline{OA} \cong \overline{OB} \quad \text{(Hy)}$$
$$\overline{OX} \cong \overline{OX} \quad \text{(Leg)}$$

By CPCTC, $\overline{AX} \cong \overline{BX}$. In addition, $\angle AOX \cong \angle BOX$.

Returning to the problem at hand, the apothem $\overline{OX}$ is the side opposite the 60° angle in a 30-60 right triangle (that is, in $\triangle AOX$):

$$OX = \tfrac{1}{2} OA \sqrt{3}$$
$$= \tfrac{1}{2}(14)\sqrt{3}$$
$$OX = 7\sqrt{3}$$

Theorem 14.9 summarizes the properties of an apothem.

THEOREM 14.9 PROPERTIES OF AN APOTHEM

- An apothem of a regular polygon is the perpendicular bisector of the side to which it is drawn.

- An apothem bisects the central angle determined by the side to which it is drawn.

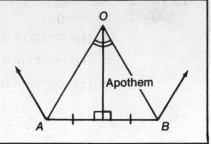

We are now in a position to develop a convenient formula for finding the area of a *regular* polygon. Interestingly, the formula is analogous to the formula for the area of a triangle. The apothem of a regular polygon corresponds to the altitude of a triangle and the perimeter of a regular polygon corresponds to the base of a triangle, so that the area of a regular polygon may be found by multiplying one-half the length of the apothem by the perimeter of the regular polygon.

THEOREM 14.10 THE AREA OF A REGULAR POLYGON

The area (A) of a regular polygon is equal to one-half the product of the length of an apothem (a) and its perimeter (p): $A = \frac{1}{2}ap$.

OUTLINE OF PROOF

GIVEN ABC . . . is a regular polygon having n sides. Let:

a = length of an apothem,
s = length of each side,
p = perimeter of ABC . . . .

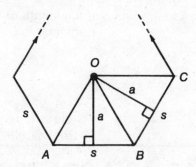

PROVE Area ABC . . . $= \frac{1}{2}ap$.

- Draw the radii of the polygon.

- Find the sum of the areas of each triangle thus formed:

$$\text{Area } ABC \ldots = \text{Area } \triangle AOB + \text{area } \triangle BOC + \cdots$$

$$= \tfrac{1}{2}as + \tfrac{1}{2}as + \cdots + \tfrac{1}{2}as$$

$$= \tfrac{1}{2}a(s + s + \cdots + s)$$

$$= \tfrac{1}{2}ap$$

EXAMPLE 14.13 The length of a side of a regular pentagon is 20. Find each of the following:

a The length of the apothem correct to the nearest hundredth.

b The area correct to the nearest whole number.

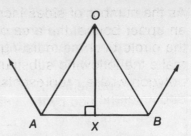

SOLUTION First draw a representative triangle of the polygon.

a Since an apothem is the perpendicular bisector of the side to which it is drawn, △ OAX is a right triangle and $AX = \frac{1}{2}(20) =$ 10. The measure of central angle $AOB = \frac{360}{5} = 72$. Apothem $\overline{AX}$ bisects central angle AOB so that $m \angle AOX = \frac{1}{2}(72) = 36$. We must find OX by using an appropriate trigonometric ratio. The arithmetic is simplest if we find the tangent of angle OAX. Since $m \angle OAX = 90 - 36 = 54$, we may write the following:

$$\tan 54° = \frac{\text{side opposite } \angle}{\text{side adjacent } \angle}$$

$$= \frac{OX}{AX}$$

$$1.3764 = \frac{OX}{10}$$

$$OX = 10(1.3764)$$

$$OX = 13.76 \quad \text{correct to the nearest hundredth}$$

b If the length of a side of a regular pentagon is 20, then its perimeter is 5 times 20 or 100.

$$A = \frac{1}{2}ap$$

$$= \frac{1}{2}(13.76)(100)$$

$$= \frac{1}{2}(1376)$$

$$= 688 \text{ square units}$$

14.5 AREA OF A CIRCLE, SECTOR, AND SEGMENT

Consider what happens if the number of sides, *n*, of an inscribed regular polygon gets larger and larger, as shown in Figure 14.13.

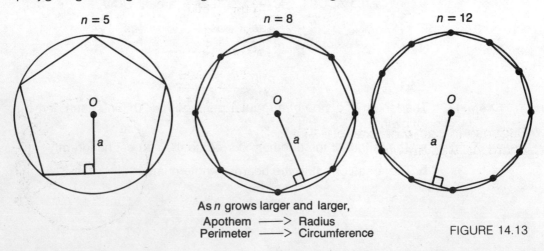

$n = 5$ $n = 8$ $n = 12$

As *n* grows larger and larger,
Apothem ———> Radius
Perimeter ———> Circumference

FIGURE 14.13

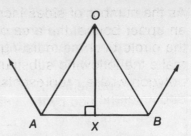

As the number of sides increases indefinitely without being constrained by an upper bound, the area of the inscribed regular polygon and the area of the circle become indistinguishable. Under these circumstances, we may make the following substitutions in the formula for the area of a regular polygon where r represents the radius of the circle and C its circumference:

As the number of sides becomes infinitely large

Area of regular polygon $= \frac{1}{2}a \quad p$

Area of circumscribed circle $= \frac{1}{2}r \quad C$

or

$= \frac{1}{2}(r)(2\pi r)$

Area of a circle $= \pi r^2$

THEOREM 14.11 AREA OF A CIRCLE

The area A of a circle is equal to the product of the constant π and the square of the length of the radius r of the circle: $A = \pi r^2$.

**EXAMPLE
14.14**

a Find the area of circle O.

b Find the area of the shaded region.

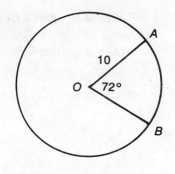

SOLUTION

a $A = \pi r^2$

$= \pi(10)^2$

$= 100\pi$ square units

b The shaded region is called a *sector* of the circle. In a manner analogous to determining the length of an arc of a circle, we form the following proportion:

$$\frac{\text{Area of sector}}{\text{Area of circle}} = \frac{\text{degree measure of sector arc}}{360°}$$

Let $x =$ the area of the sector.

$$\frac{x}{100\pi} = \frac{72°}{360°}$$

$$\frac{x}{100\pi} = \frac{1}{5}$$

$$5x = 100\pi$$

$$\boxed{x = 20\pi}$$

Suppose in part **b** of Example 14.14, chord $\overline{AB}$ was drawn.

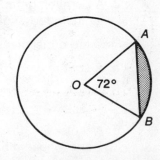

If it was known that the area of triangle *AOB* was 50, for the sake of this example, what would be the area of the region bounded by chord $\overline{AB}$ and the minor arc that it cuts off? The region described is called a *segment* of the circle. To find the area of this segment, all we need do is to subtract the area of triangle *AOB* from the area of sector. *AOB*.

$$\text{Area of segment } \overline{AB} = \text{area sector } AOB - \text{area } \triangle AOB$$

$$= 20\pi - 50$$

Unless otherwise directed, we generally leave the answer in this form.

Let us pause to summarize the concepts introduced in this last example.

DEFINITIONS OF SECTOR AND SEGMENT OF A CIRCLE

- A *sector of a circle* is a region of a circle bounded by two radii and the minor arc they determine.

- A *segment of a circle* is a region of a circle bounded by a chord and the minor arc that it cuts off.

THEOREM 14.12   AREA OF A SECTOR

The area of a sector of a circle is determined by the following proportion:

$$\frac{\text{Area of sector}}{\text{Area of circle}}$$

$$= \frac{\text{degree measure of sector arc}}{360°}$$

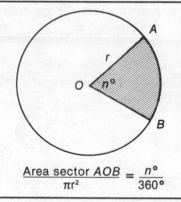

$$\frac{\text{Area sector } AOB}{\pi r^2} = \frac{n°}{360°}$$

REVIEW EXERCISES FOR CHAPTER 14

1. Find the area of each of the following:

   a A rectangle whose base is 6 and whose diagonal is 10.

   b A square whose diagonal is 8.

   c A parallelogram having two adjacent sides of 12 and 15 centimeters and an included angle of measure 60.

2. Find the dimensions of each of the following:

   a A square whose area is 144 square centimeters.

   b A rectangle whose area is 75 and whose base and altitude are in the ratio of 3:1.

   c A rectangle having an area of 135 and whose base is represented by $x + 2$ and whose altitude is represented as $2x + 1$.

For Exercises 3 to 11, find the area of each of the figures. Whenever appropriate, answers may be left in radical form.

3.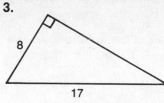

4.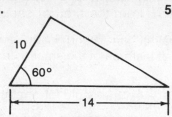

5.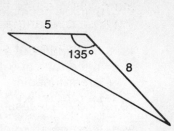

6.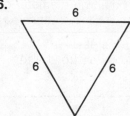

7.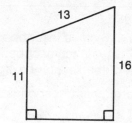

8.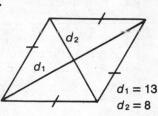

9.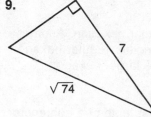

10.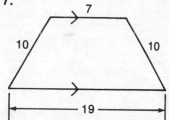

11.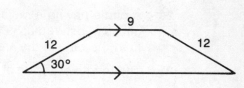

12. The area of an equilateral triangle is given. Find the length of a side of the equilateral triangle.

   a $16\sqrt{3}$ **b** $11\sqrt{3}$ **c** 4

13. Find the length of the shorter diagonal of a rhombus if:

   a The length of the longer diagonal is 15 and the area is 90.

   b The lengths of the diagonals are in the ratio of 2:3 and the area of the rhombus is 147.

14. Find the length of an altitude of a trapezoid if:

   a Its area is 72 and the sum of the lengths of the bases is 36.

   b Its area is 80 and its median is 16.

   c The sum of the lengths of the bases is numerically equal to one-third of the area of the trapezoid.

15. Find the area of a rhombus if its perimeter is 68 and the length of one of its diagonals is 16.

16. Find the area of a triangle if the length of a pair of adjacent sides are 6 and 14 and the measure of the included angle is:

   a 90   **b** 30   **c** 120 (Leave answer in radical form.)

17. $\triangle JKL \sim \triangle RST$. If $JL = 20$ and $RT = 15$, find the ratio of their areas.

18. The areas of two similar polygons are 81 and 121.

  a Find their ratio of similitude.

  b If the perimeter of the smaller polygon is 45, find the perimeter of the larger polygon.

19. The lengths of a pair of corresponding sides of a pair of similar triangles are in the ratio of 5:8. If the area of the smaller triangle is 75, find the area of the larger triangle.

20. Find the measure of a central angle for a regular octagon.

21. If the measure of an interior angle of a regular polygon is 150, find the measure of a central angle.

22. Find the area of a regular hexagon inscribed in a circle having a diameter of 20 cm.

23. Fill in the following table for a regular polygon having four sides:

	SIDE	RADIUS	APOTHEM	AREA
a	6 cm	?	?	?
b	?	?	?	49 cm²
c	?	?	5 cm	?
d	?	8 cm	?	?

24. A circle having a radius of 6 cm is inscribed in a regular hexagon. Another circle whose radius is 6 cm is circumscribed about another regular hexagon. Find the ratio of the area of the smaller hexagon to the larger hexagon.

25. The length of a side of a regular decagon is 20 cm.

  a Find the length of the apothem correct to the nearest tenth of a centimeter.

  b Using the answer obtained in part (a), find the area of the decagon.

Unless otherwise indicated, answers may be left in terms of π.

26. Fill in the following table:

	RADIUS	DIAMETER	CIRCUM-FERENCE	AREA
a	5	?	?	
b	?	9	?	?
c	?	?	?	49π
d	?	?	18π	?
e	?	?	?	$\dfrac{64}{25}\pi$

27. The ratio of the areas of two circles is 1:9. If the radius of the smaller circle is 5 cm, find the length of the radius of the larger circle.

28. The ratio of the lengths of the radii of two circles is 4:25. If the area of the smaller circle is 8 square units, find the area of the larger circle.

29. The diameters of two concentric circles are 8 and 12. Find the area of the ring-shaped region (called an annulus) bounded by the two circles.

30. Fill in the following table:

	RADIUS OF CIRCLE	AREA OF CIRCLE	DEGREE MEASURE OF SECTOR ARC	AREA OF SECTOR
a	12	?	45°	?
b	?	36π	?	12π
c	?	?	72°	20π
d	9	?	160°	?

31. In circle O, radii OA and OB are drawn such that OA is equal to the length of chord $\overline{AB}$. If $OA = 12$, find the area of:

   a Sector AOB.

   b Segment AB.

32. A square whose side is 8 cm in length is inscribed in a circle. Find the area of the segment formed by a side of the square and its intercepted arc.

33. What is the ratio of the area of the inscribed and circumscribed circles of a square having a side of 6 cm?

For each of the following figures find the area of the shaded region.

34.

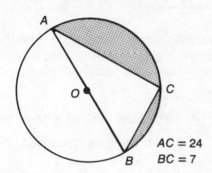

$AC = 24$
$BC = 7$

35.

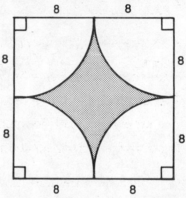

36. A regular pentagon is inscribed in a circle having a radius of 20 cm. Find the area of the sector formed by drawing radii to a pair of consecutive vertices of the pentagon.

37. An isosceles right triangle whose hypotenuse has a length of 16 is inscribed in a circle. Find the area of the segment formed by a side of the triangle and its intercepted arc.

38. A regular hexagon is inscribed in a circle. The length of the apothem is $4\sqrt{3}$.

   a Find the area of the circle.

   b Find the area of the hexagon.

   c Find the area of the segment cut off by a side of the hexagon.

39.   **GIVEN** G is the midpoint of $\overline{CV}$.

      PROVE   **a** Area $\triangle CLG$ = area $\triangle VLG$.

               b Area $\triangle BLC$ = area $\triangle BLV$.

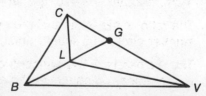

40. **GIVEN** $\square ABCD$, $\overline{BE} \cong \overline{FD}$.

PROVE Area $\triangle FAC$ = area $\square ABCD$.

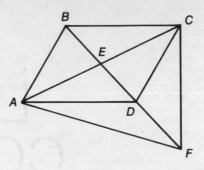

41. Prove using algebraic methods that for an equilateral triangle:

  a The length of the radius of the inscribed circle is one-third the length of the altitude of the triangle.

  b The areas of the inscribed and circumscribed circles are in the ratio of 1 : 4.

CHAPTER FIFTEEN
LOCUS AND COORDINATE GEOMETRY

15.1 SIMPLE LOCUS: DESCRIBING POINTS WHICH SATISFY A SINGLE CONDITION

Draw a point on a piece of paper and label it point K. Take a ruler and locate all points which are 3 in. from point K. How many such points can you find? How would you concisely describe the set of *all* points which satisfy the condition that the point be 3 in. from point K?

The set of all points which are 3 in. from point K forms a *circle* having point K as its center and a radius of 3 in. The circle drawn is said to represent the *locus* of all points 3 in. from point K.

DEFINITION OF LOCUS

A *locus* (plural: *loci*) is the set of all points, and only those points, which satisfy one or more stated conditions.

Think of a locus as a path consisting of one or more points such that each point along the path conforms to the given conditions. In our example, the given condition was that the points had to be 3 in. from point K. The circular path having point K as a center and a radius of 3 in. is the locus.

Figure 15.1 uses our example to summarize the steps to be followed in determining a locus. (Note that the representative points in Figure 15.1 have been labeled P_1, P_2, and P_3. It is a common practice in mathematics to use *subscripts* to name a related set of points. The subscript is the number written one-half line below the letter. The symbol P_1, for example, is read "P one" or "P sub-one." The symbol P_2 names the point "P two" and so on.)

STEP 1 Draw a diagram which includes all the given information (point K) and several representative points which satisfy the given condition (that the point be 3 in. from point K).

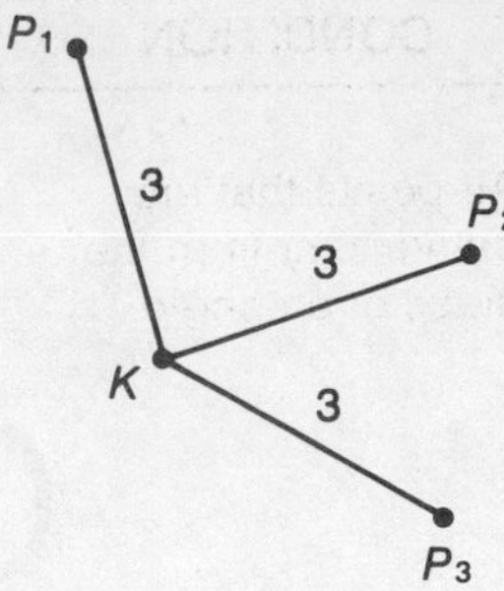

STEP 2 Keep drawing points until you discover a pattern. Connect the points using a broken curve or line.

STEP 3 Describe the locus in a complete sentence.

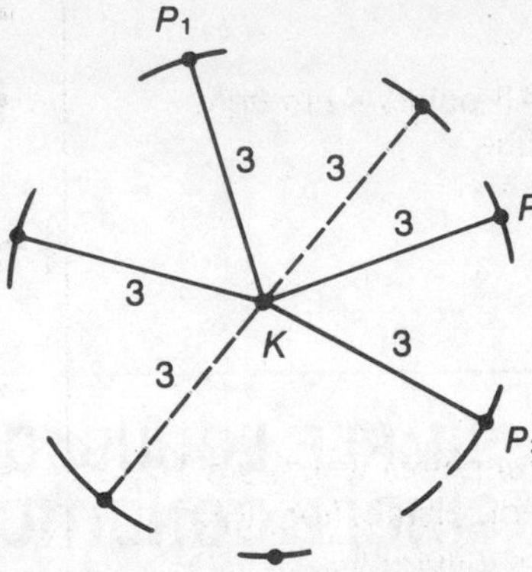

The locus is a circle having point K as its center and a radius of 3 in.

FIGURE 15.1

Some additional examples of determining a locus are provided in Table 15.1.

SUMMARY

- The locus of all points at a given distance from a point is a circle having the point as its center and the given distance as the length of its radius.

- The locus of all points equidistant from the sides of an angle is the ray that bisects the angle.

- The locus of all points at a given distance from a line is two lines parallel to the original line, on opposite sides of the original line, and each at the given distance from the line.

- The locus of all points equidistant from two parallel lines is a line parallel to the two lines and halfway between them.

- The locus of all points equidistant from the end points of a line segment is the perpendicular bisector of the segment.

TABLE 15.1

CONDITION	DIAGRAM	LOCUS
1. All points that are equidistant from the sides of an angle.		The ray that bisects the angle.
2. All points 4 cm from line *l*.	4 4 4 *l* 4 4 4	Two lines parallel to line *l*, on opposite sides of *l*, and a distance of 4 cm from line *l*.
3. All points which are equidistant from two parallel lines.	*d* *d* *d* *d* *d*	A line parallel to the two lines and halfway between them.
4. All points equidistant from the end points of a line segment.	*A* *B* *Z* — *M* — *Y* *C*	The perpendicular bisector of the segment. (*ZA* = *YA*, for example, since △*ZAM* ≅ △*YAM* by SAS.)

15.2 COMPOUND LOCI: DESCRIBING POINTS WHICH SATISFY MORE THAN ONE CONDITION

Tree *A* stands 5 meters away from tree *B*. A map indicates that treasure is buried 2 meters from tree *A* (condition 1) *and* 4 meters from tree *B* (condition 2). Where would you dig for the treasure? We can narrow down the possibilities by determining the points at which the loci of the individual conditions intersect. Since the sum of the lengths of the radii of the two circles (6 meters) is greater than the distance between the trees (5 meters), the circles which define the locus for each condition will intersect at two different points. This is illustrated in Figure 15.2. The treasure is buried at either point *X* or point *Y*.

Suppose the trees were 6 meters apart. Where would you dig? Since the circles which define the locus for each condition will be tangent to each other, the desired locus is a single point that corresponds to the point of tangency.

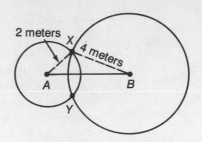

NOTE: $AB = 5$ meters

FIGURE 15.2

In general, the loci which satisfy two or more conditions are found by:

- Determining the locus that satisfies the first condition.

- Using the same diagram to determine the locus for each of the remaining conditions.

- Noting the points of intersection of the loci (if any).

EXAMPLE 15.1 Two parallel lines are 8 inches apart. Point A is located on one of the lines. Find the number of points which are the same distance from each of the parallel lines *and* which are 5 inches from point A.

SOLUTION STEP 1 Identify the given and the conditions.

GIVEN
Two parallel lines 8 inches apart and a point A located on one of them.

CONDITIONS

1. Points must be equidistant from the two lines; and
2. 5 inches from point A.

STEP 2 Draw a diagram which reflects the given information.

STEP 3 The locus of all points equidistant from two parallel lines is a line parallel to the original lines and midway between them.

STEP 4 The locus of all points 5 inches from point A is a circle having point A as its center and a radius of 5 inches.

STEP 5 Since the length of the radius of the circle is greater than the distance of the middle line from the line that contains point A, the circle

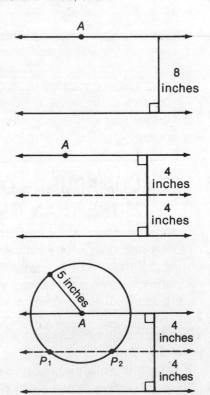

intersects this line in *two* distinct points.

STEP 6   There are two points which are the same distance from each of the parallel lines and which are 5 inches from point *A*.

In the previous example, how would the solution be affected if the second condition was changed to, "4 inches from point *A*"? The circle would now be tangent to the middle line so that there would be exactly one point which satisfied the locus conditions. If the second condition specified a distance less than 4 inches from point *A*, then the circle and the middle line would not intersect so that there would be *no* points which satisfy the locus conditions.

CONCURRENCE THEOREMS

In Figure 15.3, lines *j, k,* and *l* intersect at point *A*; lines *x, y,* and *z* do *not* meet at a common point. Lines *j, k,* and *l* are said to be *concurrent* at point *A*. Point *A* is referred to as the point of concurrency.

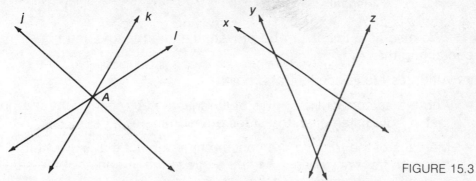

FIGURE 15.3

If the perpendicular bisectors of sides $\overline{ST}$ and $\overline{RT}$ of $\triangle RST$ are drawn, the two perpendicular bisectors will meet at a point, say point *K*. This is illustrated in Figure 15.4. Interestingly, the perpendicular bisector of $\overline{RS}$, the third side of the triangle, will intersect the other two perpendicular bisectors at point *K*. The perpendicular bisectors of the sides of a triangle are concurrent at point *K*. The line of reasoning that supports this conclusion goes as follows:

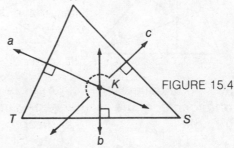

FIGURE 15.4

- Since line *a* is the perpendicular bisector of $\overline{RT}$, any point on line *a* is equidistant from points *R* and *T*. Hence, point *K* is equidistant from points *R* and *T*.

- Similarly, since line *b* is the perpendicular bisector of $\overline{ST}$, point *K* is equidistant from points *S* and *T*.

- Since point *K* is equidistant from points *R* and *S*, it must lie on the perpendicular bisector of $\overline{RS}$ which establishes that line *c* passes through point *K*.

The preceding analysis leads to the conclusions which are summarized in Theorem 15.1.

THEOREM 15.1 CONCURRENCE OF THE ⊥ BISECTORS OF △SIDES

In a triangle,

- The perpendicular bisectors of the sides are concurrent at a point.

- The point of concurrency is equidistant from the vertices of the triangle.

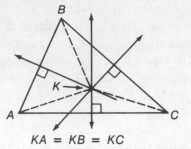

$$KA = KB = KC$$

REMARK Since the point of concurrency is equidistant from the vertices of the triangle, it represents the center of the circle which can be circumscribed about the triangle. For this reason, the point of concurrency is sometimes referred to as the *circumcenter* of the triangle.

We also note that it can be proven that in a triangle, the lines containing the:

- Altitudes are concurrent at a point.

- Angle bisectors of the angles of the triangle are concurrent at a point that is equidistant from the sides of the triangle.

- Medians of the triangle are concurrent at a point that is two-thirds of the distance from any vertex of the triangle to the midpoint of the side opposite the vertex.

See Figure 15.5.

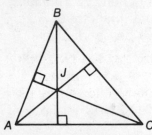
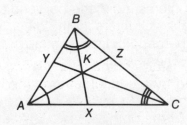
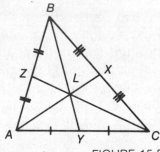

FIGURE 15.5

Point *J* is called the *orthocenter* of the triangle.

Point *K* is called the *incenter* of the triangle and represents the center of the circle which can be inscribed in the triangle. (NOTE $KX = KY = KZ$.)

Point *L* is called the *centroid* of the triangle:

$$AL = \frac{2}{3} AX$$

$$BL = \frac{2}{3} BY$$

$$CL = \frac{2}{3} CZ$$

15.3 THE COORDINATE PLANE

From previous studies in mathematics you may recall that the intersection of a horizontal line and a vertical line on a grid of square boxes can be used to represent a coordinate system in a plane. See Figure 15.6. The horizontal line is called the *x axis* and the vertical line is referred to as the *y axis*. The point at which they intersect is called the *origin*. There is a positive and negative portion of each axis. The negative part of the *x* axis is located to the left of the origin. The negative part of the *y* axis is found below the origin. The positive portions of these axes are located to the right and above the origin.

A point is located by specifying its *x* and *y* *coordinates* which are written as an *ordered pair* of numbers. The first member of the pair always represents the *x* coordinate (abscissa) of the point; the second member is the *y* coordinate (ordinate) of the ordered pair:

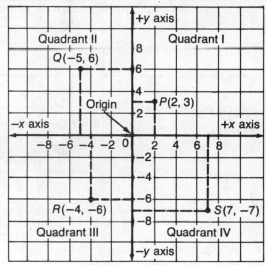

FIGURE 15.6

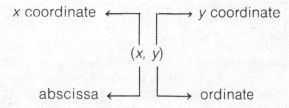

The point whose coordinates are (2, 3) is located by beginning at the origin and moving horizontally to the right along the *x* axis, stopping at $x = 2$, and then moving vertically up 3 units. Figure 15.6 illustrates the plotting of several points, one point in each of the four *quadrants*. A point on the coordinate plane is named by a capital letter immediately followed by its coordinates. Notice that the signs (positive or negative) of the coordinates determine the quadrant that the point will be located in.

SIGNS OF (x, y)	POINT IS IN QUADRANT
(+, +)	I
(−, +)	II
(−, −)	III
(+, −)	IV

Example 15.2 illustrates how to calculate the area of a triangle given the coordinates of its vertices. The approach is based on circumscribing a rectangle about the triangle. The area of the desired triangle is then found

by summing the areas of the "corner" right triangles and subtracting this value from the area of the rectangle.

EXAMPLE 15.2 The coordinates of the vertices of △ABC are A(−3, 3), B(4, 9), and C(9, 5).

    a Draw △ABC.

    b Find the area of △ABC.

SOLUTION   **a**

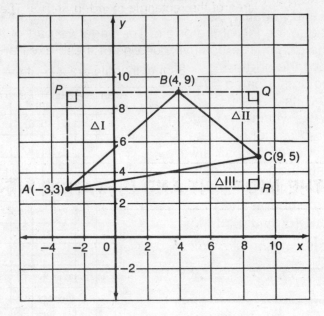

b To find the area of △ABC proceed as follows:

1. Circumscribe a rectangle about △ABC by drawing segments through the vertices of the triangle which are parallel to either the x axis or the y axis.

2. Find the length of the segments along the perimeter of the rectangle. By counting boxes we obtain: $AP = 6$ units, $PB = 7$ units, $BQ = 5$ units, $QC = 4$ units, $CR = 2$ units, and $AR = 12$ units.

3. Find the area of the rectangle:

$$\text{Area rectangle } APQR = bh$$

$$= AR \times AP$$

$$= 12 \times 6$$

$$\text{Area rectangle } APQR = 72 \text{ square units}$$

4. Find the area of each of the "corner" (right) triangles:

Area △I = $\frac{1}{2}AP \times PB$	Area △II = $\frac{1}{2}BQ \times QT$	Area △III = $\frac{1}{2}CR \cdot AR$
$= \frac{1}{2}(6)(7)$	$= \frac{1}{2}(5)(4)$	$= \frac{1}{2}(2)(12)$
Area △I = 21	Area △II = 10	Area △III = 12

5. Find the sum of the areas of triangles I, II, and III:

$$\text{Area } \triangle\text{I} = 21$$

$$\text{Area } \triangle\text{II} = 10$$

$$+ \text{ Area } \triangle\text{III} = 12$$

$$\overline{\text{Sum of areas} = 43 \text{ square units}}$$

6. Subtract the sum of the areas calculated in Step 5 from the area of the rectangle found in Step 3. The result is the area of $\triangle ABC$:

$$\text{Area rectangle } APQR = 72$$

$$- \text{ Sum of triangle areas} = 43$$

$$\overline{\text{Area } \triangle ABC = 29 \text{ square units}}$$

15.4 THE MIDPOINT AND DISTANCE FORMULAS

In Figure 15.7, what are the coordinates of the *midpoint* of $\overline{AB}$? The midpoint of $\overline{AB}$ will be located at $M(6, 5)$. In general, the coordinates of the midpoint of a segment joining two points is found by taking the average of their x coordinates and then finding the average of their y coordinates.

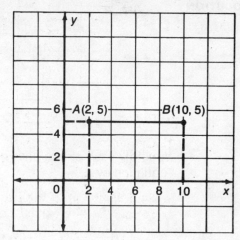

FIGURE 15.7

THEOREM 15.2 THE MIDPOINT FORMULA

The coordinates of the midpoint $M(x_m, y_m)$ of a segment whose end points are $A(x_1, y_1)$ and $B(x_2, y_2)$ may be found using the formulas:

$$x_m = \frac{x_1 + x_2}{2}$$

and

$$y_m = \frac{y_1 + y_2}{2}$$

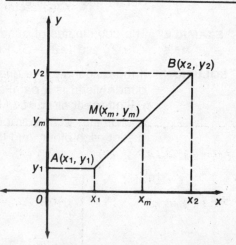

EXAMPLE
15.3

Find the coordinates of the midpoint of a segment whose end points are $H(4, 9)$ and $K(-10, 1)$.

SOLUTION

$$x_1 \rightarrow \quad \quad x_2 \rightarrow$$
$$H(4, 9) \quad \quad K(-10, 1)$$
$$\uparrow y_1 \quad \quad \uparrow y_2$$

$$x_m = \frac{x_1 + x_2}{2} \quad\quad\quad y_m = \frac{y_1 + y_2}{2}$$

$$= \frac{4 + (-10)}{2} \quad\quad\quad = \frac{9 + 1}{2}$$

$$= \frac{-6}{2} \quad\quad\quad = \frac{10}{2}$$

$$x_m = -3 \quad\quad\quad y_m = 5$$

The coordinates of the midpoint of $\overline{HK}$ are $(-3, 5)$.

EXAMPLE
15.4

$M(7, -1)$ is the midpoint of $\overline{WL}$. If the coordinates of end point W are $(5, 4)$, find the coordinates of end point L.

SOLUTION

$$x_m = \frac{5 + x}{2}$$

$$W(5, 4) \quad\quad\quad\quad L(x, y)$$

$$y_m = \frac{4 + y}{2}$$

$$x_m = 7 = \frac{5 + x}{2} \quad\quad\quad\quad y_m = -1 = \frac{4 + y}{2}$$

Multiply each side of the equation by 2:

$$2(7) = 2\left(\frac{5 + x}{2}\right) \quad\quad\quad\quad 2(-1) = 2\left(\frac{4 + y}{2}\right)$$

$$14 = 5 + x \quad\quad\quad\quad -2 = 4 + y$$

$$9 = x \quad\quad\quad\quad -6 = y$$

Hence, the coordinates of point L are $(9, -6)$.

EXAMPLE
15.5

The coordinates of quadrilateral $ABCD$ are $A(-3, 0)$, $B(4, 7)$, $C(9,2)$, and $D(2, -5)$. Prove that $ABCD$ is a parallelogram.

SOLUTION

If the diagonals of a quadrilateral bisect each other, then the quadrilateral is a parallelogram. In Figure 15.8, diagonals $\overline{AC}$ and $\overline{BD}$ intersect at point E. If the coordinates of point E are the coordinates of the midpoints of $\overline{AC}$ and $\overline{BD}$, then the diagonals bisect each other, and the quadrilateral is a parallelogram.

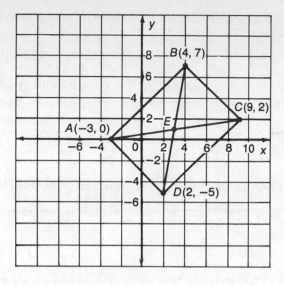

FIGURE 15.8

To find the midpoint of $\overline{AC}$:

$$x_m = \frac{-3 + 9}{2} = \frac{6}{2} = 3$$

$$y_m = \frac{0 + 2}{2} = \frac{2}{2} = 1$$

Hence, the coordinates of the midpoint of $\overline{AC}$ are (3, 1).

To find the midpoint of $\overline{BD}$:

$$x_m = \frac{4 + 2}{2} = \frac{6}{2} = 3$$

$$y_m = \frac{7 + (-5)}{2} = \frac{2}{2} = 1$$

The coordinates of the midpoint of $\overline{BD}$ are (3, 1). Since the diagonals have the same midpoint, they bisect each other which establishes that the quadrilateral is a parallelogram.

FINDING THE DISTANCE BETWEEN TWO POINTS

The distance between two points in the coordinate plane is the length of the segment that joins the two points. Figure 15.9 illustrates how to find the distance between two points which determine a horizontal or a vertical line. In each case, the distance between the two given points is found by

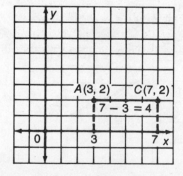
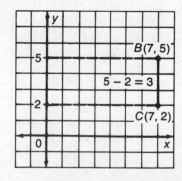

FIGURE 15.9

subtracting a single pair of coordinates. The length of a *horizontal* segment is obtained by calculating the difference in the *x* coordinates of the two points. The length of a *vertical* segment is found by subtracting the *y* coordinates of the two points.

It will prove convenient to introduce the Greek letter Δ (delta) to represent the difference or change in a pair of values. In general, if the coordinates of two points are $P(x_1, y_1)$ and $Q(x_2, y_2)$ then $\Delta x = x_2 - x_1$ and $\Delta y = y_2 - y_1$. If the coordinates of P and Q are $P(1, 7)$ and $Q(5, 9)$, then $\Delta x = 5 - 1 = 4$ and $\Delta y = 9 - 7 = 2$.

THEOREM 15.3 LENGTHS OF HORIZONTAL AND VERTICAL SEGMENTS

$$PQ = \Delta x = x_2 - x_1 \ (x_2 > x_1) \qquad RW = \Delta y = y_2 - y_1 \ (y_2 > y_1)$$

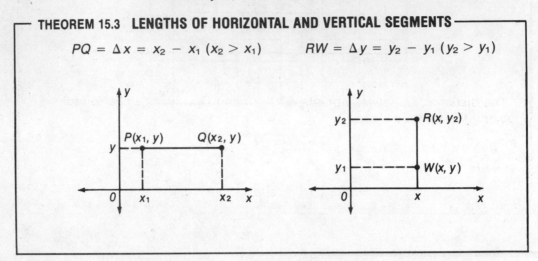

Finding the distance between two points which determine a slanted line is a bit more complicated. Our strategy will seek to capitalize on our ability to easily find the lengths of horizontal and vertical segments. Figure 15.10 demonstrates that the distance between points A and B can be found indirectly by first forming a right triangle in which $\overline{AB}$ is the hypotenuse, and horizontal and vertical segments are legs of the triangle. After determining the lengths of the legs of this right triangle, a straightforward application of the Pythagorean Theorem leads to the length of $\overline{AB}$ which represents the distance between points A and B. The general formula for determining the distance between any two given points is offered in Theorem 15.4.

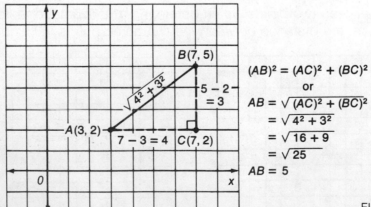

$$(AB)^2 = (AC)^2 + (BC)^2$$
$$\text{or}$$
$$AB = \sqrt{(AC)^2 + (BC)^2}$$
$$= \sqrt{4^2 + 3^2}$$
$$= \sqrt{16 + 9}$$
$$= \sqrt{25}$$
$$AB = 5$$

FIGURE 15.10

THEOREM 15.4 THE DISTANCE FORMULA

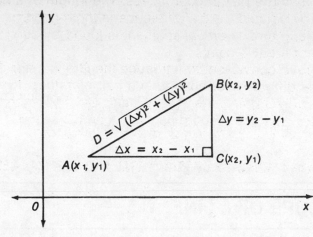

The distance, D, between points $A(x_1, y_1)$ and $B(x_2, y_2)$ may be found using the formula:

$$D = \sqrt{(\Delta x)^2 + (\Delta y)^2}$$

where

$$\Delta x = x_2 - x_1 \quad \text{and} \quad \Delta y = y_2 - y_1$$

EXAMPLE 15.6 Prove that the parallelogram whose coordinates are $A(-3, 0)$, $B(4, 7)$, $C(9, 2)$, and $D(2, -5)$ is a rectangle. (See Figure 15.8.)

SOLUTION A parallelogram is a rectangle if its diagonals are equal in length. Let us find and then compare the lengths of diagonals $\overline{AC}$ and $\overline{BD}$.

To find the length of $\overline{AC}$:

$A(-3, 0)$ $\qquad\qquad$ $C(9, 2)$

$\Delta x = x_2 - x_1$ $\qquad\qquad$ $\Delta y = y_2 - y_1$

$\quad = 9 - (-3)$ $\qquad\qquad$ $\quad = 2 - 0$

$\quad = 9 + 3$ $\qquad\qquad$ $\Delta y = 2$

$\Delta x = 12$

$$AC = \sqrt{(\Delta x)^2 + (\Delta y)^2}$$

$$= \sqrt{(12)^2 + (2)^2}$$

$$= \sqrt{144 + 4}$$

$$AC = \sqrt{148}$$

To find the length of $\overline{BD}$:

$B(4, 7)$ $\qquad\qquad$ $D(2, -5)$

$\Delta x = x_2 - x_1$ $\qquad\qquad$ $\Delta y = y_2 - y_1$

$\quad = 2 - 4$ $\qquad\qquad$ $\quad = -5 - 7$

$\Delta x = -2$ $\qquad\qquad$ $\Delta y = -12$

$$BD = \sqrt{(\Delta x)^2 + (\Delta y)^2}$$
$$= \sqrt{(-2)^2 + (-12)^2}$$
$$= \sqrt{4 + 144}$$
$$BD = \sqrt{148}$$

CONCLUSION Since AC and BD are each equal to $\sqrt{148}$, they must be equal to each other. Hence, $AC = BD$ which means parallelogram $ABCD$ is a rectangle.

15.5 SLOPE OF A LINE

If you imagine that a line represents a hill, then some lines will be more difficult to walk up than other lines.

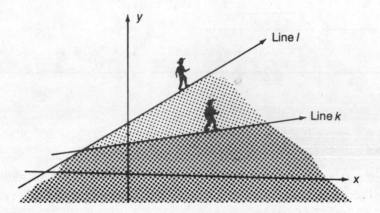

Line l would be more difficult to climb than line k since line l is *steeper* than line k. Another name for steepness is *slope*. The slope of a line may be expressed as a number. To do this, select any two different points on the line. In traveling from one point to the other, compare the change in the vertical distance (Δy) to the change in the horizontal distance (Δx). The ratio of these quantities $\left(\dfrac{\Delta y}{\Delta x}\right)$ represents the slope of the line.

DEFINITION OF SLOPE

The slope, m, of a nonvertical line which passes through the points $A(x_1, y_1)$ and $B(x_2, y_2)$ is given by the ratio in the change in the values of their y coordinates to the change in the value of their x coordinates:

$$\text{Slope} = m = \frac{\Delta y}{\Delta x} = \frac{y_2 - y_1}{x_2 - x_1}$$

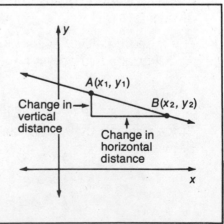

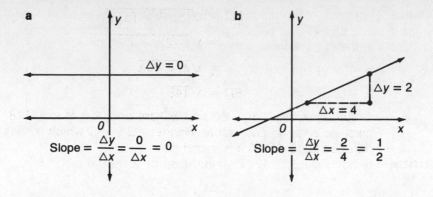

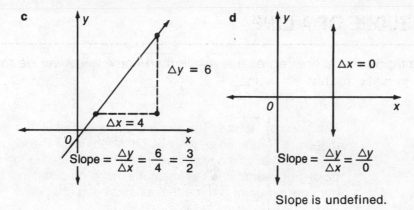

Slope is undefined.

FIGURE 15.11

As the series of graphs in Figure 15.11 illustrates, the steeper the line, the larger the value of its slope. Since a horizontal line has no steepness, its slope has a numerical value of 0. A vertical line is impossible to "walk up" so that its slope is not defined. Mathematically speaking,

■ The y coordinate of every point on a *horizontal* line is the same so that $\Delta y = 0$. For any two points on a horizontal line, the slope formula may be expressed as

$$m = \frac{\Delta y}{\Delta x} = \frac{0}{\Delta x}$$

Since 0 divided by any number is 0, the slope of a horizontal line is always 0.

■ The x coordinate of every point on a vertical line is the same so that $\Delta x = 0$. For any two points on a vertical line, the slope formula leads to

$$m = \frac{\Delta y}{\Delta x} = \frac{\Delta y}{0}$$

Since division by 0 is not defined, the slope of a vertical line is not defined.

EXAMPLE 15.7 For each of the following pairs of values, plot the points and draw a line through the points. Use the slope formula to determine the slope of each line.

a $A(4, 2)$ and $B(6, 5)$

b $P(0, 3)$ and $Q(5, -1)$

c $J(4, 3)$ and $K(8, 3)$

d $W(2, 1)$ and $C(2, 7)$

SOLUTION **a** See Figure 15.12. To find the slope, observe that:

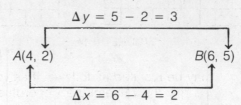

$$\text{Slope of } \overleftrightarrow{AB} \quad m = \frac{\Delta y}{\Delta x} = \frac{y_2 - y_1}{x_2 - x_1} = \frac{5 - 2}{6 - 4} = \frac{3}{2}$$

The slope of $\overleftrightarrow{AB}$ is $\frac{3}{2}$. The slope of $\overleftrightarrow{AB}$ was calculated assuming that point B was the second point. When using the slope formula either of the two points may be considered the second point. For example, let's repeat the slope calculation considering point A to be the second point (that is, $x_2 = 4$ and $y_2 = 2$):

$$m = \frac{\Delta y}{\Delta x} = \frac{y_2 - y_1}{x_2 - x_1} = \frac{2 - 5}{4 - 6} = \frac{-3}{-2} = \frac{3}{2}$$

The same value, $\frac{3}{2}$, is obtained for the slope of $\overleftrightarrow{AB}$, regardless of whether A is taken as the first or second point.

b See Figure 15.13.

$$m = \frac{\Delta y}{\Delta x} = \frac{y_2 - y_1}{x_2 - x_1} = \frac{-1 - 3}{5 - 0} = \frac{-4}{5}$$

The slope of $\overleftrightarrow{PQ}$ is $-\frac{4}{5}$. The slope of $\overleftrightarrow{AB}$ in part (a) was *positive* in value and the slope of $\overleftrightarrow{PQ}$ has a *negative* value. Compare the directions of these lines. Notice that if a line has a positive slope, then it climbs up *and* to the right; if a line has a negative slope, then it falls down *and* to the right. These observations

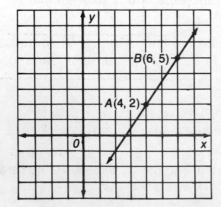

FIGURE 15.12

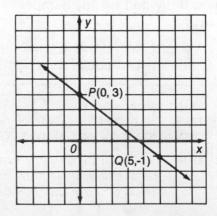

FIGURE 15.13

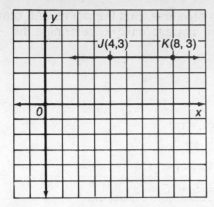

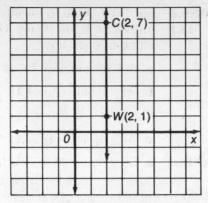

FIGURE 15.14   FIGURE 15.15

may be restated as follows: if as you move along a line from left to right, the value of the x coordinates increase while the value of the y coordinates also increase, then the line has a positive slope; if as the value of the x coordinates increase, the value of the y coordinates decrease, then the line has a negative slope.

c See Figure 15.14. Since the line is horizontal, there is no change in y, so that $\Delta y = 0$.

$$m = \frac{\Delta y}{\Delta x} = \frac{0}{\Delta x} = 0$$

d See Figure 15.15. Since the line is vertical, there is no change in x, so that $\Delta x = 0$. The slope is undefined.

A special relationship exists between the slopes of parallel lines. Since the steepness of two nonvertical lines which are parallel are the same, their slopes are equal. Conversely, if it is known that two nonvertical lines have the same slope, then the lines must be parallel.

THEOREM 15.5   SLOPES OF PARALLEL LINES

- If two nonvertical lines are parallel, then their slopes are equal.

- If two nonvertical lines have the same slope, then they are parallel.

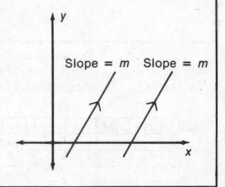

EXAMPLE 15.8   Show that the quadrilateral whose vertices are $A(-3, -8)$, $B(-2, 1)$, $C(2, 5)$, and $D(7, 2)$ is a trapezoid.

SOLUTION   Since a trapezoid is a quadrilateral that has *exactly* one pair of parallel sides, we must show that one pair of sides have the same slope and one pair of sides have different slopes.

$$\text{Slope of } \overline{AB} = \frac{\Delta y}{\Delta x} = \frac{1 - (-8)}{-2 - (-3)} = \frac{1 + 8}{-2 + 3} = \frac{9}{1} = 9$$

$$\text{Slope of } \overline{BC} = \frac{\Delta y}{\Delta x} = \frac{5 - 1}{2 - (-2)} = \frac{4}{2 + 2} = \frac{4}{4} = 1$$

$$\text{Slope of } \overline{CD} = \frac{\Delta y}{\Delta x} = \frac{2 - 5}{7 - 2} = \frac{-3}{5} = -\frac{3}{5}$$

$$\text{Slope of } \overline{AD} = \frac{\Delta y}{\Delta x} = \frac{2 - (-8)}{7 - (-3)} = \frac{2 + 8}{7 + 3} = \frac{10}{10} = 1$$

Since the slope of $\overline{BC}$ = the slope of $\overline{AD}$, $\overline{BC} \parallel \overline{AD}$ and $\overline{BC}$ and $\overline{AD}$ are the bases of the trapezoid.

Since the slope of $\overline{AB} \neq$ slope of $\overline{CD}$, $\overline{AB}$ and $\overline{CD}$ are the legs or nonparallel sides of the trapezoid.

A less obvious relationship exists between the slopes of *perpendicular* lines. If two lines are perpendicular, then the slope of one line will be the negative reciprocal of the slope of the other line. For example, suppose line *l* is perpendicular to line *k*. If the slope of line *l* is $\frac{2}{3}$ then the slope of line *k* would be $-\frac{3}{2}$. Conversely, if it is known that the slopes of two lines are negative reciprocals of one another (or, equivalently, that the product of their slopes is negative), then the lines must be perpendicular.

THEOREM 15.6 SLOPES OF PERPENDICULAR LINES

- If two lines are perpendicular, then their slopes are negative reciprocals of one another.

- If the slopes of two lines are negative reciprocals of one another, then the lines are perpendicular.

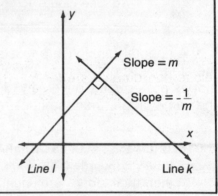

EXAMPLE 15.9   Show by means of slope that the triangle whose vertices are $A(2, 0)$, $B(11, 8)$, and $C(6, 10)$ is a right triangle.

SOLUTION   A triangle is a right triangle if two sides are perpendicular. Let us find and then compare the slopes of each side of the triangle.

$$\text{Slope of } \overline{AB} = \frac{\Delta y}{\Delta x} = \frac{8 - 0}{11 - 2} = \frac{8}{9}$$

$$\text{Slope of } \overline{AC} = \frac{\Delta y}{\Delta x} = \frac{10 - 0}{6 - 2} = \frac{10}{4} = \frac{5}{2}$$

$$\text{Slope of } \overline{BC} = \frac{\Delta y}{\Delta x} = \frac{10 - 8}{6 - 11} = \frac{2}{-5} = -\frac{2}{5}$$

Notice that the slopes of $\overline{AC}$ and $\overline{BC}$ are negative reciprocals. Hence, $\overline{AC} \perp \overline{BC}$, and angle C of $\triangle ABC$ is a right angle.

15.6 THE EQUATION OF A LINE

For each point that lies on a given line, the same relationship between the values of their x and y coordinates must hold. As an illustration, the coordinates for several representative points of line l in Figure 15.16 are (0, 1), (1, 3), and (2, 5). In comparing the x and y coordinates of each of these points, do you see a pattern? In each case the y coordinate is arrived at by multiplying the corresponding x coordinate by 2 and then adding 1. A general rule which expresses this relationship is the equation $y = 2x + 1$, which is referred to as *the equation of line l*. The x and y coordinates for *every* point on line l in Figure 15.16 must satisfy this equation. Without looking at the figure, can you predict the y coordinate of a point which lies on line l if its x coordinate is -2? All we need do is substitute -2 for x in the equation $y = 2x + 1$:

$$y = 2x + 1$$
$$= 2(-2) + 1$$
$$= -4 + 1$$
$$y = -3$$

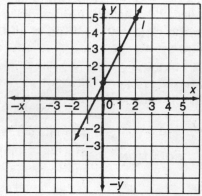

FIGURE 15.16

Now check the figure. Notice that the point whose coordinates are $(-2, -3)$ lies on line l.

In the equation $y = 2x + 1$, let's see if any special significance can be attached to the numbers 2 and 1. Referring to Figure 15.16, notice the line crosses the y axis at the y value of 1. We call this point the y *intercept* of a line. The slope of the line can be found by considering any two points on the line, say (1, 3) and (2, 5). For these points, let $x_1 = 1$, $y_1 = 3$, $x_2 = 2$, and $y_2 = 5$, so that:

$$m = \frac{\Delta y}{\Delta x} = \frac{5 - 3}{2 - 1} = \frac{2}{1} = 2$$

The slope of the line is 2. The coefficient of x in the equation $y = 2x + 1$ is also 2. Hence,

$$y = 2x + 1$$

slope of y intercept
the line   of the line

THEOREM 15.7   THE EQUATION OF A LINE

The set of all points which lie on the graph of a nonvertical line can be represented by an equation of the form $y = mx + b$, where m represents the slope of the line and b is the y coordinate of the point at which the line crosses the y axis. (This form of the equation of a line is referred to as the *slope-intercept* form.)

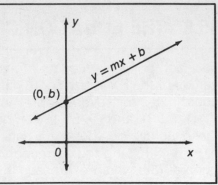

The slope-intercept form of the equation of a line requires the following *two* items: the slope of the line and y intercept. If it is known that the slope of a line is 7 and its y intercept is -9, then the equation of the line is $y = 7x - 9$. Sometimes the slope and y intercept of a line are not explicitly stated. Instead, information is given which permits these features of a line to be determined. This is illustrated in the next example.

EXAMPLE 15.10   Write the equation of a line given the following information.

a The line is parallel to the line whose equation is $y = 4x - 1$ and passes through the point (2, 15).

b The line contains the points (10, -1) and (5, 1).

SOLUTION   **a** The slope of the line $y = 4x - 1$ is 4. Since parallel lines have the same slope, the slope of the desired line must also be 4. Its equation takes the form $y = 4x + b$. Since the point (2, 15) is a point on the line, it must satisfy the equation. This means that if we replace x by 2 and y by 15, a true statement results:

$$y = 4x + b$$

$$15 = 4(2) + b$$

This substitution allows us to solve for the value of b:

$$15 = 8 + b$$

$$7 = b$$

Since $m = 4$ and $b = 7$, the equation of the desired line is $y = 4x + 7$.

b Our initial concern is to find the value of the slope of the line by using the formula $m = \dfrac{\Delta y}{\Delta x}$. The value of b can then be found using the technique illustrated in part (a) of this example.
To find slope:

$$\Delta y = 1 - (-1)$$

(10, -1)   and   (5, 1)

$$\Delta x = 5 - 10$$

$$m = \frac{\Delta y}{\Delta x} = \frac{1 - (-1)}{5 - 10} = \frac{1 + 1}{-5} = -\frac{2}{5}$$

To find the y intercept:

$$y = -\frac{2}{5}x + b$$

Choose either of the two given points, say (5, 1). Substitute 5 for *x* and 1 for *y*:

$$1 = -\frac{2}{\cancel{5}}(\cancel{5}) + b$$

Solve for *b*:

$$1 = -2 + b$$
$$3 = b$$

The equation of the line whose slope is $-\frac{2}{5}$ and *y* intercept is 3, is $y = -\frac{2}{5}x + 3$.

EXAMPLE 15.11 The equation of line *l* is $2y - 3x = 12$. Determine the slope of a line that is perpendicular to line *l*.

SOLUTION Rewrite the equation of line *l* in slope-intercept form:

$$2y - 3x = 12$$
$$2y = 3x + 12$$
$$y = \frac{3}{2}x + \frac{12}{2}$$

The slope of line *l* is $\frac{3}{2}$. The slopes of perpendicular lines are negative reciprocals. The negative reciprocal of $\frac{3}{2}$ is $-\frac{2}{3}$ which must represent the slope of a line that is perpendicular to line *l*.

EXAMPLE 15.12 Write the equation of a line which passes through the point $A(-1, 2)$ and is parallel to the given axis.

a *x* axis

b *y* axis

SOLUTION

a

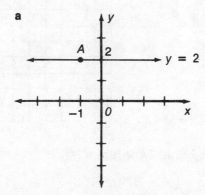

b

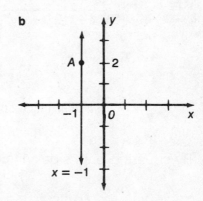

Example 15.12 illustrates that the equation of a *horizontal* line (a line parallel to the *x* axis) takes the form $y = b$, where *b* is the *y* coordinate of the point at which the line intersects the *y* axis. The general form of the equation of a *vertical* line (a line parallel to the *y* axis) is $x = a$, where *a* is the *x* coordinate of the point at which the line crosses the *x* axis.

EXAMPLE 15.13   Find the locus of all points 3 units from the *y* axis.

SOLUTION   The locus of all points 3 units from the *y* axis (a line) may be determined by drawing two vertical lines on either side of the *y* axis and 3 units from the *y* axis. The locus of all points 3 units from the *y* axis are the lines whose equations are $x = -3$ and $x = 3$.

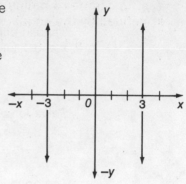

15.7   EQUATION OF A CIRCLE

Until now we have considered drawing only lines in a coordinate plane. Let us describe the locus of all points which are 4 units from the point $A(2, 3)$. We know that the locus is a circle having A as its center and a radius of 4 units. Suppose we wished to find the *equation* which defines this locus. We proceed by choosing a representative point, $P(x, y)$, that satisfies the locus condition. The distance between point A and point P must therefore be 4. A straightforward application of the distance formula where $\Delta x = x - 2$ and $\Delta y = y - 3$ yields the following result:

$$PA = \sqrt{(\Delta x)^2 + (\Delta y)^2}$$
$$PA = \sqrt{(x - 2)^2 + (y - 3)^2}$$

Since $PA = 4$, we may write:

$$\sqrt{(x - 2)^2 + (y - 3)^2} = 4$$

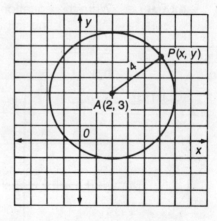

By squaring both sides of this equation, we obtain

$$(x - 2)^2 + (y - 3)^2 = 16$$

coordinates of the center of the circle

the square of the radius

A circle whose center is located at
$A(p, q)$ and has a radius r units in length
is defined by the equation:

$$(x - p)^2 + (y - q)^2 = r^2$$

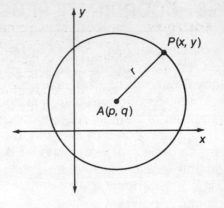

**EXAMPLE
15.14**

Write the equation for the following circles.

a Center is at $(3, -5)$ and has a radius 7 units in length.

b Center is at the origin and has a radius of 5 units.

SOLUTION

a Here, $p = 3$, $q = -5$, and $r = 7$.

$$(x - p)^2 + (y - q)^2 = r^2$$
$$(x - 3)^2 + y - (-5)^2 = 7^2$$
$$(x - 3)^2 + (y + 5)^2 = 49$$

b At the origin, $p = 0$ and $q = 0$. Since $r = 5$,

$$(x - p)^2 + (y - q)^2 = r^2$$
$$(x - 0)^2 + (y - 0)^2 = 5^2$$
$$x^2 + y^2 = 25$$

Notice in part **b** of the previous example that an equation of the form
$x^2 + y^2 = r^2$ defines a circle whose center is at the origin and has a radius
r units in length.

**EXAMPLE
15.15**

Determine the coordinates of the center of a circle and the length
of its radius given the following equations.

a $x^2 + y^2 = 100$

b $(x + 1)^2 + (y - 4)^2 = 39$

SOLUTION

a *Center:* origin.   *Radius:* $\sqrt{100} = 10$.

b Rewrite equation as follows:

$$[x - (-1)]^2 + (y - 4)^2 = 39$$
$$\text{Center: } (-1, 4) \quad \text{Radius: } \sqrt{39}$$

15.8 COORDINATE PROOFS

By placing a polygon in a convenient position in a coordinate plane, and then representing the coordinates of its vertices using letters and zeros, many theorems can be easily proven. We will illustrate this by showing how the methods of coordinate geometry can be used to establish that the diagonals of a rectangle are congruent.

Our first concern is to decide where in the coordinate plane a representative rectangle should be placed. Figure 15.17 gives some possible positionings of a rectangle in a coordinate plane. One of these has a clear advantage over the other two placements. We generally look to position a polygon so that:

- The origin is a vertex of the polygon.

- At least one of the sides of the polygon coincides with a coordinate axis.

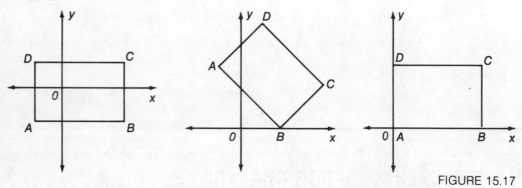

FIGURE 15.17

As we shall see, this tends to simplify matters as it introduces zeros for some of the coordinates of the vertices. Figure 15.17(*c*) conforms to these guidelines. Since the sides of a rectangle are perpendicular to each other, a pair of adjacent sides of the rectangle can be made to coincide with the coordinate axes such that the origin is a vertex of the rectangle. The actual proof follows.

GIVEN Rectangle *ABCD* with diagonals $\overline{AC}$ and $\overline{BD}$.

PROVE $AC = BD$.

PLAN Use the distance formula to show $AC = BD$.

PROOF STEP 1 Find *AC*.

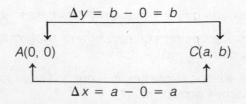

Use the distance formula:

$$AC = \sqrt{(\Delta x)^2 + (\Delta y)^2}$$
$$= \sqrt{(a)^2 + (b)^2}$$
$$AC = \sqrt{a^2 + b^2}$$

STEP 2 Find BD.

$$\Delta y = b - 0 = b$$

$B(a, 0)$ $D(0, b)$

$$\Delta x = 0 - a = -a$$

Use the distance formula:

$$BD = \sqrt{(\Delta x)^2 + (\Delta y)^2}$$
$$= \sqrt{(-a)^2 + (b)^2}$$
$$BD = \sqrt{a^2 + b^2}$$

Since AC and BD are both equal to $\sqrt{a^2 + b^2}$, they are equal to each other. Since $AC = BD$, $\overline{AC} \cong \overline{BD}$.

REVIEW EXERCISES FOR CHAPTER 15

In Exercises 1 to 4, 6, 8, and 9, find the loci.

1. The locus of all points 1 cm from a circle whose radius is 8 cm.

2. The locus of the centers of all circles tangent to each of two parallel lines which are 10 inches apart.

3. The locus of the center of a circle that rolls along a flat surface.

4. The locus of all points equidistant from two concentric circles having radii of 7 cm and 11 cm.

5. Two circles are externally tangent. One circle remains stationary while the other circle rolls around it so that the two circles remain externally tangent to each other. What is the locus of the center of the moving circle?

6. The locus of the vertices of all isosceles triangles having a common base.

7. Two circles are internally tangent. The inside circle rolls around the outside circle so that the two circles remain internally tangent. What is the locus of the center of the moving circle?

8. The locus of all points equidistant from points A and B and 4 inches from $\overleftrightarrow{AB}$.

9. The locus of all points equidistant from two intersecting lines and 3 inches from their point of intersection.

10. Point A is 4 inches from line k. What is the locus of all points 1 inch from line k and 2 inches from point A?

11. Point X is the midpoint of $\overline{PQ}$. If $PQ = 10$ cm, what is the locus of all points equidistant from points P and Q and 5 cm from point X?

12. Lines p and q are parallel and 10 cm apart. Point A is between lines p and q and 2 cm from line q. What is the locus of all points equidistant from lines p and q and d cm from point A, if d equals the given number of centimeters.

    a 6 cm   **b** 1 cm   **c** 3 cm

13. Points A and B are d units apart. What is the locus of all points equidistant from points A and B and r units from point A given the following conditions.

    a $r < \dfrac{d}{2}$ **b** $r = \dfrac{d}{2}$ **c** $r > \dfrac{d}{2}$

In Exercises 14 to 17, draw and then find the area of the triangles whose vertices are given.

14. $A(0, 0)$, $B(5, 0)$, and $C(0, 8)$.   **15.** $R(-6, 0)$, $S(0, 0)$, and $T(-6, -4)$.

16. $A(-5, 2)$, $B(-3, 6)$, and $C(3, 1)$.   **17.** $J(2, -7)$, $K(11, 7)$, and $L(8, -3)$.

18. Fill in the following table:

	POINT A	POINT B	MIDPOINT OF AB	LENGTH OF AB
a	(3, 8)	(1, 2)	?	?
b	(−1, 3)	(−6, −9)	?	?
c	(2, 4)	?	(3, −2)	?
d	?	(−5, 1)	(−1, −4)	?
e	(0, 2a)	(2b, 0)	?	?
f	(a, b)	(c, d)	?	?

19. $P(2, -3)$ and $Q(-4, 5)$ are the end points of diameter $\overline{PQ}$ of circle O.

    a Find the coordinates of the center of the circle.

    b Find the length of a radius of the circle.

    c Determine whether the point $K(-5, -2)$ lies on the circle.

20. The coordinates of the vertices of quadrilateral $MATH$ are $M(-3, 2)$, $A(4, 8)$, $T(15, 5)$, and $H(8, y)$. If $MATH$ is a parallelogram, find the value of y.

    a by using midpoint relationships

    b by using slope relationships

21. The coordinates of the vertices of $\triangle RST$ are $R(4, -4)$, $S(-1, 5)$, and $T(13, 9)$. Find the length of the median drawn to side $\overline{ST}$.

22. The coordinates of the vertices of quadrilateral $JKLM$ are $J(-7, 6)$, $K(-4, 8)$, $L(-2, 5)$, and $M(-5, 3)$. Prove that $JKLM$ is a square. (*Hint:* Show that $JKLM$ is a rectangle having a pair of congruent adjacent sides.)

23. The coordinates of the vertices of a triangle are $J(-10, -6)$, $K(0, 6)$, and $L(6, 1)$.

    a By means of slope, show that $\triangle JKL$ is a right triangle.

    b Verify your conclusion by showing that the lengths of the sides of $\triangle JKL$ satisfy the converse of the Pythagorean Theorem.

24. The coordinates of the vertices of a triangle are $A(2, -3)$, $B(5, 5)$, and $C(11, 3)$.

   a Find the slope of a line that is parallel to $\overline{AB}$.

   b Find the slope of the altitude drawn to side $\overline{AC}$.

   c Find the slope of the median drawn to side $\overline{BC}$.

25. Three points are *collinear* if the slopes of two of the segments determined by selecting any two pairs of points are the same. For example, points A, B, and C are collinear if any one of the following relationships are true: (1) slope of $\overline{AB}$ = slope of $\overline{BC}$; or (2) slope of $\overline{AB}$ = slope of $\overline{AC}$; or (3) slope of $\overline{BC}$ = slope of $\overline{AC}$.

     Determine whether each of the following sets of points are collinear.

   a $A(-4, -5)$, $B(0, -2)$, $C(8, 4)$.

   b $R(2, 1)$, $S(10, 7)$, $T(-4, -6)$.

   c $J(1, 2)$, $K(5, 8)$, $L(-3, -4)$.

26. Show by means of slope that the quadrilateral whose vertices are $S(-3, 2)$, $T(2, 10)$, $A(10, 5)$, and $R(5, -3)$ is a:

   a parallelogram   **b** rectangle

27. Write the equation of the line which passes through the point $A(-4, 3)$ and which is parallel to the line $2y - x = 6$.

28. Write the equation of the line which passes through the point $(3, 1)$ and which is perpendicular to the line $3y + 2x = 15$.

29. Write the equation of the line which passes through points A and B with the given coordinates.

   a $A(-2, 7)$ and $B(4, 7)$ **b** $A(0, 3)$ and $B(2, 1)$

   c $A(6, 8)$ and $B(6, -3)$ **d** $A(1, -3)$ and $B(-1, 5)$

30. Write the equation(s) which describe the following loci.

   a The locus of all points which are 5 units from the x axis.

   b The locus of all points which are 4 units from the y axis.

   c The locus of all points such that their ordinates are twice their abscissas.

   d The locus of all points such that the sum of twice their ordinates and three times their abscissas is 6.

   e The locus of all points such that their ordinates exceeds their abscissas by 5.

31. Write the equation of the line which contains the point $(-5, 2)$ and which is parallel to the given axis.

   a x axis   **b** y axis

32. The coordinates of the vertices of a triangle are $A(-1, -4)$, $B(7, 8)$, and $C(9, 6)$. Write the equation of the following lines.

   a The line which passes through point B and is parallel to $\overline{AC}$.

   b The line which contains the median drawn to side $\overline{BC}$ from vertex A.

   c The line which contains the altitude drawn to side $\overline{AB}$ from vertex C.

33. Write the equation of the perpendicular bisector of the segment which joins the points $A(3, -7)$ and $B(5, 1)$.

34. The equation of the line which contains the perpendicular bisector of the segment that joins the points $R(-8, t)$ and $S(2, -3)$ is $2y + 5x = k$. Find the values of k and t.

35. Fill in the following table:

	CENTER	RADIUS	EQUATION OF CIRCLE
a	(0, 0)	4	?
b	(3, 4)	6	?
c	?	?	$(x - 1)^2 + y^2 = 81$
d	?	?	$x^2 + y^2 = 49$
e	?	?	$(x - 2)^2 + (y + 5)^2 = 51$

36. The line whose equation is $x = 5$ is tangent to a circle whose center is at the origin. Write the equation of the circle.

37. The center of a circle is located at $O(1, h)$. The line whose equation is $y = kx + 1$ is tangent to circle O at the point $P(3, 6)$. Find the values of k and h.

38. Two circles are tangent externally at point P. The equation of one of the circles is $x^2 + y^2 = 16$. If the other circle has its center on the positive y axis and has a radius of 5 units, find its equation.

Prove each of the following theorems using the methods of coordinate geometry.

39. The opposite sides of a parallelogram are congruent.

40. The length of the median drawn to the hypotenuse of a right triangle is equal to one-half the length of the hypotenuse.

41. The segment joining the midpoints of two sides of a triangle is parallel to the third side and one-half of its length.

42. The diagonals of an isosceles trapezoid are congruent.

CHAPTER SIXTEEN

COMPUTER EXPERIENCES IN GEOMETRY*

Computer programming and the study of geometry have much in common in that both depend on logical thinking. For example, the If . . . Then . . . sentence structure is not only encountered frequently in geometry but also represents a widely used and powerful programming construction. It is the IF . . . THEN . . . program statement which gives computers their decision-making ability. The following BASIC program accepts as input the measure of angle A. If the measure of angle A is equal to 90, then the message "ANGLE A IS A RIGHT ANGLE" is printed; otherwise, the message "ANGLE A IS NOT A RIGHT ANGLE" is displayed.

```
10   INPUT "ENTER THE MEASURE OF ANGLE A"; A
20   IF A = 90 THEN 50
30   PRINT "ANGLE A IS NOT A RIGHT ANGLE"
40   GOTO 60
50   PRINT "ANGLE A IS A RIGHT ANGLE"
60   END
```

Suppose the number 75 is entered as data. In line number 20, the data value of 75 is compared with 90. If it is equal to 90, then the computer will jump to line number 50. Since it is not equal to 90, the program "drops through" and the statement having the next highest line number (that is, line number 30) is executed. This would result in the message contained in the PRINT statement in line 30 to be executed. The GOTO statement transfers the control of the program to the END statement in line 60 so that the execution of the program stops.

Suppose the program is run a second time and the value of 90 is entered as the value of A. When line 20 is executed, a true condition results and the control of the program jumps to line 50. After printing the appropriate message, the execution of the program terminates at line 60.

If you think about it, a geometric proof and a computer program have much in common. The Given of a proof corresponds to the *input* in a computer program; the set of statements and reasons which make up a proof correspond to the *processing* required in a computer program; what

---

* This chapter assumes a familiarity with the elements of the BASIC programming language.

is to be proved in a proof is analogous to the desired *output* of the program. Furthermore, both a geometric proof and a computer program represent a *generalized* solution to a given problem. For example, the theorem which states that the sum of the measures of the angles of a triangle is 180, was proved to be true for *any* triangle—not for a particular or special type of triangle (for example, acute, right, or obtuse). In designing a computer solution to a problem, program *variables* are used so that the solution is not tied to a particular choice of data values. The computer then "solves" a problem by following the appropriate set of program instructions and replacing the program variables with specific numbers provided by the user of the program.

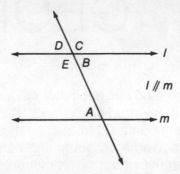

FIGURE 16.1

To illustrate, consider Figure 16.1. The problem is to find the measures of angles *B, C, D,* and *E,* given the measure of angle *A.* The programmer must first determine the general relationships which exist between angle *A* and each of the remaining angles. Based upon this analysis, a programmer solves the problem in the general case by preparing a list of instructions using program variable *A* rather than a specific number that corresponds to the measure of angle *A.* Upon execution of the program, variable *A* is replaced by an actual data value that is supplied by the user of the program and, by following the list of program instructions, the computer arrives at corresponding values for variables *B, C, D,* and *E.* Here's the program.

```
10   INPUT "THE MEASURE OF ANGLE A"; A
20   LET B = A
30   LET C = 180 - B
40   LET D = A
50   LET E = 180 - A
60   PRINT "THE MEASURE OF ANGLES B, C, D, AND E ARE:"
70   PRINT B, C, D, E
80   END
```

The following program accepts as input four values (*A, B, C,* and *D*) which represent the measures of the angles of a quadrilateral. The program determines whether the quadrilateral is a parallelogram by testing if opposite angles are equal in measure. In line 20, a logical AND is used since a quadrilateral is a parallelogram only if *both* pairs of opposite angles are equal in measure. If the quadrilateral is a parallelogram, the program then determines whether the parallelogram is a rectangle by comparing one of the four angles to 90.

```
10   INPUT A, B, C, D
20   IF A = C AND B = D THEN 50
30   PRINT "QUAD ABCD IS NOT A PARALLELOGRAM"
40   GOTO 90
50   IF A = 90 THEN 80
60   PRINT "QUAD ABCD IS A PARALLELOGRAM"
70   GOTO 90
80   PRINT "QUAD ABCD IS A RECTANGLE"
90   END
```

Suppose you entered the following four data values for *A*, *B*, *C*, and *D*: 80, 110, 80, 110. What would be the output of the program? The message "QUAD ABCD IS A PARALLELOGRAM" would be printed. But is this true? Notice that the sum of the input values is 380. The sum of the measures of the angles of any quadrilateral must always be exactly 360. In order to prevent such misleading results, always "protect" your program by including program statements which validate the input data *before* processing begins. For example, the previous program could be suitably modified by inserting the following statements:

```
  .
  .
  .
15   IF A + B + C + D < > 360 THEN 89
  .
  .
  .
85   GOTO 90
89   PRINT A; B; C; D, "BAD DATA"
90   END
```

Line number 15 determines whether the sum of the angles is *not* equal to 360. If this is the case, then the four angles cannot represent the angles of a quadrilateral. Program control is then transferred to line 89 which prints the values and the diagnostic message BAD DATA. This alerts the user that there is an error in the data which were supplied to the program.

Pythagoras discovered that if *p* represents an odd integer greater than 1, then the set of numbers,

$$p, \frac{p^2 - 1}{2}, \frac{p^2 + 1}{2}$$

form a Pythagorean triple. For example, if $p = 3$ then

$$\frac{p^2 - 1}{2} = \frac{3^2 - 1}{2} = \frac{9 - 1}{2} = \frac{8}{2} = 4$$

$$\frac{p^2 + 1}{2} = \frac{3^2 + 1}{2} = \frac{9 + 1}{2} = \frac{10}{2} = 5$$

This leads to the familiar 3, 4, 5 Pythagorean triple. Using a computer to find Pythagorean triples with this formula is easy. The program that follows uses a FOR statement to produce values of *p* from 3 to 25, in increments of 2. The calculations of the second and third members of the triple, as well as their printing, are performed within the body of the FOR/NEXT loop. Before processing begins, column headings are printed in line 10.

```
10   PRINT "     A", "     B", "     C"
20   FOR P = 3 TO 25 STEP 2
30      LET B = (P ↑ 2 − 1) / 2
40      LET C = (P ↑ 2 + 1) / 2
50      PRINT P, B, C
60   NEXT P
70   END
```

Now it's your turn. Here are some computer programs for you to write which will require you to use your knowledge of geometry.

REVIEW EXERCISES FOR CHAPTER 16

Write a BASIC program to accomplish each of the following tasks.

1. The volume of a rectangular box is given by the formula $V = LWH$, where L = length, W = width, and H = height. The surface area of the box is given by the formula $S = 2(LW + LH + WH)$. Enter L, W, and H, and print the volume and surface area of the box.

2. Input the measure of angle X $(0 < X < 180)$ and print whether angle X is acute, right, or obtuse.

3. Input the lengths of a pair of adjacent sides of a parallelogram and the lengths of its diagonals. Classify the parallelogram as a rectangle, rhombus, square, or none of these.

4. Input the lengths of the radii of two circles and the length of their line of centers. Determine the number of common internal tangents and the number of common external tangents which can be drawn. (Assume that if the circles are not congruent, then one circle does *not* lie completely in the interior of the other circle.)

5. Euclid's relationship between the numbers which form a Pythagorean triple can be expressed as follows:

$$r^2 - s^2, \ 2rs, \ r^2 + s^2$$

where r and s are positive integers and $r > s$. Generate Pythagorean triples for all values of $r \le 5$.

6. Input three positive numbers A, B, and C such that C is the largest of the three values. Determine whether the numbers can represent the lengths of the sides of a triangle. If they can, then classify the triangle as acute $(C^2 < A^2 + B^2)$, right $(C^2 = A^2 + B^2)$, or obtuse $(C^2 > A^2 + B^2)$.

7. Input 6 values $X1$, $Y1$, $X2$, $Y2$, $X3$, $Y3$ which correspond to the coordinates of three points, $A(X1, Y1)$, $B(X2, Y2)$, and $C(X3, Y3)$. Determine whether points A, B, and C are collinear. Provide for the possibility that two of the points may determine a vertical line.

8. Revise the program written in Exercise 7 so that if the points are *not* collinear, then the area of the triangle determined by the three points is printed. Accomplish this by finding the lengths of each of the three sides of the triangle and then apply Hero's formula:

$$\text{Area} = \sqrt{s(s - a)(s - b)(s - c)}$$

where *a*, *b*, and *c* represent the lengths of the three sides of the triangle and *s* (semiperimeter) equals one-half the perimeter $\left(s = \dfrac{a + b + c}{2}\right)$

9. An approximation for the value of π may be obtained by evaluating the following formula:

$$\pi = 4\left(1 - \frac{1}{3} + \frac{1}{5} - \frac{1}{7} + \frac{1}{9} - \frac{1}{11} + \cdots\right)$$

As the number of terms inside the parentheses of the formula increases, the approximation of π becomes more accurate. Print an approximation for π by using the sum of the first 25 terms. Continue the summation process until two successive terms differ by less than 0.0001. Print the result and the number of terms which was added.

10. An approximation for π may be obtained by simulating the throwing of darts onto a bull's-eye. The bull's-eye is that part of a circle having a radius of 1 unit, centered at the origin, and falling within the first quadrant. To complete the bull's-eye, the circle is framed by a square having a side 1 unit in length. The diagram is as follows:

NOTE: The area of the circular region shown is $\frac{1}{4}(\pi r^2)$. Since $r = 1$, the area is $\pi/4$ square units.

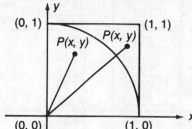

Darts are "thrown" by using the computer's built-in random number generating function to produce values for the *x* and *y* coordinates of a point *P(x, y)*. Since the values manufactured by the random number function always fall between 0 and 1, some points will lie within the circle, while others will lie outside the circle, but fall within the boundaries of the square. Design a program that:

1. Randomly generates the *x* and *y* coordinates for *P(x, y)*.

2. Counts the number of times the point *P(x, y)* falls within the circle.

3. Repeats the first two steps 100 times.

4. Calculates the ratio of the number of points accumulated in Step 2 to the total number of points generated (100 in our example). This ratio should be approximately equal to $\pi/4$, which is the ratio of the area of the circular region ($\pi/4$ square units) to the area of the square (1 square unit).

5. Multiplies the ratio calculated in Step 4 by 4 and then prints the result which should be an approximation for π.

TABLE OF TRIGONOMETRIC FUNCTIONS

Angle	Sin	Cos	Tan	Angle	Sin	Cos	Tan
0°	.0000	1.0000	.0000				
1°	.0175	.9998	.0175	46°	.7193	.6947	1.0355
2°	.0349	.9994	.0349	47°	.7314	.6820	1.0724
3°	.0523	.9986	.0524	48°	.7431	.6691	1.1106
4°	.0698	.9976	.0699	49°	.7547	.6561	1.1504
5°	.0872	.9962	.0875	50°	.7660	.6428	1.1918
6°	.1045	.9945	.1051	51°	.7771	.6293	1.2349
7°	.1219	.9925	.1228	52°	.7880	.6157	1.2799
8°	.1392	.9903	.1405	53°	.7986	.6018	1.3270
9°	.1564	.9877	.1584	54°	.8090	.5878	1.3764
10°	.1736	.9848	.1763	55°	.8192	.5736	1.4281
11°	.1908	.9816	.1944	56°	.8290	.5592	1.4826
12°	.2079	.9781	.2126	57°	.8387	.5446	1.5399
13°	.2250	.9744	.2309	58°	.8480	.5299	1.6003
14°	.2419	.9703	.2493	59°	.8572	.5150	1.6643
15°	.2588	.9659	.2679	60°	.8660	.5000	1.7321
16°	.2756	.9613	.2867	61°	.8746	.4848	1.8040
17°	.2924	.9563	.3057	62°	.8829	.4695	1.8807
18°	.3090	.9511	.3249	63°	.8910	.4540	1.9626
19°	.3256	.9455	.3443	64°	.8988	.4384	2.0503
20°	.3420	.9397	.3640	65°	.9063	.4226	2.1445
21°	.3584	.9336	.3839	66°	.9135	.4067	2.2460
22°	.3746	.9272	.4040	67°	.9205	.3907	2.3559
23°	.3907	.9205	.4245	68°	.9272	.3746	2.4751
24°	.4067	.9135	.4452	69°	.9336	.3584	2.6051
25°	.4226	.9063	.4663	70°	.9397	.3420	2.7475
26°	.4384	.8988	.4877	71°	.9455	.3256	2.9042
27°	.4540	.8910	.5095	72°	.9511	.3090	3.0777
28°	.4695	.8829	.5317	73°	.9563	.2924	3.2709
29°	.4848	.8746	.5543	74°	.9613	.2756	3.4874
30°	.5000	.8660	.5774	75°	.9659	.2588	3.7321
31°	.5150	.8572	.6009	76°	.9703	.2419	4.0108
32°	.5299	.8480	.6249	77°	.9744	.2250	4.3315
33°	.5446	.8387	.6494	78°	.9781	.2079	4.7046
34°	.5592	.8290	.6745	79°	.9816	.1908	5.1446
35°	.5736	.8192	.7002	80°	.9848	.1736	5.6713
36°	.5878	.8090	.7265	81°	.9877	.1564	6.3138
37°	.6018	.7986	.7536	82°	.9903	.1392	7.1154
38°	.6157	.7880	.7813	83°	.9925	.1219	8.1443
39°	.6293	.7771	.8098	84°	.9945	.1045	9.5144
40°	.6428	.7660	.8391	85°	.9962	.0872	11.4301
41°	.6561	.7547	.8693	86°	.9976	.0698	14.3007
42°	.6691	.7431	.9004	87°	.9986	.0523	19.0811
43°	.6820	.7314	.9325	88°	.9994	.0349	28.6363
44°	.6947	.7193	.9657	89°	.9998	.0175	57.2900
45°	.7071	.7071	1.0000	90°	1.0000	.0000	

SOLUTIONS TO EXERCISES

CHAPTER 1

1. **a** $\overrightarrow{BA}$, $\overrightarrow{BE}$, $\overrightarrow{BC}$, $\overrightarrow{BD}$.

   b $\overleftrightarrow{BD}$, $\overleftrightarrow{BE}$, $\overleftrightarrow{DE}$.

   c $\overleftrightarrow{AB}$, $\overleftrightarrow{BC}$, $\overleftrightarrow{AC}$.

   d Angles *ABE*, *EBC*, *CBD*, and *DBA*.

   e $\overrightarrow{BE}$ and $\overrightarrow{BD}$; $\overrightarrow{BA}$ and $\overrightarrow{BC}$.

2. **a** *D*   **b** *A*   **c** *B*

3. **a** ∡*AEF*   **b** ∡*BCA*   **c** ∡*EFA*

4. *D*, *N*, *R*, and *W*.

5. $\overline{NB}$ and $\overline{NW}$.

6. Triangles *ABN*; *NKR*, *TKR*, *TRW*, *NTR*, and *NTW*.

7. Angles *NRK*, *KRT*, *TRW*, *KRW*, *NRW*, and *NRT*.

8. Angle *ANT*.

9. $\overrightarrow{NB}$ and $\overrightarrow{NW}$; $\overrightarrow{RB}$ and $\overrightarrow{RW}$.

10. $\overline{TR}$ is a side of △*TKR* and △*TWR*. Also, $\overline{KR}$ is a side of △*NKR* and △*TKR*.

11. $\overline{RN}$ and $\overline{RW}$.

12. **a** An angle whose measure is less than 90 degrees is an acute angle.

    b A triangle having three sides equal in length is an equilateral triangle.

    c A ray (or segment) which divides an angle into two congruent angles is the bisector of the angle.

13. **a** Inductive.   **b** Deductive.

    c Inductive.   **d** Deductive.

14. The length of the median drawn to the hypotenuse of a right triangle is equal to one-half the length of the hypotenuse.

15. The medians intersect at the same point.

16. **a** Henry has green eyes.

    b The measure of the third angle of the triangle is 80.

17. The formula generates prime numbers for integer values of *N* from $n = 0$ to $n = 15$. For $n = 16$, the formula produces the value 289 which is not a prime number since it is divisible by 17.

CHAPTER 2

1. **a** Acute.   **b** Straight.   **c** Right.   **d** Acute.

   e Right.   **f** Obtuse.   **g** Acute.   **h** Obtuse.

2. 8

3. *Z*.

4. *m*∡*RPL* = 10.

5. 12

6. 42

7. Right.

8. 20

9. $\overline{EY} \cong \overline{AS}$; ∡*PYS* ≅ ∡*IAE*; ∡*YEP* ≅ ∡*ASI*.

10. $\overline{AF} \cong \overline{CF}$.

11. ∡*STP* ≅ ∡*OTP*.

12. $\overline{BE} \cong \overline{DE}$; ∡*ADB* ≅ ∡*CDB*.

13. Transitive.

14. Reflexive.

15. Transitive.

16. Substitution.

17. Addition.

18. Subtraction.

19. Addition.

20. Addition.

21. Halves of equals are equal.

22. Subtraction.

23. Subtraction.

24. Addition.

25. 3. Reflexive.
    4. Addition.

26. 1. Given.
    2. Congruent angles are equal in measure.
    3. Subtraction.
    4. Substitution.
    5. Angles equal in measure are congruent.

27. 1. Given.
    2. Reflexive.
    3. Subtraction.
    5. Given.
    6. Congruent segments are equal in measure.
    7. Reflexive.
    8. Subtraction.
    10. Segments equal in measure are congruent.

28. 2. Congruent angles are equal in measure.
    3. Given.
    4. Definition of angle bisector.
    5. Halves of equals are equal.
    6. Angles equal in measure are congruent.

CHAPTER 3

1. Adjacent angle pairs: 1 and 4; 2 and 3; 4 and 3; 5 and 6; 7 and 8. Vertical angle pairs: 1 and 2; 9 and 10.

2. 75

3. 136

4. 45

5. The measure of the angle is 52 and its complement has measure 38.

6. 56

7. **a** 9 **b** 40 **c** 18

8. 143

9. True (provided the measure of the angles which form the vertical pair are less than 90).

10. True.

11. True.

12. True.

13. False. Consider triangle ABC in which $AC = BC$. Point C is equidistant from points A and B, but point C is not the midpoint of side $\overline{AB}$.

14. (1) $\angle 1 \cong \angle 4$ (Given); (2) $\angle 2 \cong \angle 1$ and $\angle 3 \cong \angle 4$ (Vertical angles are congruent); (3) $\angle 2 \cong \angle 3$ (Transitive property).

15. (1) $\overline{BD}$ bisects $\angle ABC$ (Given); (2) $\angle 3 \cong \angle 4$ (Definition of angle bisector); (3) $\angle 1 \cong \angle 2$ (Supplements of congruent angles are congruent).

16. (1) $\angle 3$ is complementary to $\angle 1$; $\angle 4$ is complementary to $\angle 2$ (Given); (2) $\angle 1 \cong \angle 2$ (Vertical angles are congruent); (3) $\angle 3 \cong \angle 4$ (Complements of congruent angles are congruent).

17. (1) $\overline{AB} \perp \overline{BD}$, $\overline{CD} \perp \overline{BD}$, $\angle 2 \cong \angle 4$ (Given); (2) $\angle 1$ is complementary to $\angle 2$ and $\angle 3$ is complementary to $\angle 4$ (Adjacent angles whose exterior sides are perpendicular are complementary); (3) $\angle 1 \cong \angle 3$ (Complements of congruent angles are congruent).

18. (1) $\overline{KL} \perp \overline{JM}$, $\overline{KL}$ bisects $\angle PLQ$ (Given); (2) $\angle 1$ is complementary to $\angle 2$ and $\angle 4$ is complementary to $\angle 3$ (Adjacent angles whose exterior sides are perpendicular are complementary); (3) $\angle 2 \cong \angle 3$ (Definition of angle bisector); (4) $\angle 1 \cong \angle 4$ (Complements of congruent angles are congruent).

19. (1) $\overline{NW} \perp \overline{WT}$, $\overline{WB} \perp \overline{NT}$, $\angle 4 \cong \angle 6$ (Given); (2) $\angle 4 \cong \angle 3$ (Vertical angles are congruent); (3) $\angle 3 \cong \angle 6$ (Transitive); (4) $\angle 5$ is complementary to $\angle 6$ and $\angle 2$ is complementary to $\angle 3$ (Adjacent angles whose exterior sides are perpendicular are complementary); (5) $\angle 5 \cong \angle 2$ (Complements of congruent angles are congruent).

20. (1) $\overline{MT}$ bisects $\angle ETI$, $\overline{KI} \perp \overline{TI}$, $\overline{KE} \perp \overline{TE}$, $\angle 3 \cong \angle 1$, $\angle 5 \cong \angle 2$ (Given); (2) $\angle 1 \cong \angle 2$ (Definition of angle bisector); (3) $\angle 3 \cong \angle 5$ (Transitive); (4) $\angle 4$ is complementary to $\angle 3$ and $\angle 6$ is complementary to $\angle 5$ (Adjacent angles whose exterior sides are perpendicular are complementary); (5) $\angle 4 \cong \angle 6$ (Complements of congruent angles are congruent).

CHAPTER 4

1. Alternate interior.

2. Corresponding.

3. Corresponding.

4. **a** $x = 112$, $y = 68$. **b** $x = 104$, $y = 104$.

  c $x = 48$, $y = 80$. **d** $x = 45$, $y = 117$.

5. **a** 22 **b** 55 **c** 60

6. **a** 138 and 42 **b** 144 and 36

7. 55

8. **a**

$m\angle 2 = 80$	$m\angle 8 = 100$
$m\angle 3 = 100$	$m\angle 9 = 80$
$m\angle 4 = 80$	$m\angle 10 = 130$
$m\angle 5 = 80$	$m\angle 11 = 50$
$m\angle 6 = 50$	$m\angle 12 = 130$
$m\angle 7 = 50$	$m\angle 13 = 50$

  b

$m\angle 1 = 127$	$m\angle 8 = 127$
$m\angle 2 = 53$	$m\angle 9 = 53$
$m\angle 3 = 127$	$m\angle 10 = 116.5$
$m\angle 4 = 53$	$m\angle 11 = 63.5$
$m\angle 5 = 53$	$m\angle 12 = 116.5$
$m\angle 6 = 63.5$	$m\angle 13 = 63.5$
$m\angle 7 = 63.5$	

9. **a** If I live in the United States, then I live in New York. (False.)

  b If two angles are equal in measure, then the angles are congruent. (True.)

  c If two angles are congruent, then they are vertical angles. (False.)

  d If the sum of the measures of two angles is 90, then the angles are complementary. (True.)

  e If two angles have the same vertex, then they are adjacent. (False.)

  f If two lines are parallel, then they are perpendicular to the same line. (False.)

10. (1) $\overline{BA} \parallel \overline{CF}$, $\overline{BC} \parallel \overline{ED}$ (Given); (2) $\angle 1 \cong \angle C$ (Postulate 4.1); (3) $\angle C \cong \angle 2$ (Theorem 4.2); (4) $\angle 1 \cong \angle 2$ (Transitive).

11. (1) $\overline{LT} \parallel \overline{WK} \parallel \overline{AP}$, $\overline{PL} \parallel \overline{AG}$ (Given); (2) $\angle 1 \cong \angle KWP$ (Theorem 4.2); (3) $\angle KWP \cong \angle WPA$ (Postulate 4.1); (4) $\angle 1 \cong \angle WPA$ (Transitive); (5) $\angle WPA \cong \angle 2$ (Postulate 4.1); (6) $\angle 1 \cong \angle 2$ (Transitive).

12. (1) $\overline{QD} \parallel \overline{UA}$, $\overline{QU} \parallel \overline{DA}$ (Given); (2) $m\angle 1 = m\angle 4$, $m\angle 2 = m\angle 3$ (Postulate 4.1); (3) $m\angle 1 + m\angle 2 = m\angle 3 + m\angle 4$ (Addition); (4) $m\angle QUA = m\angle ADQ$ (Substitution); (5) $\angle QUA \cong \angle ADQ$ (Angles equal in measure are congruent).

13. (1) $\overline{AT} \parallel \overline{MH}$, $\angle M \cong \angle H$ (Given); (2) $\angle A$ is supplementary to $\angle M$ and $\angle T$ is supplementary to $\angle H$ (Theorem 4.4); (3) $\angle A \cong \angle T$ (Supplements of congruent angles are congruent).

14. (1) $\overline{IB} \parallel \overline{ET}$, $\overline{IS}$ bisects $\angle EIB$, $\overline{EC}$ bisects $\angle TEI$ (Given); (2) $m\angle BIS = \frac{1}{2}m\angle EIB$ and $m\angle TEC = \frac{1}{2}m\angle TEI$ (Definition of angle bisector); (3) $m\angle EIB = m\angle TEI$ (Postulate 4.1); (4) $m\angle BIS = m\angle TEC$ (Halves of equals are equal); (5) $\angle BIS \cong \angle TEC$ (Angles equal in measure are congruent).

15. (1) $\angle B \cong \angle D$, $\overline{BA} \parallel DC$ (Given); (2) $\angle B \cong \angle C$ (Postulate 4.1); (3) $\angle C \cong \angle D$ (Transitive); (4) $\overline{BC} \parallel \overline{DE}$ (Postulate 4.2).

16. (1) $k \parallel l$, $\angle 5 \cong \angle 8$ (Given); (2) $\angle 5 \cong \angle 1$ (Theorem 4.2); (3) $\angle 1 \cong \angle 8$ (Transitive); (4) $j \parallel l$ (Postulate 4.2).

17. (1) $\angle K = \angle P$, $m\angle J + m\angle P = 180$ (Given); (2) $m\angle J + m\angle K = 180$ (Substitution); (3) $\overline{KL} \parallel \overline{JP}$ (Theorem 4.7).

18. (1) $\overline{AB} \perp \overline{BC}$, $\angle ACB$ is complementary to $\angle ABE$ (Given); (2) $m\angle ABE + m\angle ABC + m\angle CBD = 180$ (A straight angle has measure 180); (3) $m\angle ABE + 90 + m\angle CBD = 180$ (Substitution); (4) $m\angle ABE + m\angle CBD = 90$ (Subtraction); (5) $\angle CBD$ is complementary to $\angle ABE$ (Definition of complementary angles); (6) $\angle ACB \cong \angle CBD$ (Complements of the same angle are congruent); (7) $\overleftrightarrow{AC} \parallel \overleftrightarrow{EBD}$ (Postulate 4.2).

19. (1) $\overline{AG} \parallel \overline{BC}$, $\overline{KH} \parallel \overline{BC}$, $\angle 1 \cong \angle 2$ (Given); (2) $\overline{AG} \parallel \overline{KH}$ (Segments parallel to the same segment are parallel to each other—see exercise 20); (3) Angles 1 and 2 are supplementary (Theorem 4.4); (4) Angles 1 and 2 are right angles (If two angles are supplementary and congruent, then each is a right angle); (5) $\overline{HK} \perp \overline{AB}$ (Segments which intersect to form a right angle are perpendicular).

20. GIVEN $l \parallel p$ and $m \parallel p$.
PROVE $l \parallel m$
PLAN Show $\angle 1 \cong \angle 2$.
$\angle 1 \cong \angle 3$ and $\angle 2 \cong \angle 3$. Since $\angle 1 \cong \angle 2$, $l \parallel m$ by Theorem 4.5.

21. GIVEN $\overrightarrow{XL}$ bisects $\angle AXY$,
$\overrightarrow{YM}$ bisects $\angle DYX$, $\overleftrightarrow{XL} \parallel \overleftrightarrow{YM}$.
PROVE $\overleftrightarrow{AXB} \parallel \overleftrightarrow{CYD}$
PLAN Show $m\angle AXY = m\angle DYX$.
$m\angle 1 = m\angle 2$; $m\angle AXY = 2m\angle 1$ and $m\angle DYX = 2m\angle 2$. Since doubles of equals are equal, $m\angle AXY = m\angle DYX$ and $\overleftrightarrow{AXB} \parallel \overleftrightarrow{CYD}$ by Postulate 4.2.

22. GIVEN $\overrightarrow{RAS} \parallel \overrightarrow{PCQ}$,
$\overrightarrow{AB}$ bisects $\angle HAS$,
$\overrightarrow{CD}$ bisects $\angle QCA$.
PROVE $\overrightarrow{AB} \parallel \overrightarrow{CD}$
PLAN Show $m\angle 1 = m\angle 2$.
$m\angle HAS = m\angle QCA$; $m\angle 1 = \frac{1}{2}m\angle HAS$ and $m\angle 2 = \frac{1}{2}m\angle QCA$. Hence, $m\angle 1 = m\angle 2$ and $\overrightarrow{AB} \parallel \overrightarrow{CD}$ by Theorem 4.5.

CHAPTER 5

1. **a** 52 **b** 67 **c** 47 **d** 22 **e** 127

2. $m\angle 1 = 70$ $m\angle 4 = 40$ $m\angle 7 = 70$
$m\angle 3 = 70$ $m\angle 5 = 140$ $m\angle 8 = 40$

3. **a** $x = 18$. Right triangle since $m\angle C = 90$.
b $x = 22$. Obtuse triangle since $m\angle B = 105$.
c $x = 30$. Acute triangle.

4. 40

5. 50

6. 90

7. 113

8. **a** 360 **b** 720 **c** 1,260 **d** 1,980

9. **a** 12 **b** 17 **c** 5 **d** 14

10. **a** 125 **b** 104 **c** 170

11. **a** 108 **b** 165 **c** 135 **d** 156

12. **a** 20 **b** 10 **c** 9 **d** 30

13. **c** 27

14. 10

15. **a** 8 **b** 4 **c** 15

16–18 PLAN Show two angles of one triangle are congruent to two angles of the other triangle. Then apply Corollary 5.2.4.

16. $\angle B \cong \angle D$ and $\angle ACB \cong \angle ECD$ so that $\angle A \cong \angle E$.

17. $\angle C \cong \angle DEB$ and $\angle B \cong \angle B$ so that $\angle 1 \cong \angle 2$.

18. $\angle P \cong \angle K$ and $\angle KML \cong \angle PRJ$ so that $\angle J \cong \angle L$. Hence, $\overline{KL} \parallel \overline{PJ}$ by Postulate 4.2.

19. Angles 1 and 2 are each complementary to $\angle DCE$ and are therefore congruent to each other.

20. Angles CAX and ACY are supplementary. Hence, $\frac{1}{2}m\angle CAX + \frac{1}{2}m\angle ACY = \frac{1}{2}(180) = 90$. By substitution, the sum of the measures of angles BAC and BCA is equal to 90. This implies that $m\angle ABC = 90$ and $\angle ABC$ is a right angle.

CHAPTER 6

1. Use SAS. $\overline{BM} \cong \overline{BM}$; right angle $BMA \cong$ right angle BMC; $\overline{AM} \cong \overline{CM}$.

2. Use ASA. $\angle SRT \cong \angle WRT$; $\overline{RT} \cong \overline{RT}$; $\angle STR \cong \angle WTR$.

3. Use ASA. $\angle F \cong \angle A$; by addition, $\overline{AC} \cong \overline{FD}$; $\angle EDF \cong \angle BCA$.

4. Use ASA. $\angle R \cong \angle T$; $\overline{SR} \cong \overline{ST}$; $\angle S \cong \angle S$.

5. Use ASA. Right angle $REW \cong$ right angle THW; $\overline{EW} \cong \overline{HW}$; $\angle EWR \cong \angle HWT$.

6. Use AAS. $\angle UQX \cong \angle DAX$; $\angle UXQ \cong \angle DXA$; $\overline{QU} \cong \overline{DA}$.

7. Use Hy-Leg. $\overline{JL} \cong \overline{EV}$ (Hyp) and by addition, $\overline{KL} \cong \overline{TV}$ (Leg).

8. Use ASA. Right angle $JKL \cong$ right angle ETV; $\overline{KL} \cong \overline{TV}$; by taking supplements of congruent angles, $\angle EVT \cong \angle JLK$.

9. Use AAS. $\angle ARF \cong \angle ARI$; by subtraction, $\angle RFA \cong \angle RIA$; $\overline{AR} \cong \overline{AR}$.

10. Use SAS. $\overline{TS} \cong \overline{RS}$; by subtraction, $\angle TSW \cong \angle RSP$; $\overline{SW} \cong \overline{SP}$.

11. Use AAS. Since $\overline{RP} \parallel \overline{SW}$, $\angle TSW \cong \angle SRP$ (Corresponding angles). Also, since $\overline{SP} \parallel \overline{TW}$, $\angle STW \cong \angle RSP$. $\overline{TW} \cong \overline{SP}$.

12. Use SSS. $\overline{AB} \cong \overline{DE}$; $\overline{BM} \cong \overline{EM}$; $\overline{AM} \cong \overline{DM}$ by transitivity.

13. Use AAS. $\angle 1 \cong \angle 2$. $\angle A \cong \angle 1$ and $\angle D \cong \angle 2$, so that $\angle A \cong \angle D$ (Angle); since complements of congruent angles are congruent, $\angle AMB \cong \angle DME$ (Angle); $\overline{BM} \cong \overline{EM}$ (Side).

14. Use ASA. Right angle AFC = right angle BDC; $\angle C \cong \angle C$; $\overline{FC} \cong \overline{DC}$.

15. Use SAS. $\overline{AD} \cong \overline{BF}$; $\angle BAD \cong \angle ABF$; $\overline{AB} \cong \overline{AB}$.

16. Use Hy-Leg. $\overline{AB} \cong \overline{DC}$ (Hyp); by subtraction, $\overline{AF} \cong \overline{CE}$ (Leg).

17. Use AAS. Right angle KLG = right angle JRG; $\angle G \cong \angle G$; $\overline{KL} \cong \overline{JR}$.

18. Use AAS. Right angle KRO = right angle JLO; $\angle ROK \cong \angle LOJ$; by halves of equals are equal, $\overline{KR} \cong \overline{JL}$.

19. Use ASA. $\angle AEB \cong \angle DEC$; $\overline{EB} \cong \overline{EC}$; by halves of equals are equal, $\angle 1 \cong \angle 2$.

20. Use SSS (all other methods, except Hy-Leg, can also be used).

> **GIVEN** $\triangle ABC \cong \triangle XYZ$ and $\triangle RST \cong \triangle XYZ$.
>
> **PROVE** $\triangle ABC \cong \triangle RST$.
>
> **PLAN** Using the reverse of the definition of congruent triangles, $\overline{AB} \cong \overline{XY}$ and $\overline{RS} \cong \overline{XY}$ so that $\overline{AB} \cong \overline{RS}$. Similarly, show that each of the remaining pairs of corresponding sides are congruent.

CHAPTER 7

1. a 76   **b** 20   **c** 40

    d 32   **e** 16   **f** 36

2. Show $\triangle HGF \cong \triangle HJF$ by SAS. By CPCTC, $\angle GFH \cong \angle JFH$. $\angle 1 \cong \angle 2$ since supplements of congruent angles are congruent.

3. Show $\triangle ABC \cong \triangle DCB$ by Hy-Leg. $\angle 1 \cong \angle 2$ by CPCTC.

4. Show $\triangle JPY \cong \triangle KPX$ by ASA. $\overline{JY} \cong \overline{KX}$ by CPCTC.

5. $\overline{PJ} \cong \overline{PK}$ by addition. Show $\triangle KXP \cong \triangle JYP$ by SAS where $\overline{PK} \cong \overline{PJ}$, $\angle P \cong \angle P$, and $\overline{PX} \cong \overline{PY}$. By CPCTC, $\overline{KX} \cong \overline{JY}$.

6. Show $\triangle UTQ \cong \triangle DWA$ by SAS where $\overline{UT} \cong \overline{DW}$, $\angle QTU \cong \angle AWD$ (Alternate exterior angles); $\overline{QT} \cong \overline{AW}$ (Subtraction). By CPCTC, $\angle UQT \cong \angle DAW$ which implies $\overline{UQ} \parallel \overline{AD}$.

7. Show $\triangle SRT \cong \triangle HWT$ by AAS. By CPCTC, $\overline{RT} \cong \overline{WT}$ and T is the midpoint of $\overline{RW}$.

8. Show $\triangle ADB \cong \triangle CDB$ by SSS. By CPCTC, $\angle ADB \cong \angle CDB$ and $\overline{DB}$ bisects $\angle ADC$.

9. Show $\triangle BAM \cong \triangle CDM$ by AAS. By CPCTC, $\overline{AM} \cong \overline{DM}$ and $\overline{BC}$ bisects $\overline{AD}$.

10. $\angle RHK \cong \angle NHK$ and $\angle HRK \cong \angle HNK$ (Supplements of congruent angles are congruent). If two angles of one triangle are congruent to corresponding angles of another triangle, then the third pair of angles are congruent. Hence, $\angle HKR \cong \angle HKN$ which implies $\overline{HK} \perp \overline{RN}$.

11. Show $\triangle RTS \cong \triangle ACB$ by SSS. By CPCTC, $\angle 1 \cong \angle 2$. If two angles are supplementary (see Given) and congruent, then each is a right angle. Hence, $\overline{ST} \perp \overline{TR}$ and $\overline{BC} \perp \overline{AC}$.

12. (1) Show $\triangle RLS \cong \triangle RLT$ by Hy-Leg where $\overline{RL} \cong \overline{RL}$ (Hyp) and $\overline{RS} \cong \overline{RT}$ (Leg).

     (2) By CPCTC, $\overline{SL} \cong \overline{TL}$ and $\angle RLS \cong \angle RLT$. Taking supplements of congruent angles, $\angle SLW \cong \angle TLW$. $\triangle SLW \cong \triangle TLW$ by SAS: $\overline{SL} \cong \overline{TL}$ (Side), $\angle SLW \cong \angle TLW$ (Angle), and $\overline{LW} \cong \overline{LW}$ (Side). By CPCTC, $\angle SWL \cong \angle TWL$ so that $\overline{WL}$ bisects $\angle SWT$.

13. Show $\triangle PMK \cong \triangle PML$ by Hy-Leg. By CPCTC, $\overline{KM} \cong \overline{LM}$ and M is the midpoint of $\overline{KL}$.

14. Show $\triangle PMK \cong \triangle PML$ by SSS. By CPCTC, $\angle KMP \cong \angle LMP$ which implies $\overline{PM} \perp \overline{KL}$.

15. Show $\triangle PSL \cong \triangle LTP$ by AAS. By CPCTC, $\overline{PS} \cong \overline{LT}$.

16. Show $\triangle XLR \cong \triangle XPS$ by Hy-Leg where $\overline{XR} \cong \overline{XS}$ (Hyp) and $\overline{XL} \cong \overline{XP}$ (Leg). By CPCTC, $\angle R \cong \angle S$ which implies $\triangle RTS$ is isosceles.

17. $\angle 1 \cong \angle 3$. Since parallel lines form congruent corresponding angles, $\angle 1 \cong \angle 2$ and $\angle 3 \cong \angle 4$. By transitivity, $\angle 2 \cong \angle 4$ which implies $\triangle PQR$ is isosceles.

18. Show $\triangle TWL \cong \triangle PXF$ by Hy-Leg where by addition $\overline{TL} \cong \overline{PF}$ (Hyp) and $\overline{WL} \cong \overline{XF}$ (Leg). By CPCTC, $\angle 1 \cong \angle 2$ which implies $\triangle FML$ is isosceles.

19. Show $\triangle KLE \cong \triangle ABW$ by AAS: $\angle E \cong \angle W$ (Base Angles Theorem); $\angle BAW \cong \angle LKE$ (Supplements of congruent angles are congruent); $\overline{EL} \cong \overline{WB}$. By CPCTC, $\overline{KL} \cong \overline{AB}$.

20. By the Base Angles Theorem, $\angle 1 \cong \angle 2$ and $\angle 3 \cong \angle 4$ since supplements of congruent angles are congruent. $\triangle VOK \cong \triangle VLZ$ by SAS. By CPCTC, $\overline{VK} \cong \overline{VZ}$ which implies that $\triangle KVZ$ is isosceles.

21. Show $\triangle BJH \cong \triangle KJL$ by ASA where right $\angle BHJ$ is congruent to right $\angle KLJ$, $\overline{HJ} \cong \overline{LJ}$ (Converse of the Base Angles Theorem), and $\angle HJB \cong \angle LJK$. By CPCTC, $\overline{JB} \cong \overline{JK}$.

22–26. These exercises are double congruence proofs requiring that one pair of triangles be proven congruent in order to obtain congruent pairs of parts which may be used in proving a second pair of triangles congruent.

22. First show $\triangle ABC \cong \triangle ADC$ by SAS in order to obtain $\overline{BC} \cong \overline{DC}$ and $\angle ACB \cong \angle ACD$. Taking supplements of congruent angles yields $\angle BCE \cong \angle DCE$. Since $\overline{CE} \cong \overline{CE}$, $\triangle BCE \cong \triangle DCE$ by SAS.

23. First show $\triangle BEC \cong \triangle DEC$ by SSS in order to obtain $\angle BCE \cong \angle DCE$ so that $\angle BCA \cong \angle DCA$. Show $\triangle ABC \cong \triangle ADC$ by SAS. By CPCTC, $\angle BAC \cong \angle DAC$ which implies that $\overline{CA}$ bisects $\angle DAB$.

24. First show $\triangle ABL \cong \triangle CDM$ by SAS in order to obtain $\overline{BL} \cong \overline{DM}$ and $\angle ALB \cong \angle CMD$. Taking supplements of congruent angles gives $\angle BLC \cong \angle DMA$. Using the addition property, $\overline{AM} \cong \overline{CL}$. Hence, $\triangle CLB \cong \triangle AMD$ by SAS. By CPCTC, $\angle CBL \cong \angle ADM$.

25. Show $\triangle AFD \cong \triangle CFE$ by ASA where right $\angle ADF$ = right $\angle CEF$, $\overline{AF} \cong \overline{CF}$ (Converse of the Base Angles Theorem), $\angle AFD \cong \angle CFE$. By CPCTC, $\overline{FD} \cong \overline{FE}$. $\triangle BDF \cong \triangle BEF$ by Hy-Leg. By CPCTC, $\angle DBF \cong \angle EBF$ which implies that $\overline{BF}$ bisects $\angle DBE$.

26. Show $\triangle BDF \cong \triangle BEF$ by SSS so that $\angle BDF \cong \angle BEF$. Show $\triangle AFD \cong \triangle CFE$ by ASA where $\angle DFA \cong \triangle EFC$; $\overline{DF} \cong \overline{EF}$; $\angle ADF \cong \angle CEF$. By CPCTC, $\overline{AF} \cong \overline{CF}$ which implies that $\triangle AFC$ is isosceles.

27. **GIVEN** $\triangle ABC \cong \triangle RST$ and $\overline{BH}$ and $\overline{SK}$ are altitudes.

PROVE $\overline{BH} \cong \overline{SK}$.

PLAN Show $\triangle BAH \cong \triangle SRK$ by AAS where right $\angle BHA$ = right $\angle SKR$, and using the reverse of the definition of congruent triangles, $\angle A \cong \angle S$ and $\overline{AB} \cong \overline{RS}$. By CPCTC, $\overline{BH} \cong \overline{SK}$.

28. **GIVEN** Equilateral $\triangle ABC$.

PROVE $\angle A \cong \angle B \cong \angle C$.

PLAN Since $\overline{AB} \cong \overline{BC}$, $\angle A \cong \angle C$. Since $\overline{AC} \cong \overline{BC}$, $\angle A \cong \angle B$. Hence, $\angle B \cong \angle C$ so that $\angle A \cong \angle B \cong \angle C$.

29. **GIVEN** $\triangle PEG$, $\overline{PE} \cong \overline{GE}$, altitudes $\overline{GH}$ and $\overline{PK}$ are drawn.

PROVE $\overline{GH} \cong \overline{PK}$.

PLAN Show $\triangle GHP \cong \triangle PKG$ by AAS where right $\angle PHG$ = right $\angle GKP$ $\angle HPG \cong \angle KGP$ (Base Angles Theorem); $\overline{PG} \cong \overline{PG}$. By CPCTC, $\overline{GH} \cong \overline{PK}$.

30. **GIVEN** $\triangle ART$, $\overline{AR} \cong \overline{TR}$, medians $\overline{AY}$ and $\overline{TX}$ are drawn.

PROVE $\overline{AY} \cong \overline{TX}$.

PLAN Show $\triangle AXT \cong \angle TYA$ by SAS where $\overline{AT} \cong \overline{AT}$; $\angle XAT \cong \angle YTA$ (Base Angles Theorem); $\overline{AX} \cong \overline{TY}$ (Halves of equals are equal). By CPCTC, $\overline{AY} \cong \overline{TX}$.

CHAPTER 8

1. *Converse:* If I live in the United States, then I live in Los Angeles. (False.)
Inverse: If I do not live in Los Angeles, then I do not live in the United States. (False.)
Contrapositive: If I do not live in the United States, then I do not live in Los Angeles. (True.)

2. *Converse:* If tomorrow is Tuesday, then today is Monday. (True.)
Inverse: If today is not Monday, then tomorrow is not Tuesday. (True.)
Contrapositive: If tomorrow is not Tuesday, then today is not Monday. (True.)

3. *Converse:* If the sides of a triangle are congruent, then the angles opposite these sides are congruent. (True.)
Inverse: If two angles of a triangle are not congruent, then the sides opposite these angles are not congruent. (True.)

Contrapositive: If the sides of a triangle are not congruent, then the angles opposite these sides are not congruent. (True.)

4. $\overline{BC}$.

5. $\angle S$.

6. a True. **b** False. **c** True.
d True. **e** True.

7. $\angle B$.

8. a $\angle R$ **b** $\angle S$

9. a Yes. **b** No. **c** No. **d** Yes.

10. $m\angle 1 = m\angle 2$ and $m\angle 3 > m\angle 2$. By substitution, $m\angle 3 > m\angle 1$ which implies $AB > BD$.

11. $m\angle 1 > m\angle A$. By substitution, $m\angle 2 > m\angle A$ which implies $AD > ED$.

12. $m\angle 4 > m\angle 3$ and $m\angle 3 > m\angle AEC$. Hence, $m\angle 4 > m\angle AEC$.

13. Since $AC > BC$, $m\angle 3 > m\angle EAC$. Since $m\angle 2 < m\angle EAC$, $m\angle 3 > m\angle 2$ which implies $AD > BD$.

14. Since $AD > BD$, $m\angle 3 > m\angle 2$. $m\angle 1 = m\angle 2$ and by substitution $m\angle 3 > m\angle 1$. But $m\angle 4 > m\angle 3$. Hence, $m\angle 4 > m\angle 1$ which implies $AC > DC$.

15–20. Use an indirect proof.

15. Assume $\angle 1 \cong \angle 2$. By transitivity, $\angle 2 \cong \angle 3$. But this contradicts the exterior angle of a triangle inequality theorem. Hence, $\angle 1 \not\cong \angle 2$.

16. Assume $\overline{AB} \cong \overline{BC}$. Then $\angle 2 \cong \angle C$ and, by transitivity, $\angle 1 \cong \angle C$. But this contradicts the exterior angle of a triangle inequality theorem. Hence, $\overline{AB} \not\cong \overline{BC}$.

17. Assume $RW = WL$. Then $m\angle WRL = m\angle WLR$. $m\angle WLR > m\angle T$. By substitution, $m\angle WRL > m\angle T$. Since $m\angle SRT > m\angle WRL$, $m\angle SRT > m\angle T$ which implies that $TS > RS$. This contradicts the given. Hence, $RW \ne WL$.

18. Assume $\triangle ADC$ is isosceles. Then $AD = CD$. $\triangle ADB \cong \triangle CDB$ by SAS. By CPCTC, $\overline{AB} \cong \overline{CB}$, but this contradicts the given ($\triangle ABC$ is *not* isosceles). Hence, $\triangle ADC$ is *not* isosceles.

19. Assume $\overline{AB} \parallel \overline{DE}$. Then $\angle B \cong \angle D$. Since $\overline{DE} \cong \overline{CE}$, $\angle D \cong \angle DCE$. By transitivity, $\angle B \cong \angle DCE$. Since $\angle ACB \cong \angle DCE$, $\angle B \cong \angle ACB$ which implies $\overline{AC} \cong \overline{AB}$. This contradicts the given ($AC > AB$). Hence $\overline{AB}$ is *not* parallel to $\overline{DE}$.

20. Assume $\overline{BD} \perp \overline{AC}$. $\triangle ABD \cong \triangle CBD$ by ASA. By CPCTC, $\overline{AB} \cong \overline{CB}$ which contradicts the assumption that $\triangle ABC$ is scalene. Hence, $\overline{BD}$ is *not* perpendicular to $\overline{AC}$.

21. **GIVEN** Equilateral $\triangle RST$. Point X is any point on $\overline{RT}$ and $\overline{SX}$ is drawn.

PROVE $SX < RS$.

PROOF $m\angle SXR > m\angle T$. Since an equilateral triangle is also equiangular, $m\angle R = m\angle T$ so that $m\angle SXR > m\angle R$ which implies that $RS > SX$ or, equivalently, $SX < RS$. Since $RS = ST = RT$, $SX < ST$ and $SX < RT$.

22.

GIVEN △BET with $\overline{BE} \cong \overline{TE}$. Angle E is obtuse.

PROVE $BT > BE$.

PROOF $m\angle T < 90$ since a triangle may have at most one nonacute angle. Hence, $m\angle E > m\angle T$ which implies $BT > BE$. Since $BE = TE$, $BT > TE$.

23. Suppose point P is any point not on line l. Draw a segment from point P and perpendicular to l, intersecting l at point A. Choose any other point on l, say point X, so that A and X are different points. Draw $\overline{PX}$. $\overline{PX}$ is the hypotenuse of right triangle PAX. Since $\overline{PX}$ lies opposite the greatest angle of the triangle, it must be the longest side of the triangle. Hence, $PA < PX$.

CHAPTER 9

1. 36

2. $m\angle T = m\angle M = 105$.
$m\angle H = m\angle A = 75$.

3. $m\angle R = m\angle G = 55$.
$m\angle T = m\angle I = 125$.

4. 23

5. $EF = 19$, $RT = 38$.

6. **a** $YT = 28.5$ and $BE = 41.5$.
b $YT = 35$, $LM = 41$, $BE = 47$.

7. 22

8. 45

9. In $\square ABCD$, $\overline{AB} \cong \overline{CD}$ and $\overline{AB} \parallel \overline{CD}$ so that $\angle BAE \cong \angle DCF$. Show △AEB ≅ △CFD by SAS. By CPCTC, $\angle ABE \cong \angle CDF$.

10. In $\square ABCD$, $\overline{AD} \parallel \overline{BC}$ so that $\overline{BH} \parallel \overline{DE}$ and $\angle H \cong \angle E$. $\angle BAD \cong \angle BCD$. Taking supplements of congruent angles, $\angle FAE \cong \angle GCH$. △FAE ≅ △GCH by AAS. By CPCTC, $\overline{AF} \cong \overline{CG}$.

11. In $\square ABCD$, $\overline{AB} \cong \overline{CD}$ and $\overline{AB} \parallel \overline{DC}$. $\overline{AB} \cong \overline{BE}$ (Definition of midpoint). By transitivity, $\overline{CD} \cong \overline{BE}$; $\angle EBF \cong \angle DCF$; $\angle EFB \cong \angle DFC$. △BEF ≅ △CDF by AAS. By CPCTC, $\overline{EF} \cong \overline{FD}$.

12. Right △ABM ≅ right △DCM by SAS. By CPCTC, $\overline{AM} \cong \overline{DM}$, which implies that △AMD is isosceles.

13. Since the diagonals of a rhombus are the perpendicular bisectors of each other, $\overline{AR} \cong \overline{CR}$ and angles ARS and CRS are right angles. Show △SRA ≅ △SRC by SAS. By CPCTC, $\overline{SA} \cong \overline{SC}$ which implies that △ASC is isosceles.

14. $\angle EBC \cong \angle ECB$. Since $\overline{AD} \parallel \overline{BC}$, $\angle AFB \cong \angle EBC$ and $\angle DGC \cong \angle ECB$. By transitivity, $\angle AFB \cong \angle DGC$. $\angle A \cong \angle D$ and $\overline{AB} \cong \overline{DC}$ so that △FAB ≅ △GDC by AAS. By CPCTC, $\overline{AF} \cong \overline{DG}$.

15. Since $AD > DC$, $m\angle ACD > m\angle DAC$. Since $\overline{AB} \parallel \overline{DC}$, $m\angle BAC = m\angle ACD$. By substitution, $m\angle BAC > m\angle DAC$.

16. Since ABCD is a parallelogram, $m\angle ABC = m\angle ADC$. $m\angle RBS = m\angle SDR$ since halves of equals are equal. In triangles BAR and DCS, $\angle A \cong \angle C$ and $\angle ABR \cong \angle CDS$. This implies that

angles CSD and ARB, the third pair of angles, are congruent. Since supplements of congruent angles are congruent, $\angle BRD \cong \angle DSB$. BRDS is a parallelogram since both pairs of opposite angles are congruent.

17. $\overline{BL} \cong \overline{DM}$ and $\overline{BL} \parallel \overline{DM}$. Taking supplements of congruent angles, $\angle ALB \cong \angle CMD$. △ALB ≅ △CMD by SAS. By CPCTC, $\overline{AB} \cong \overline{CD}$ and $\angle BAL \cong \angle DCM$ so that $\overline{AB} \parallel \overline{CD}$. Hence, ABCD is a parallelogram by Theorem 9.9.

18. $\overline{AB} \cong \overline{CD}$ and $\overline{AB} \parallel \overline{CD}$ so that $\angle BAL \cong \angle DCM$. △ALB ≅ △CMD by ASA, $\overline{BL} \cong \overline{DM}$ and $\angle ALB \cong \angle CMD$ by CPCTC. Taking supplements of congruent angles, $\angle BLM \cong \angle DML$ so that $\overline{BL} \parallel \overline{DM}$. Hence, BLDM is a parallelogram by Theorem 9.9.

19. $\overline{AB} \cong \overline{BC}$ since a rhombus is equilateral. △ABL ≅ △BCM by SSS. By CPCTC, $\angle B \cong \angle C$. Since $\overline{AB} \parallel \overline{DC}$, angles B and C are supplementary. Hence, each is a right angle and ABCD is a square.

20. Use an indirect proof. Assume ABCD is a rectangle. Then △BAD ≅ △CDA by SAS. By CPCTC, $\angle 1 \cong \angle 2$ which contradicts the given.

21. $DE = \frac{1}{2}BC$ and $DF = \frac{1}{2}AB$. Since $DE = DF$, $\frac{1}{2}BC = \frac{1}{2}AB$ implies that $BC = AB$. Hence, △ABC is isosceles.

22. If $\overline{BC} \parallel \overline{WT}$, then $\overline{AC} \parallel \overline{WT}$ (Extensions of parallel segments are parallel). $\overline{AW} \parallel \overline{TC}$ (Segments of parallel lines are parallel). Hence, WACT is a parallelogram.

23. Show △RSW ≅ △WTR by SAS. By CPCTC, $\angle TRW \cong \angle SWR$ so that $\overline{RP} \cong \overline{WP}$ (Converse of the Base Angles Theorem).

24. By addition, $AG = DF$. $\overline{EF} \cong \overline{EG}$ (Given) and by the Converse of the Base Angles Theorem, $\angle EGF \cong \angle EFG$. $\overline{BG} \cong \overline{CF}$ (Given). △BAG ≅ △CDF by SAS. By CPCTC, $\overline{AB} \cong \overline{DC}$. Hence, trapezoid ABCD is isosceles.

25. By the Base Angles Theorem, $\angle PLM \cong \angle PML$. Since $\overline{LM}$ is a median, $\overline{LM} \parallel \overline{AD}$ so that $\angle PLM \cong \angle APL$ and $\angle PML \cong \angle DPM$. By transitivity, $\angle APL \cong \angle DPM$. Show △LAP ≅ △MDP by SAS. By CPCTC, $\angle A \cong \angle D$ which implies trapezoid ABCD is isosceles.

26. Since ABCD is a trapezoid, $\overline{BC} \parallel \overline{KD}$. Show $\overline{BK} \parallel \overline{DC}$ as follows: $\angle BAK \cong \angle CDA$ and by transitivity $\angle BKA \cong \angle CDA$; since corresponding angles are congruent, $\overline{BK} \parallel \overline{DC}$. BKDC is a parallelogram since both pairs of opposite sides are parallel.

27. See the hint in the statement of the problem. Since the diagonals of a rhombus bisect its angles, $m\angle EDA > m\angle EAD$. This implies that $AE > DE$. Equivalently, $2AE > 2DE$. Since the diagonals of a rhombus bisect each other, AC may be substituted for 2AE and BD may be substituted for 2DE so that $AC > BD$.

28. See the hint in the statement of the problem. By dropping perpendiculars, a parallelogram is formed. Since the opposite sides of a parallelogram are equal in length, the perpendicular segments have the same length.

29.
GIVEN ▱$ABCD$. Points X, Y, Z, and W are the midpoints of sides $\overline{AB}$, $\overline{BC}$, $\overline{CD}$, and $\overline{DA}$, respectively.

PROVE Quadrilateral $WXYZ$ is a parallelogram.

PLAN Draw diagonal $\overline{BD}$. In $\triangle BAD$, $\overline{XW}$ is parallel to $\overline{BD}$. In $\triangle BCD$, $\overline{YZ}$ is parallel to $\overline{BD}$. Hence, $\overline{XW} \parallel \overline{YZ}$. Similarly, by drawing diagonal $\overline{AC}$, show that $\overline{XY} \parallel \overline{WZ}$. Since both pairs of opposite sides are parallel, quadrilateral $WXYZ$ is a parallelogram.

30.
GIVEN Rectangle $ABCD$. Points E, F, G, and H are the midpoints of sides $\overline{AB}$, $\overline{BC}$, $\overline{CD}$, and $\overline{DA}$, respectively.

PROVE Quadrilateral $EFGH$ is a rhombus.

PLAN First prove $EFGH$ is a parallelogram (see exercise 29). Next, show that an adjacent pair of sides are congruent. Show $\overline{HE} \cong \overline{HG}$ by proving $\triangle EAH \cong \triangle GDH$ by SAS.

CHAPTER 10

1. 84

2. 72

3. 75, 105, 75, 105

4. **a** 24 **b** 10 **c** $\frac{47}{8}$

d 5 **e** −3 or 7

5. **a** 8 **b** $6e^2$ **c** $\frac{1}{4}$ **d** $\sqrt{54}$

6. **a** Yes. **b** Yes. **c** No. **d** No.

7. $\frac{AL}{AB} = \frac{AM}{AG}$; $\frac{LB}{AB} = \frac{MG}{AG}$; $\frac{AL}{LB} = \frac{AM}{MG}$.

8. **a** Yes. **b** Yes. **c** No.

d No. **e** Yes.

9. **a** 15 **b** 4 **c** 7

d 24 **e** $KE = 9$, $KH = 3$.

10. ∡$G \cong$ ∡S, ∡$A \cong$ ∡H, ∡$L \cong$ ∡E
$\frac{GA}{SH} = \frac{AL}{HE} = \frac{GL}{SE}$

11. 40

12. $MY = 5$, $MX = 35$.

13. 44

14. 12 and 27

15. **a** 12 and 8 **b** 17.5 and 14

c 10 and 16 **d** 5

16. 2. Parallel Postulate.
3. Postulate 10.1.
4. Substitution.
5. If two lines are parallel, then the corresponding angles are congruent.
6. Given.
7. Transitive property.

8. If two angles of one triangle are congruent to two angles of another triangle, then the third pair of angles are congruent.
9. ASA Postulate.
10. CPCTC.
11. Substitution.

17–29 are based on the application of the AA Theorem of Similarity.

17. ∡$W \cong$ ∡Y and right ∡$HAW \cong$ right ∡KBY implies $\triangle HWA \sim \triangle KYB$.

18. ∡$XBC \cong$ ∡YST and ∡$C \cong$ ∡T (Reverse of the definition of similar triangles). Hence, $\triangle BXC \sim \triangle SYT$.

19. ∡$CTM \cong$ ∡AWB (Reverse of the definition of similar triangles). Since ∡$AWB \cong$ ∡CWB (Definition of angle bisector), ∡$CTM \cong$ ∡CWB by transitivity. ∡$MCT \cong$ ∡BCW. Hence, $\triangle MCT \sim \triangle BCW$.

20. ∡$S \cong$ ∡MWB (Angle) and ∡$A \cong$ ∡WCA (Base Angles Theorem). Since $\overline{AW} \parallel \overline{ST}$, ∡$A \cong$ ∡BTS and by transitivity, ∡$BTS \cong$ ∡WCA (Angle). Hence, $\triangle BCW \sim \triangle BTS$.

21. ∡$WBC \cong$ ∡WCB and taking supplements of these angles, ∡$ABW \cong$ ∡TCW (Angle). Since $\overline{AT}$ bisects ∡STW, ∡$STA \cong$ ∡WTC. ∡$A \cong$ ∡STA since $\overline{ST} \parallel \overline{AW}$. By transitivity, ∡$A \cong$ ∡WTC (Angle). Hence, $\triangle ABW \sim \triangle TCW$.

22. Show $\triangle TAX \sim \triangle WHY$. ∡$W \cong$ ∡ATX and ∡$Y \cong$ ∡AXT since $\parallel$ line segments have congruent corresponding angles.

23. Show $\triangle RMN \sim \triangle RAT$. ∡$RMN \cong$ ∡A and ∡$RNM \cong$ ∡T. Write $\frac{MN}{AT} = \frac{RN}{RT}$. Using the Converse of the Base Angles Theorem, $NT = MN$. Substitute NT for MN in the proportion.

24. Show $\triangle SQP \sim \triangle WRP$. ∡$SPQ \cong$ ∡WPR (Angle). Since $\overline{SR} \cong \overline{SQ}$, ∡$SRQ \cong$ ∡SQR. ∡$SRQ \cong$ ∡WRP (Definition of angle bisector). By transitivity, ∡$SQR \cong$ ∡WRP (Angle).

25. Show $\triangle PMQ \sim \triangle MKC$. Right ∡$MCK \cong$ right ∡PMQ. Since $\overline{TP} \cong \overline{TM}$, ∡$TPM \cong$ ∡TMP.

26. Show $\triangle PMT \sim \triangle JKT$. Since $\overline{MP} \cong \overline{MQ}$, ∡$MPQ \cong$ ∡MQP. Since $\overline{JK} \parallel \overline{MQ}$, ∡$J \cong$ ∡MQP so that by transitivity, ∡$J \cong$ ∡MPQ (Angle). ∡$K \cong$ ∡QMT (Congruent alternate interior angles). ∡$QMT \cong$ ∡PMT (since $\triangle MTP \cong \triangle MTQ$ by SSS). By transitivity, ∡$K \cong$ ∡PMT (Angle). Write the proportion $\frac{PM}{JK} = \frac{PT}{JT}$. Substitute TQ for PT (see Given) in the proportion.

27. Show $\triangle EIF \sim \triangle HIG$. Since $\overline{EF}$ is a median, $\overline{EF} \parallel \overline{AD}$ so that ∡$FEI \cong$ ∡GHI and ∡$EFI \cong$ ∡HGI.

28. Rewrite the product as $ST \times ST = TW \times RT$. Show $\triangle SWT \sim \triangle RST$. Right ∡$RST =$ right ∡SWT and ∡$T \cong$ ∡T.

29. Show $\triangle AEH \sim \triangle BEF$. ∡$BEF \cong$ ∡HEA and ∡$EAH \cong$ ∡EBF (Halves of equals are equal).

30. Apply Postulate 10.1. Since $\overline{XY} \parallel \overline{LK}$, $\frac{JY}{YK} = \frac{JX}{XL}$.

Since $\overline{XZ} \parallel \overline{JK}$, $\dfrac{JX}{XL} = \dfrac{KZ}{ZL}$ By transitivity, $\dfrac{JY}{YK} =$

$\dfrac{KZ}{ZL}$ $YK = XZ$ since $YKZX$ is a parallelogram.

Using substitution, $\dfrac{JY}{XZ} = \dfrac{KZ}{ZL}$ Cross-multiplying

yields the desired product.

CHAPTER 11

1. $r = 8$, $s = 8\sqrt{3}$, $t = 4\sqrt{3}$.
2. $r = 25$, $s = 5\sqrt{5}$, $t = 10\sqrt{5}$.
3. $r = 16$, $s = 9$, $t = 15$.
4. 8
5. 6
6. $4\sqrt{6}$
7. $9\sqrt{5}$
8. 7
9. $4\sqrt{10}$
10. **a** 4 and 16

    b $8\sqrt{5}$
11. 8
12. $6\sqrt{2}$
13. 24
14. $100\sqrt{7}$
15. $\frac{20}{3}$
16. 80
17. 18
18. $5\sqrt{3}$
19. $\sqrt{74}$
20. **a** 4   **b** $8\sqrt{3}$
21. $6\sqrt{2}$
22. $36\sqrt{2}$
23. **a** 4   **b** $4\sqrt{3}$ **c** $4\sqrt{2}$
24. **a** Altitude $= 8/\sqrt{3}$, leg $= 16/\sqrt{3}$.

    b Altitude $= 8$, leg $= 8\sqrt{2}$.

    c Altitude $= 8\sqrt{3}$, leg $= 16$.
25. $x = 4\sqrt{3}$, $y = 4$, $z = 2\sqrt{37}$.
26. $2\sqrt{37}$
27. $10\sqrt{3}$
28. $\frac{24}{25}$
29. $\cos R = \frac{40}{41}$, $\tan R = \frac{9}{40}$.
30. 32°
31. Base $= 15$, leg $= 12$.
32. **a** 57.7   **b** 112.9
33. $x = 20$, $y = 20.1$.
34. $AD = 6.2$, $CD = 21.9$.

CHAPTER 12

1. 66
2. 52.5
3. 25
4. 66
5. 12.5
6. 38
7. 66
8. 61
9. 24
10. 75
11. 36
12. 26
13. 106
14. 25
15. **a** 74

    b 16

    c 74

    d 16

    e 32

    f 90

    g 16

    h 16

    i 74
16. **a** $m\angle DEB = 90$, $m\angle ABC = 31$, $m\angle AED = 90$.

    b $m\overset{\frown}{AC} = 56$, $m\angle DEB = 79$, $m\angle AED = 101$.

    c $m\overset{\frown}{AC} = 51$, $m\angle ABC = 25.5$, $m\angle AED = 97$.

    d $m\overset{\frown}{BD} = 150$, $m\overset{\frown}{AC} = 82$, $m\angle DEB = 116$.
17. $x = 82$, $y = 59$, $z = 39$.
18. **a** $m\angle CAQ = 60$, $m\angle DBC = 40$, $m\angle AEB = 60$, $m\angle CPQ = 40$.

    b $m\overset{\frown}{AB} = 58$, $m\angle CAQ = 66$, $m\angle DBC = 37$, $m\angle AEB = 66$.

    c $m\overset{\frown}{ADC} = 164$, $m\angle CAQ = 82$, $m\angle DBC = 50$, $m\angle CPQ = 50$.

    d $m\overset{\frown}{ADC} = 142$, $m\overset{\frown}{AB} = 44$, $m\angle DBC = 49$, $m\angle AEB = 71$.
19. **a** $m\overset{\frown}{WA} = 30$, $m\overset{\frown}{AT} = 90$, $m\overset{\frown}{ST} = 150$, $m\overset{\frown}{SW} = 90$.

    b $m\angle WTS = 45$.

    c $m\angle TBS = 90$.

    d $m\angle TWP = 120$.

    e $m\angle WPT = 15$.
20. **a** $m\overset{\frown}{KF} = 76$, $m\overset{\frown}{FM} = 104$, $m\overset{\frown}{JK} = 100$, $m\overset{\frown}{JM} = 80$, $m\overset{\frown}{HM} = 28$.

    b $m\angle KPF = 76$.

c $m\angle KJH = 76$.

d $m\angle KLJ = 12$.

21. Show $\triangle AOM \cong \triangle BOM$ by Hy-Leg. By CPCTC, central $\angle AOX \cong$ central $\angle BOX$ which implies $\overset{\frown}{AX} \cong \overset{\frown}{BX}$.

22. By subtraction, $\overset{\frown}{RS} \cong \overset{\frown}{WT}$. $\angle RTS \cong \angle WST$ since inscribed angles which intercept congruent arcs are congruent.

23. Since $OA > AC$, $m\angle OCA > m\angle AOC$. $m\angle BOC > m\angle OCA$ so that $m\angle BOC > m\angle AOC$. Since $m\angle BOC = m\overset{\frown}{BC}$ and $m\angle AOC = m\overset{\frown}{AC}$, by substitution, $m\overset{\frown}{BC} > m\overset{\frown}{AC}$.

24. Since $m\overset{\frown}{BC} > m\overset{\frown}{AC}$, $m\angle BOD > m\angle AOD$. $m\angle ADO > m\angle BOD$ so that $m\angle ADO > m\angle AOD$. This implies that $OA > AD$.

25. Show $\triangle XPM \cong \triangle YQM$ by AAS where right $\angle XPM \cong$ right $\angle YQM$, $\angle XMP \cong \angle YMQ$, $\overline{XP} \cong \overline{YQ}$.

26. Since $\overline{FE} \cong \overline{FG}$, $m\angle G = m\angle E$. $m\overset{\frown}{FG} = 2m\angle E$ so that $m\angle BFG = m\angle E$. By transitivity, $m\angle G = m\angle BFG$ so that $\overline{AB} \parallel \overline{EG}$.

27. Show $\triangle AMB \cong \triangle AMD$ by SAS. $\overline{BM} \cong \overline{DM}$; $\angle M$ is inscribed in a semicircle so that right $\angle AMB \cong$ right $\angle AMD$; $\overline{AM} \cong \overline{AM}$. By CPCTC, $\angle BAM \cong \angle DAM$ which implies that $\overset{\frown}{BM} \cong \overset{\frown}{CM}$.

28. **a** $\angle W \cong \angle RSW$ since inscribed angles which intercept congruent arcs are congruent. $\angle RSW \cong \angle NSW$ (Definition of angle bisector). By transitivity, $\angle W \cong \angle NSW$ which implies that $\triangle NWS$ is isosceles.

    b $\overset{\frown}{WT} \cong \overset{\frown}{RW}$ since congruent inscribed angles intercept congruent arcs. Since $\overset{\frown}{RW} \cong \overset{\frown}{SK}$ (Given), $\overset{\frown}{WT} \cong \overset{\frown}{SK}$. $\angle STK \cong \angle TKW$ (Inscribed angles which intercept congruent arcs are congruent). Since the base angles of $\triangle NTK$ are congruent, the triangle is isosceles.

29. $m\overset{\frown}{LM} > m\overset{\frown}{KL}$ implies that $m\angle MKL > m\angle LMK$. Since $\overline{JK} \parallel \overline{LM}$, $m\angle JKM = m\angle LMK$. By substitution, $m\angle MKL > m\angle JKM$.

30. Since $\overline{MA} \cong \overline{MT}$, $\angle MAT \cong \angle ATM$. $\angle SRA \cong \angle ATM$ (Inscribed angles which intercept the same arc are congruent). By transitivity, $\angle SRA \cong \angle MAT$ which implies that $\overline{SR} \parallel \overset{\frown}{ATW}$. Quadrilateral $RSTW$ is a parallelogram since $\overline{SR} \cong \overline{TW}$ (Given) and $\overline{SR} \cong \overline{TW}$.

CHAPTER 13

1. **a** 2 common internal tangents.
   2 common external tangents.

   b None.

   c 1 common external tangent.

2. **a** 11

   b 3

3. 3

4. 9

5. 14

6. 1 or 21

7. 18

8. 2

9. 20

10. 1

11. 5

12. 27

13. 5

14. 12

15. 12

16. **a** 13

    b $JW = 48$, $OA = 7$.

    c 12

17. $2\sqrt{5}$

18. 6

19. **a** 66

    b 314

    c 34π

20. **a** 16

    b 7.5

    c $\dfrac{11.5}{\pi}$

21. 9π

22. $9\pi\sqrt{2}$

23. **a** 6π

    b $45°$

    c 10

24. $3\pi\sqrt{2}$

25. **a** $120°$

    b 12π

    c $24\pi + 36\sqrt{3}$

26. Triangles OAR and OBR are right triangles since angles S and T are right angles and $\overline{OA} \parallel \overline{MS}$ and $\overline{OB} \parallel \overline{MT}$ making corresponding angles congruent. $\triangle OAR \cong \triangle OBR$ by Hy-Leg. By CPCTC, $OA = OB$. $\overline{SR} \cong \overline{TR}$ by Theorem 13.2.

27. $\overline{OA} \perp \overline{SR}$ and $\overline{OB} \perp \overline{TR}$ (See exercise 26). By Theorem 13.1, $OA = OB$ so that $\triangle OAR \cong \triangle OBR$ by Hy-Leg. By CPCTC, $\angle AOR \cong \angle BOR$. Taking supplements of these congruent angles gives $\angle AOM \cong \angle BOM$.

28. Since a square contains 4 right angles and is equilateral, $\overline{OX} \perp \overline{QT}$, $\overline{OY} \perp \overline{PJ}$, and $OX = OY$. By Theorem 13.2, $\overline{QT} \cong \overline{PJ}$ so that $\overset{\frown}{QT} \cong \overset{\frown}{PJ}$. By arc subtraction, $\overset{\frown}{QP} \cong \overset{\frown}{JT}$.

29. Show $\triangle HBW \sim \triangle MBL$. $\angle ABW \cong \angle H$ since they are measured by one-half of the same arc measure. Since $ABLM$ is a parallelogram, $\overline{AB} \parallel \overline{ML}$ so that $\angle ABW \cong \angle BML$. By transitivity, $\angle H \cong \angle BML$ (Angle). $\angle MBL \cong \angle HBW$ (Angle).

30. Show $CD:BC = BC:AB$ by proving $\triangle ABC \sim \triangle BCD$. Right $\angle ACB \cong$ right $\angle CDB$. $\angle CAB \cong \angle CBD$ since they are measured by one-half the same arc measure.

31. Show $\triangle KLP \sim \triangle KJM$. $\angle LKP \cong \angle JKM$ and right $\angle KLP \cong$ right $\angle KJM$.

32. Show $\triangle KLP \sim \triangle KJM$. Right $\angle KLP \cong$ right $\angle KJM$ (Angle). Since $\overline{JP} \cong \overline{JM}$, $\angle JPM \cong \angle JMK$. $\angle KPL \cong \angle JPM$. By transitivity, $\angle KPL \cong \angle JMK$ (Angle).

33. **a** $\angle NTK \cong \angle WTK$. Since $\overline{KW} \cong \overline{TW}$. $\angle WKT \cong \angle WTK$. By transitivity, $\angle NTK \cong \angle WKT$ which implies that $\overrightarrow{NTP} \parallel \overrightarrow{KW}$.
   b Rewrite product as $TW \cdot TW = JT \cdot TK$. Show $\triangle JTW \sim \triangle WTK$. $\angle T \cong \angle T$ (Angle). $\angle JWT \cong \angle NTK$ and $\angle K \cong \angle NTK$ so that $\angle K \cong \angle JWT$ (Angle).

34. **GIVEN** Circles A and B are tangent externally at point P. $\overleftrightarrow{XY}$ is tangent to $\odot A$ at point X and tangent to $\odot B$ at point Y. $\overrightarrow{PQ}$ intersects $\overleftrightarrow{XY}$ at point Q.

    PROVE $\overline{QX} \cong \overline{QY}$.

    PLAN By Theorem 13.3, $\overline{QX} \cong \overline{QP}$ and $\overline{QP} \cong \overline{QY}$. By transitivity, $\overline{QX} \cong \overline{QY}$.

35. **GIVEN** $\overleftrightarrow{XY}$ is tangent to $\odot A$ at point X and tangent to $\odot B$ at point Y. $\overrightarrow{PQ}$ is tangent to $\odot A$ at point P and tangent to $\odot B$ at point Q.

    PROVE $\overline{XY} \cong \overline{PQ}$.

    PLAN *Case 1.* Assume circles A and B are not congruent. $\overleftrightarrow{XY}$ cannot be parallel to $\overleftrightarrow{PQ}$. Therefore, $\overleftrightarrow{XY}$ and $\overleftrightarrow{PQ}$ will intersect, extended if necessary, say at point C. By Theorem 13.3, $\overline{CX} \cong \overline{CP}$ and $\overline{CY} \cong \overline{CQ}$. By subtraction, $\overline{XY} \cong \overline{PQ}$. *Case 2.* Assume circles A and B are congruent. Then $\overleftrightarrow{XY} \parallel \overleftrightarrow{PQ}$. Draw the line of centers $\overline{AB}$ and radii $\overline{AX}$, $\overline{BY}$, $\overline{AP}$, and $\overline{BQ}$. Quadrilateral $AXYB$ is a parallelogram since $\overline{AX} \cong \overline{BY}$ (Congruent circles have congruent radii) and $\overline{AX} \parallel \overline{BY}$ (Segments perpendicular to the same line are parallel). Similarly, $APQB$ can be shown to be a parallelogram. Hence, $\overline{XY} \cong \overline{AB}$ and $\overline{AB} \cong \overline{PQ}$ so that $\overline{XY} \cong \overline{PQ}$.

36. **GIVEN** Lines l, m, and k are tangent to circle P at points C, D, and E, respectively, such that $l \parallel m$ and line k intersects line l at point A and line m at point B.

    PROVE $\angle APB$ is a right angle.

    PLAN Since $l \parallel m$, angles CAB and DBA are supplementary. $\overline{PA}$ bisects $\angle CAB$ and PB bisects $\angle DBA$ (Refer back to the discussion of the proof of Theorem 13.3). Hence, the sum of the measures of angles PAB and PBA is 90 which implies that the remaining angle of $\triangle APB$, namely

$\angle APB$, must have a measure of 180 − 90 or 90.

CHAPTER 14

1. **a** 48
   b 32
   c $90\sqrt{3}$

2. **a** Side = 12 cm.
   b Altitude = 5, base = 15.
   c Altitude = 15, base = 9.

3. 60

4. $35\sqrt{3}$

5. $10\sqrt{2}$

6. $9\sqrt{3}$

7. 104

8. 52

9. 17.5

10. 162

11. $54 + 36\sqrt{3}$

12. **a** 8
    b $2\sqrt{11}$
    c $\dfrac{4}{\sqrt[3]{3}}$

13. **a** 12
    b 14

14. **a** 4
    b 5
    c 6

15. 240

16. **a** 42
    b 21
    c $21\sqrt{3}$

17. $\frac{1}{6}$

18. **a** $\frac{9}{11}$
    b 55

19. 192

20. 45

21. 30

22. $150\sqrt{3}$

23. **a** Radius = $3\sqrt{2}$, apothem = 3, area = 36.
    b Side = 7, radius = $3.5\sqrt{2}$, apothem = 3.5.
    c Side = 10, radius = $5\sqrt{2}$, area = 100.
    d Side = $8\sqrt{2}$, apothem = $4\sqrt{2}$, area = 128.

24. 3:4

25. **a** 30.8
    b 3,080

26. **a** $D = 10$, $C = 10\pi$, $A = 25\pi$

    b $R = 4.5$, $C = 9\pi$, $A = \frac{81}{4}\pi$

    c $R = 7$, $D = 14$, $C = 14\pi$

    d $R = 9$, $D = 18$, $A = 81\pi$

    e $R = \frac{8}{5}$, $D = \frac{16}{5}$, $C = \frac{16}{5}\pi$

27. 15 cm

28. 312.5

29. 20π

30. **a** Area of circle = 144π, area of sector = 18π.

    b Radius = 6, degree measure of sector arc = 120°.

    c Radius = 10, area of circle = 100π.

    d Area of circle = 81π, area of sector = 36π.

31. **a** 24π

    b $24\pi - 36\sqrt{3}$

32. $8\pi - 16$

33. $1:2$

34. $\frac{625}{8}\pi - 84$

35. $256 - 64\pi$

36. 80π

37. $16\pi - 32$

38. **a** 64π

    b $96\sqrt{3}$

    c $\frac{32}{3}\pi - 16\sqrt{3}$

39. **a** From point L draw the altitude to side $\overline{CV}$, intersecting $\overline{CV}$ at point H. Area $\triangle CLG = \frac{1}{2}CG \cdot LH$ and area $\triangle VLG = \frac{1}{2}VG \cdot LH$. Since $CG = VG$, area $\triangle CLG$ = area $\triangle VLG$. (*Note: In $\triangle CLV$, $\overline{LG}$ is a median. We have just proven that a median divides a triangle into two triangles having the same area.*)

    b In $\triangle BCV$, $\overline{BG}$ is a median so that area $\triangle BCG$ = area $\triangle BVG$ (see part **a** of this exercise). Next, subtract:

$$\text{Area } \triangle BCG = \text{area } \triangle BVG$$
$$-\text{Area } \triangle CLG = \text{area } \triangle VLG$$
$$\text{Area } \triangle BLC = \text{area } \triangle BLV$$

40. The area of $\square ABCD$ is equal to the sum of the areas of triangles AEB, CEB and ADC. Since $\triangle CED \cong \triangle AEB$ and $\triangle AED \cong \triangle CEB$, the area of $\square ABCD$ is equal to the sum of the areas of triangles CED, AED, and ADC. $\overline{BE} \cong \overline{ED}$ so that $\overline{ED} \cong \overline{FD}$. Since $\overline{CD}$ and $\overline{AD}$ are medians of triangles ECF and EAF, respectively, area $\triangle CED$ = area $\triangle CDF$ and area $\triangle AED$ = area $\triangle FAD$. By substitution, the area of $\square ABCD$ is equal to the sum of the areas of triangles CDF, FAD, and ADC which is equivalent to $\triangle FAC$.

41. **a** Circle O is inscribed in equilateral $\triangle ABC$ where apothem $\overline{OX} \perp \overline{AC}$. In right $\triangle AOX$, $\overline{OA}$

represents the radius of the circumscribed circle, while $\overline{OX}$ is a radius of the inscribed circle. Since $m\angle OAX = 30$ and $AX = \frac{1}{2}AC$, $OX = \frac{1}{\sqrt{3}}AX = \frac{1}{2\sqrt{3}}AC = \frac{\sqrt{3}}{6}AC$. The altitude to side $AC = \frac{\sqrt{3}}{2}AB = \frac{\sqrt{3}}{2}AC$. Hence, OX is one-third the length of the altitude.

    b $AX = \frac{\sqrt{3}}{2}OA$ or $OA = \frac{2}{\sqrt{3}}AX = \frac{2}{\sqrt{3}}\left(\frac{1}{2}AC\right)$

    OA is therefore equal to $\frac{1}{\sqrt{3}}AC$. Since OX may be represented by $\frac{1}{2\sqrt{3}}AC$ (see part **a**), $OX = \frac{1}{2}OA$. That is, the radius of the inscribed circle is one-half the length of the radius of the circumscribed circle. Since their areas have the same ratio as the square of the ratio of their radii, the ratio of the areas of the circle is $1:4$.

CHAPTER 15

1. Two circles concentric with the original circle having radii of 7 cm and 9 cm.

2. A line midway between the 2 parallel lines, parallel to each line, and 5 inches from each line.

3. A line parallel to the surface at a distance equal to the radius length.

4. A concentric circle having a radius of 9 cm.

5. A circle concentric with the stationary circle whose radius length is the sum of the lengths of the radii of the two given circles.

6. The perpendicular bisector of the base.

7. A circle concentric with the stationary circle whose radius length is difference of the lengths of the radii of the two given circles.

8. Two points determined by the intersection of the perpendicular bisector of $\overline{AB}$ and the two lines which are parallel to and on either side of $\overleftrightarrow{AB}$ and 4 inches from $\overleftrightarrow{AB}$.

9. Four points determined by the angle bisector of each pair of vertical angles formed by the intersecting lines and a circle whose center is their point of intersection.

10. No points.

11. The two points determined by the intersection of the perpendicular bisector of $\overline{PQ}$ and the circle whose center is at point X and which has a radius of 5 cm.

12. **a** 2   **b** 0   **c** 1

13. **a** 0   **b** 1   **c** 2

14. 20

15. 12

16. 17

17. 24

18. **a** $M(2, 5)$, $AB = \sqrt{40}$.

   b $M(-3.5, -3)$, $AB = 13$.

   c $B(4, -8)$, $AB = \sqrt{148}$.

   d $A(3, -9)$, $AB = \sqrt{164}$.

   e $M(b, a)$, $AB = 2\sqrt{a^2 + b^2}$.

   f $M\left(\dfrac{a + c}{2}, \dfrac{b + d}{2}\right)$, $AB = \sqrt{(a - c)^2 + (b - d)^2}$.

19. **a** $O(-1, 1)$.

   b 5

   c Yes, since $OK = 5$.

20. $y = -1$.

21. $\sqrt{125}$

22. First show $JKLM$ is a parallelogram. Since slope of $JK = \frac{2}{3}$ and the slope of $KL = -\frac{3}{2}$, $\overline{JK} \perp \overline{KL}$. Since $JK = KL = \sqrt{13}$, $JKLM$ is a square.

23. **a** Slope of $JK = \frac{6}{5}$ and slope of $KL = -\frac{5}{6}$ so that $\overline{JK} \perp \overline{KL}$.

   b $JL = \sqrt{305}$, $KL = \sqrt{61}$, and $JK = \sqrt{244}$ so that $(JL)^2 = (KL)^2 + (JK)^2$.

24. **a** $\frac{8}{3}$ **b** $\frac{3}{2}$ **c** $\frac{7}{6}$

25. **a** Collinear.

   b Noncollinear.

   c Collinear.

26. Slope of $\overline{ST} = \frac{6}{5}$, slope of $\overline{TA} = -\frac{5}{6}$, slope of $\overline{AR} = \frac{6}{5}$ and slope of $\overline{SR} = -\frac{5}{6}$. Hence, $\overline{ST} \parallel \overline{AR}$ and $\overline{TA} \parallel \overline{SR}$. Also, $\overline{ST} \perp \overline{TA}$ and so on.

27. $y = \frac{1}{2}x + 5$.

28. $y = \frac{3}{2}x - \frac{7}{2}$.

29. **a** $y = 7$.

   b $y = -x + 3$.

   c $x = 6$.

   d $y = -4x + 1$.

30. **a** $y = 5$, $y = -5$.

   b $x = 4$, $x = -4$.

   c $y = 2x$.

   d $2y + 3x = 6$.

   e $y = x + 5$.

31. **a** $y = 2$.

   b $x = -5$.

32. **a** $y = x + 1$.

   b $y = \frac{11}{9}x - \frac{25}{9}$.

   c $y = -\frac{2}{3}x + 12$.

33. $y = -\frac{1}{4}x - 2$.

34. $k = 4$, $t = 22$.

35. **a** $x^2 + y^2 = 16$,

   b $(x - 3)^2 + (y - 4)^2 = 36$.

   c $(1, 0)$, 9.

d $(0, 0)$, 7.

e $(2, -5)$, $\sqrt{51}$.

36. $x^2 + y^2 = 25$.

37. $k = \frac{5}{3}$, $h = \frac{36}{9}$.

38. $x^2 + (y - 9)^2 = 25$.

39. Let the coordinates of $\square ABCD$ be represented by $A(0, 0)$, $B(s, t)$, $C(r + s, t)$, and $D(r, O)$. Use the distance formula to verify $AB = DC = \sqrt{S^2 + t^2}$ and $BC = AD = r$.

40. Let the coordinates of right $\triangle ABC$ be represented by $A(0, 0)$, $B(0, 2r)$, and $C(2s, 0)$. If M is the midpoint of $\overline{BC}$, then its coordinates are $M(s, r)$. $AM = \sqrt{s^2 + r^2}$ and $BC = 2\sqrt{s^2 + r^2}$ so that $AM = \frac{1}{2}BC$.

41. Let the coordinates of $\triangle ABC$ be represented by $A(0, 0)$, $B(2r, 2s)$, and $C(2t, 0)$. Let L and M be the midpoints of $\overline{AB}$ and $\overline{BC}$, respectively. Then $L(r, s)$ and $M(r + t, s)$. Show slope of $\overline{LM}$ = slope of $\overline{AC}$ and that $LM = \frac{1}{2}AC$.

42. Let the coordinates of isosceles trapezoid $ABCD$ be represented by $A(0, 0)$, $B(r, s)$, $C(t, s)$, and $D(r + t, 0)$. $AC = \sqrt{t^2 + s^2} = BD$.

CHAPTER 16

These program solutions were tested on an Apple IIe computer.

```
5   REM *****EXERCISE 1*****
10  INPUT L,W,H
20  V = L * W * H
30  S = 2 * (L * W + L * H + W * H)
40  PRINT "VOLUME = ";V
50  PRINT "SURFACE AREA = ";S
60  END
```

```
5   REM *****EXERCISE 2*****
10  INPUT X
20  IF X < 90 THEN 60
30  IF X > 90 THEN 80
40  PRINT X;" IS A RIGHT ANGLE"
50  GOTO 90
60  PRINT X;" IS AN ACUTE ANGLE"
70  GOTO 90
80  PRINT X;" IS AN OBTUSE ANGLE"
90  END
```

```
5   REM   *****EXERCISE 3*****
8   INPUT "LENGTHS OF ADJACENT SIDES ? ";A,B
10  INPUT "LENGTHS OF DIAGONALS ? ";D1,D2
20  IF A = B AND D1 = D2 THEN 70
30  IF A < > B AND D1 = D2 THEN 90
40  IF A = B THEN 110
50  PRINT "NONE OF THESE"
60  GOTO 120
70  PRINT "SQUARE"
80  GOTO 120
90  PRINT "RECTANGLE"
100  GOTO 120
110  PRINT "RHOMBUS"
120  END
```

```
5   REM *****EXERCISE 4*****
8   INPUT "RADII ? ";R1,R2
10    INPUT "LINE OF CENTERS ? ";D
15    PRINT : PRINT
20    IF D > R1 + R2 THEN 70
30    IF D < R1 + R2 THEN 100
40    PRINT "TWO EXTERNAL AND ONE INTERNAL"
50    PRINT "COMMON TANGENTS"
60    GOTO 120
70    PRINT "TWO EXTERNAL AND TWO INTERNAL"
80    PRINT "COMMON TANGENTS"
90    GOTO 120
100   PRINT "TWO EXTERNAL AND NO INTERNAL"
110   PRINT "COMMON TANGENTS"
120   END


5   REM *****EXERCISE 5*****
10    FOR R = 2 TO 5
20    FOR S = 1 TO R - 1
30  A = R * R - S * S
40  B = 2 * R * S
50  C = R * R + S * S
60    PRINT A,B,C
70    NEXT S
80    NEXT R
90    END


5   REM *****EXERCISE 6*****
10    INPUT A,B,C
20    IF C < A OR C < B THEN  PRINT "CHECK DATA": END
30    IF A < B + C AND B < A + C AND C < A + B THEN 60
40    PRINT "NO TRIANGLE CAN BE FORMED"
50    GOTO 999
60    IF C * C < A * A + B * B THEN 100
70    IF C * C = A * A + B * B THEN 120
80    PRINT "TRIANGLE IS OBTUSE"
90    GOTO 999
100   PRINT "TRIANGLE IS ACUTE"
110   GOTO 999
120   PRINT "TRIANGLE IS RIGHT"
999   END


5   REM *****EXERCISE 7*****
10    INPUT X1,Y1,X2,Y2,X3,Y3
20    IF X2 - X1 = 0 OR X3 - X2 = 0 THEN 999
30    IF (Y2 - Y1) / (X2 - X1) = (Y3 - Y2) / (X3 - X2) THEN 60
40    PRINT "POINTS ARE NOT COLLINEAR"
50    GOTO 999
60    PRINT "POINTS ARE COLLINEAR"
999   END


41    REM ***SUBROUTINE FOR EXERCISE 8***
43    REM ***MAKE THE FOLLOWING CHANGES***
45    REM ***CHANGE LINE 50 AS FOLLOWS:
50    GOSUB 800
55    GOTO 999
790   STOP
800   REM ***SUBROUTINE FOR AREA***
810 A = ((X2 - X1) ^ 2 + (Y2 - Y1) ^ 2) ^ .5
820 B = ((X3 - X1) ^ 2 + (Y3 - Y1) ^ 2) ^ .5
830 C = ((X3 - X2) ^ 2 + (Y3 - Y2) ^ 2) ^ .5
840 S = (A + B + C) / 2
850 AR = (S * (S - A) * (S - B) * (S - C)) ^ .5
860   RETURN


5   REM *****EXERCISE 9*****
10 PI = 0
20 K = 0
30    FOR N = 1 TO 10000 STEP 2
40 K = K + 1
45 X = PI
50 PI = PI + 4 * ((1 / N) * ( - 1) ^ (K + 1))
60    IF K = 25 THEN  PRINT PI
70    IF  ABS (X - PI) < .001 THEN 90
80    NEXT N
90    PRINT "PI APPROXIMATELY = ";PI
100   PRINT "NUMBER OF TERMS = ";N
110   END


5   REM *****EXERCISE 10*****
8 K = 0
10    FOR N = 1 TO 100
20 X =  RND (1):Y =  RND (1)
30    IF X ^ 2 + Y ^ 2 < = 1 THEN K = K + 1
40    NEXT N
50 PI = 4 * (K / 100)
60    PRINT "PI APPROXIMATELY = ";PI
70    END
```

Index

SAS Postulate 86
SSS Postulate 85
Contrapositive 127
Converse 57; 127
Convex polygon 65
Coordinate proofs 316—317
Coplanar 49
Corollary 70
Correspondence 81—82
Corresponding parts 82; 89
Cosine ratio 198
CPCTC 99

Deductive reasoning 8
Defined terms 1, 3
Definition 6
Degree 13; 214
Determined line 104—105
Diameter 210—211
  perpendicular to a chord 221
Distance
  between a point and a
    line 43
  between two parallel
    lines 156
  between two points 42—43
  equidistant 42
  formula in the coordinate plane
    305
  shortest 43
Double congruence proofs
  112—113
Drawing conclusions from
  diagrams 20—21

Equality 21
  addition, subtraction,
    multiplication, and
    division 24—26
  halves of equals 26
  reflexive, symmetric, and
    transitive 22—23
  substitution 23
Equation
  of a circle 315
  of a line 311
Euclid's Fifth Postulate 60

45-45 right triangle 196—197

Geometric mean (see mean
  proportional)

HL (Hy-Leg) Postulate of
  Congruence 93
Hypotenuse 92

IF...THEN...structure
  in BASIC 321
  in geometry 9
Included angle and side 83—84
Indirect measurement 202—203
Indirect method of proof 124
Inductive reasoning 8
Inequalities
  angles of a triangle 122
  exterior angle of a
    triangle 74
  properties of 119—120
  sides of a triangle 121
Inscribed Angle Theorem 229
  corollaries 241
Inscribed polygon 227
Inverse 127
Isosceles triangle 107—111
  base 104
  base angles 104
  base angles theorem 107
  converse of base angles
    theorem 107
  legs 104
  vertex angles 104

Line of centers 253
Line segments 3
  addition of measures 25—26
  altitude 106
  bisector of 19
  congruent 18
  corresponding sides of
    congruent triangles 82
  measure of 13
  median 106
  midpoint formula 301
  midpoint of 19
  subtraction of measures
    25—26
Lines 2
  equation of 311
  parallel 49
  perpendicular 41
  slope 306
Lobachevsky 60
Locus
  definition 293
  determining a locus 294
  determining compound
    loci 296

Mean proportional 162
Measure 13
  of a central angle 215
  of a line segment 13
  of an angle 13—14
  of an angle formed by a
    tangent and a chord
    233

of an angle formed by two
    chords intersecting in
    the interior of a
    circle 234—235
  of an angle whose vertex lies
    in the exterior of a
    circle 236—237
  of an arc 214
  of an inscribed angle 229
Median
  in similar triangles 177
  of a trapezoid 146
  of a triangle 11; 106—107; 147
  to the hypotenuse 146
Midpoint formula 301
Midpoint of a line segment 19
Midpoints of a Triangle Theorem
  147

Non-Euclidean geometries 60

Obtuse
  angle 15
  triangle 78
Ordered pair 299
Ordinate 299
Overdetermined line 105

Parallel lines
  properties of 51—56
  proving lines are parallel
    57—59
Parallel Postulate 59—60
Parallelograms
  properties of 135—138
  properties of special
    139—141
  proving a quadrilateral is a
    parallelogram
    142—143
Perpendicular lines 41
  bisectors 41
  existence of 41—42
  proving lines are perpendicular
    43—44; 103
Perimeters of similar polygons
  171
Pi (π) 262
Plane 2, 7, 49
Playfair, John 60
Point 2
  collinear 5
  coordinates of 299
  midpoint 19
  of tangency 224
Polygons
  circumscribed 227
  classifying by the number of
    sides 67
  congruent 83
  convex 65—66
  diagonals 67
  equiangular 68

DISCARD/SOLD
FRIENDS MLS